COAL
CONVERSION

by
E. J. HOFFMAN
Energy Consultant
Laramie, Wyoming

THE ENERGON COMPANY
Laramie, Wyoming

Copyright © by E.J. Hoffman

All Rights Reserved

This book or any part thereof
must not be reproduced in any form
without the written permission of the publisher.

Library of Congress Catalog Number: 77-93533
International Standard Book Number: 0-9601552-1-X
Printed in the United States of America

Modern Printing Company
Laramie, Wyoming

PREFACE

It is the purpose of this volume to detail in orderly fashion the more common chemical conversions of coal and their applications, along with more esoteric considerations. These conversions include combustion and partial combustion, hydrogenation and reaction with steam. Certain other reactions and transformations are presented for completeness. Emphasis is on the fundamentals behind the technology, rather than on the technology itself in all its variations.

The so-called indirect hydrogenation of coal — involving the further reaction of carbon monoxide and hydrogen as intermediates — is of sufficient importance to be presented as well, and leads naturally to the discussion of catalysis.

At the same time, this work represents a survey of the literature and practices for the chemical conversion of coal, and for investigations underway or proposed. It is in part based on secondary sources since the original literature has been surveyed to such an extent that much of the pertinent information has by now appered in review-type articles. Otherwise, the original sources have been consulted. In other instances, new information is drawn upon.

There is an element of discernment aimed at the selective inclusion of features and data which are basic to the subject. In so doing, the body of the work provides a manageable reservoir of condensed and concise information by which further study and experiment may be facilitated, and which hopefully will indicate new directions.

As a special addition, calculations are appended delineating methods and methodology for the estimation of overall net efficiency or net energy yield, determinations which have become of relevance with the new stress on net energy analysis. Thermodynamic changes in the heats of reaction provide the most elemental or abstract basis for the determinations.

The literature on coal conversion is growing exponentially, as befits a topic of such immediacy. Much of this information is fragmented, occurring as reports or presented at formal or informal meetings and in the proceedings of conferences and symposia. The further sifting out and consolidation has not yet been undertaken. This monograph is one such effort.

Recognition is in order for the several works on coal which have appeared in the past and for specialized compilations on various applications, particularly in processing. The pre-eminent treatise on coal in its many aspects is *Chemistry of Coal Utilization*.* Flow diagrams and descriptions for may of the coal conversion processes which have surfaced are provided by Tom Hendrickson's *Synthetic Fuels Data Handbook*** and by *Coal Conversion Technology 1976*.*** Additionally, there are the annual reports and other publications of the Energy Research and Development Administration (now part of the Department of Energy).

The modern practices of coal combustion pertain to pulverized fuels in entrained systems. The notable work in this area is *Combustion of Pulverized Coal*.[†] Research and development in combustion, however, is increasingly aimed at fluidized systems and may be a portent of the future, though so far not realized.

The subject of catalysis in coal-related reactions is of growing significance — though somewhat diffused throughout the literature — and the attempt is made to treat catalysts and catalysis in considerable depth in order to define more closely this emerging approach.

The pollution-free combustion of solid fuels touches upon several other topical areas, notably off-gas cleanup, even the manufacture and use of "smokeless" briquettes or pellets. (The latter comprises a lesser-known art and science, the raison d'etre of the Institute for Briquetting and Agglomeration.) Still other aspects of coal conversion will interface with the problems and alternatives of energy transport: rail use, slurry-pipelining of coal and/or gas piepelining vs high-voltage electrical transmission. Etc. While it is not the intent of this work to cover all aspects of coal conversion and its impact, at least note is made.

Work in coal conversion is by no means confined to the United States. Activity in other countries is attested to by the many references appearing in Chemical Abstracts. In citing much of this literature from

---

* *Chemistry of Coal Utilization*, H. H. Lowry Ed., Wiley, New York. Volume I and II (1945). Supplementary Volume (1963).

** *Synthetic Fuels Data Handbook*, Cameron Engineers, Denver, Colorado (1975).

*** I. Howard-Smith and G. J. Werner, *Coal Conversion Technology 1976*, Noyes Data Corporation, Park Ridge, New Jersey (1976).

[†] M. A. Field, D. W. Gill, B. B. Morgan and P. G. W. Hawksley, *Combustion of Pulverized Coal*, British Coal Utilization Research Association, Leatherhead, Surrey (1967).

non-English journals, the abstract appearing in Chemical Abstracts is provided with the reference designator "CA" followed by the volume and page number. Reference to the British Coal Utilization Research Association is provided by the designator "BCURA".

Acknowledgement is made to present and former associates and to Robert G. Morrison for material on carbon monoxide. While the utilization of coal is of long standing, the re-dedication of coal in alleviating our energy problems has provided the encouragement for producing this volume.

Laramie, Wyoming　　　　　　　　　　　　　　　　　　　E. J. Hoffman
March, 1978

Contents

1. Introduction   1

    1.1 Chemical Conversions   3
        Combustion and Partial Combustion, Hydrogenation, Indirect Hydrogenation, the Coal-Steam Reaction, Direct Conversion, Catalysis

    1.2 Reacting Systems   9
        Fixed Versus Suspended Beds

    1.3 State of the Technology   11

    1.4 Trends in Research and Development   15
        Energy Conversion

Supplement:
    Miscellaneous Coal Liquefaction and Gasification Processes   24

2. Physical and Chemical Changes in Coals   47

    2.2 The Characteristics of Coal   48
        Analysis, Rank, Petrography, Physical and Thermodynamic Properties

    2.2 Moisture in Coals   58
        Analysis, Drying

    2.3 The Action of Solvents   60

    2.4 Pyrolysis   64
        Carbonization, Effects During Firing, Effect of Moisture

    2.5 Addition and Substitution Reactions   73
        Hydrolysis, Alkalies, Organic Solvents, Other Effects

Contents

3. Reactions with Air or Oxygen (Combustion) — 81

    3.1 Influence of Coal Quality — 84

    2.3 Low-Temperature Reactions — 85
Atmospheric Oxidation, Moisture, Pyrites, Spontaneous Combustion

    3.3 The Energetics of Combustion — 90
Heat Balance, Soot Formation, Convection and Radiation, Fouling, Excess Air

    3.4 Combustion Mechanisms — 102
Stages of Reaction, Homogeneous Behavior and Ignition, Surface Effects, Combustion Rates, Coal/Air Transport, Additives and Catalysis; Experimental Systems

    3.5 Effects of Moisture on Combustion — 119
Steam-Carbon Reaction, CO Oxidation, Furnace Performance

    3.6 Coal/Water Systems — 126
Transport of Coal/Water Suspensions, Combustion of Coal/Water Slurries, Integrated Systems, Conclusions

    3.7 Combustion Systems — 133
Fixed or Slowly-Moving Beds, Fluidized Combustion, Entrained Systems

4. Partial Combustion — 153

    4.1 Reactions Producing Carbon Monoxide — 154
Reaction Times, The Boudouard Reaction

    4.2 Producer Gas — 169
Variables, Sulfur

    4.3 *In Situ* Gasification — 173
Fracture, Problems of In Situ Gasification, Conversion

5. Reactions with Hydrogen 183

   5.1 Hydrogenation Without Carrier   185

   5.2 Liquid-Phase Hydrogenation   189

   5.3 Vapor-Phase Hydrogenation   194

   5.4 Mild Hydrogenation   195
      Bitumens, Reaction Conditions, The Makeup and Properties of Synthetic Bitumens

   5.5 Dehydration   200

6. Reactions with Steam 205

   6.1 Water-Gas Reactions   205

   6.2 Catalytic Effects   208
      Process Applications

   6.3 Endothermicity   213

   6.4 Steam-Gasification Systems   214

7. Reactions of Carbon Monoxide and Hydrogen 223

   7.1 Products and Conditions   224

   7.2 Catalyst Compositions and Properties   233

   7.3 Catalyst Preparation and Activity   238

   7.4 Effects of Operating Variables   245
      Temperature, Pressure, Reaction Time, Composition

   7.5 Experimental Studies   251

8. Catalysis and Heterogeneous Reactions 265

   8.1 Catalyst Activation and Regeneration   272
      Activation and Reaction, The Surface Reaction

   8.2 The Premise of Direct Conversion   277

Contents

8.3	The Hoffman Process	280
	Description of Reactions, Experimental Systems, Catalyst Behavior, Experimental Results, Other Factors	

9. The Production of Hydrogen and Carbon Monoxide as Fuels and Reaction Intermediates — 287

9.1	Hydrogen From Coal-Steam Systems	288
9.2	Carbon Monoxides Properties and Sources	292
	Properties	
9.3	Generation of Carbon Monoxide	294
	Gas Producers, Novel Gasification Schemes, Carbon Monoxide from Basic Oxygen Furnaces (BOF), Hot-Gas Cleanup, Conversion of Carbon Dioxide to Carbon Monoxide, Other Methods for the Production of Carbon Monoxide	
9.4	Maximizing Carbon Monoxide Production	315
9.5	Separation and Recovery of Carbon Monoxide	317
	Copper Aluminum Chloride, Low Concentrations of Carbon Monoxide, Low Temperature Separation of Carbon Monoxide	

10. Ancillary Considerations — 329

10.1	Water Requirements for Coal Conversion	331
	Other Figures	
10.2	Removal of Sulfur Oxides and Particulates	338
10.3	Acid Gas Absorption Systems	345
	Alkalies, Hot Potassium Carbonate Solutions, Solvents, Amines, Selective Absorption, Heats of Solution, Thermal Imput, Adsorptive Methods, Novel Methods for CO_2 Removal and Recovery, NO_x Absorption	
10.4	Conversion of Hydrogen Sulfide	358
10.5	Drying and Compression	359
10.6	Other Compounds	360

APPENDIX.		The Determination of Net Energy Efficiencies for Coal Conversion Processes	365
		Net Energy Analysis	
	I.	Comparison of Reaction Sequences For the Production of High-Btu Gas (Methane)	371
		Direct Methanation, Indirect Methanation, Direct Hydrogenation	
	II.	Stoichiometry Based on Coal Anaylsis	375
		Analysis of the Conversion, Conversion to Benzene, Conversion to Oxygenated Compounds, Pyrolysis	
	III.	Carbon Balances and Heats of Reaction	383
		Direct Methanation, Indirect Methanation (O-BPG), Indirect Methanation (External Combustion), Direct Methanation	
		Supplement to Section III	391
	IV.	Thermodynamic Efficiencies for Methane Production	395
		Conversion, Efficiency, Feed Carbon, Total Actual Carbon Requirement, Oxygen, Latent Heat, Reaction Heat Requirements, Size Reduction, Acid Gas Removal and Recovery, Compression Requirements, Additional Plant Energy Requirements, Results	
	V.	Production of Low- or Medium-Btu Gas	425
		Assumptions, Reaction Stoichiometry, Energy Inputs for Oxygen or Air, Expansion, Additional Requirements, Results, Combustion of Hot Gas	
SUBJECT INDEX			451

Chapter 1

INTRODUCTION

The continued downturn in known domestic reserves of oil and natural gas, signified by falling reserve/production ratios, emphasizes what is suspected by many: that coal will become the immediate substitute energy rersource of the near future.

Perhaps the most striking overview of the inevitable depletion of finite resources of petroleum has been provided by King Hubbert (1), formerly of the U. S. Geological Survey. His theoretical Gaussian distribution curves are shown in Figure 1.1, where the difference between the cumulative ogives representing discoveries and production provides the bell-shaped normal curve that denotes reserves. The experimental curve in Figure 1.2 is a history of known U. S. reserves. The similarity between theory and data is cause for apprehension.

If it is conceded, therefore, that the utilization of coal is to complement or supplement and eventually replace oil and natural gas, then in what manner becomes the question.

The most efficient use from an energy standpoint is to combust the coal and utilize the heat directly, a path requiring apparently sophisticated pollution controls which so far can be accommodated in only very large operations such as electrical power generation—which, incidentally, is a more indirect conversion of energy, with attendant efficiency losses.

It gets around to the use of more convenient forms of fuel—forms which are both desirable and necessary—and brings up the chemical conversion of coal to non-polluting liquid and gaseous fuels.

DEFINITIONS

Broadly speaking, conversion of coal involves changes from one form or manifestation of energy to another, whether chemical, mechanical or electrical. For the most part we will be here concerned with chemical changes, notably combustion and the transformation of coal to liquid and gaseous fuels.

More specifically, the chemical conversion of coal delineates those thermo/chemical processes by which the coal molecular structure is broken down and/or rearranged in varying degree. It involves destructive

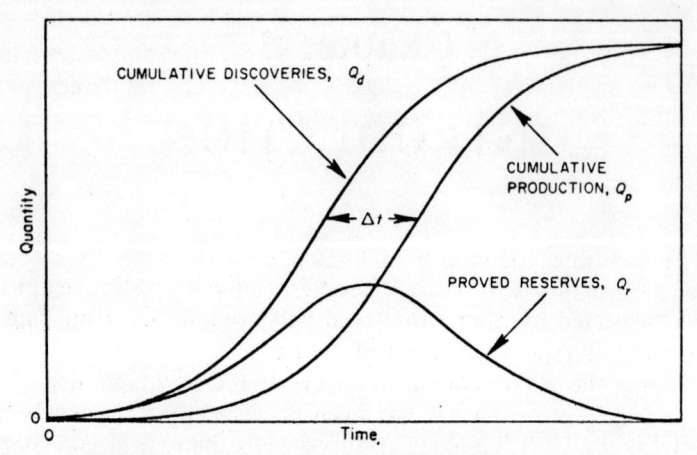

Fig. 1.1. Generalized form of curves of cumulative discoveries, cumulative production, and proved reserves for a petroleum component during a full cycle of production. Δt indicates the time lapse between discovery and production. [From Hubbert (1), with permission of National Academy of Sciences.]

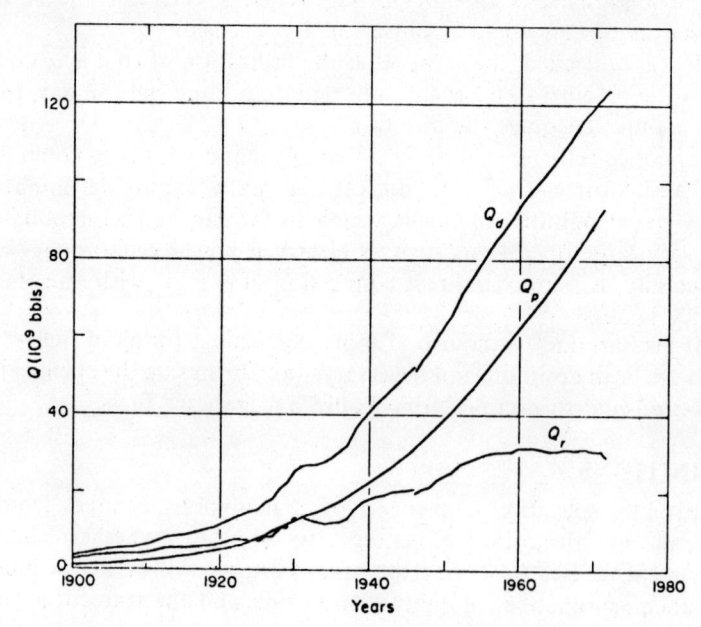

Fig. 1.2. Data for the United States, exclusive of Alaska, on cumulative production (Q_p), cumulative proved discoveries (Q_d), and proved reserves (Q_r) of crude oil. [From Hubbert (1). Also consult "U.S. Energy Resources, a Review as of 1972," Committee on Interior and Insular Affairs, U.S. Senate, June, 1974.]

Introduction

reactions with free of bound oxygen and/or free of bound water (steam), and may be abetted by other organic and inorganic compounds, present or added. Gasification is the more usual embodiment, ranging from pyrolysis through combustion, with heavier product components appearing as condensable liquids.

Combustion is a chemical change of the first magnitude and is used to generate heat energy for direct use or for further conversion to mechanical and/or electrical energy. Conversion to gaseous and liquid fuels is predominately a chemical change from one form of chemical energy to another.

Heat energy, however, is also rejected in chemical conversions, in one way or another—and is evidenced by an efficiency loss in the processing, a matter examined in the Appendix.

By complete combustion there is generally implied the full use of the calorific value of the fuel with carbon dioxide and water (or water vapor) as the principal reaction products. This is as opposed to partial combustion and/or gasification where the reaction products have further calorific value and may be regarded as fuels *per se* or chemical intermediates for further conversion. As will be stressed subsequently from time to time, it is all a matter of degree.

1.1 CHEMICAL CONVERSIONS

For the purposes here the chemical conversion of coal is categorized as follows:

1) Combustion, the reaction with oxygen (or air) to yield carbon dioxide and water as the principal products.

2. Partial combustion, the reaction with oxygen (or air) to yield carbon monoxide and hydrogen as the principal products.

3) Direct hydrogenation, the reaction with hydrogen to produce semi-solid, liquid and gaseous fuel products.

4) Indirect hydrogenation, the reaction with steam to produce carbon monoxide and hydrogen, which may in turn be catalytically reacted to produce hydrocarbons and/or oxygenated compounds (the Fischer-Tropsch synthesis).

To this categorization may be added pyrolysis—a complicated chemical change involving decomposition and further reactions with water or steam, present or formed, plus the distilling effect of the steam. In many ways pyrolysis is the precursor to the other reactions.

Additionally, there is the action of solvents (some of which for instance are hydrogen donors), and reactions that occur with still other compounds, present or introduced.

For the present study, however, the four categories enumerated are the principal conversions covered. A few further introductory remarks are in order.

COMBUSTION AND PARTIAL COMBUSTION

Interestingly, combustion for example—when examined in more detail—requires qualification. Thus with sufficient excess air, carbon dioxide and water tend to be the principal reaction products; however, carbon monoxide and hydrogen may also co-exist, the concentrations depending upon reaction conditions, rates, and temperature levels. With lower air-to-fuel ratios, increasing concentrations of carbon monoxide and hydrogen may appear, gases that are themselves combustibles (producer gas), and the condition of partial combustion is approached.

On the other hand, higher temperatures as encountered in complete combustion favor the existence of CO over CO_2 at equilibrium. And the presence of steam also will favor hydrogen. The matter becomes increasingly circumspect.

The stoichiometric reaction between carbon and steam yields an equimolal mixture of hydrogen and carbon monoxide called water gas or blue gas. By using an excess of steam, particularly with catalytic activity, a conversion of CO occurs by the CO-shift reaction to yield increasing concentrations of H_2. Such gas, of a lower heating value per volume (but not on a weight basis), is used as an intermediate for further reaction, and referred to as synthesis gas. In the limit, pure hydrogen is obtainable, subject to the requirements of equilibrium.

It becomes in final analysis a matter of both rate and equilibrium and the relative concentrations of reactants and products, along with system conditions. System conditions in turn relate to flame temperatures achieved, rate of heat transfer, and other elusive factors. In practice, the combustion process is very complex.

HYDROGENATION

Direct hydrogenation of coal occurs at different conditions of temperature and pressure, with or without the action of a solvent, depen-

Introduction

ding upon the sort of product desired. The reactions are enhanced by the action of catalysts. The severity of hydrogenation determines to what extent solid, liquid, or gaseous products will be produced.

Hydrogenation results in a change in the carbon/hydrogen ratio. This may be accomplished by producing two products: one of a lower C/H ratio, and one of a higher, such as coke. Or primarily one product of a lower C/H ratio may be the objective — i.e., complete upgrading.

Typical atomic C/H ratios for carbonaceous and related materials and derived products are:

Material	C/H Atomic Ratio
Coals	1.08 to 2.65
Colorado coal	1.16
Gilsonite	0.71
Tar sands	0.58
Shale oil	0.55 to 0.67
Crude oil	0.5 to 0.65
Pa. crude	0.51
Texas crude	0.64
Coal tar	0.75
Synthetic fuels	0.5
C_8H_{18} (isooctane)	0.44
C_6H_6 (benzene)	1.00
CH_4 (methane)	0.25

Hence, demonstrably, the conversion of coals to products with lower C/H ratios—that is, higher H/C ratios—involves an upgrading in hydrogen content.

INDIRECT HYDROGENATION

Indirect hydrogenation of coal first involves the conversion of coal and steam to some mixture of CO and H_2 (synthesis gas) and then the reaction of the synthesis gas to form hydrocarbons and oxygenated compounds. The production of the synthesis gas is in itself a major controlling factor in indirect hydrogenation, requiring the use of air or oxygen for oxidation to maintain the endothermic carbon-steam reaction. The nature and proportions of the product yield will depend upon the catalyst, the H_2/CO ratio, and the operating conditions. If liquid hydrocarbons are the predominately important product, the reaction of the synthesis gas is referred to as the Fischer-Tropsch process. The stoichiometric conversion of CO and H_2 to methane is called methanation. A wide variety of reactions and products may be proposed, many of which evidently can occur.

The reactions of CO and H_2 to form hydrocarbons and/or oxygenated compounds, using catalysts of Group VIII transition metals, are highly exothermic. This exothermicity requires sometimes novel means for heat removal and temperature control.

While pyrolysis of coal and the simultaneous cracking and hydrogenation (hydrocracking) involved in direct hydrogenation lead both to fuel products and valuable hydrocarbon chemicals, particularly of ring structures, the greater variety of chemicals valuable *per se* or as intermediates appears to be in the oxygenated compounds obtainable via indirect hydrogenation.

In the light of the philosophy that the most valuable utilization of fossil fuel sources may ultimately prove to be non-fuel uses, it is well to examine the conversion of coal from the standpoint of what will result in, at present, fuel usage, but can result in, at a later date, non-fuel or chemical usage. It stands that if a conversion is economically attractive for the production of a fuel, then a similar process will more than likely possess an advantage for non-fuel purposes.

THE COAL-STEAM REACTION

In order to produce hydrogen for direct hydrogenation of CO-H_2 mixtures for indirect hydrogenation, coal or other carbonaceous materials must be reacted with steam. That is, in simple terms

$$C + H_2O(g) = CO + H_2$$
$$CO + H_2O(g) = CO_2 + H_2$$

The equimolal mixture of CO and H_2 produced via the first reaction is referred to as water gas. As the proportions change due to the effect of the second reaction, the mixture is more generally referred to as synthesis gas—though it may also be used as a fuel gas.

The first reaction—here considered the principal steam-gasification reaction—poses the outstanding difficulty in coal gasification due to the very great endothermicity of the reaction. Heat has to be added directly or indirectly, in one way or another, in order to sustain the reaction. Cyclic air-blowing is the classic method. The continuous addition of oxygen to the reaction system is the more modern approach. More will be said of this later.

The second reaction, called the CO-shift or water-gas shift, is used to adjust the proportions of CO and H_2. It may occur in varying degree during gasification, depending upon conditions and concentrations, or may be conducted in a second stage, usually with a catalyst.

The further reaction of CO and H_2 to yield mixtures of hydrocarbons and/or oxygenated compounds is in turn catalyzed by catalyst mixtures predominant with members of the Group VIII transition elements. Of these, iron and its oxides form the most common, available, and cheapest varieties of catalyst, though nickel is more reactive and more specific.

The foregoing introduces some of the reasons for the premise that catalysis may provide the ultimate key for the conversion of coal or other hydrogen-deficient carbonaceous materials to more convenient and valuable fuels and chemicals, a development that marks the "third generation" of process approaches to coal conversion.

DIRECT CONVERSION

Aside from the specialty chemicals available only, say, by pyrolysis or direct hydrogenation of coal, the direct conversion of coal and steam to hydrocarbons and oxygenated compounds is, theoretically at least, potentially the simplest procedure for the large scale production of both fuels and chemicals.

For reactions involving direct hydrogenation, the hydrogen is first produced by a coal-steam reaction (or from natural gas reforming). For reactions involving indirect hydrogenation, the synthesis gas mixture (CO and H_2) must also first be produced by a coal-steam reaction (or from natural gas). Thus the shortest possible path of reaction would be a one-stage overall conversion of a coal-steam mixture to hydrocarbon and/or oxygenated product. Such a route possesses a double advantage. The coal-steam reaction is endothermic, while the CO-H_2 reactions are exothermic. Apart, they create problems associated with heat generation and removal — together, the heat effects tend to cancel.

While the desirability of such a step may be obvious, it has not been pursued for the simple reason that the first reaction step, the conversion of steam and carbon to CO and H_2, is assumed to be only thermodynamically possible at conditions unfavorable to the further conversion of CO and H_2 to the final products — or vice versa.

Consequently, the overall chemical conversion of coal has always been done in at least two stages. The first step or steps involves the production of H_2 or synthesis gas (CO and H_2) at one set of conditions, and the second step performed apart from the first — involving the reaction of the coal and H_2 or the reaction of the synthesis gas at a new set of conditions.

CATALYSIS

The key to such a successful conversion lies in the role of the catalysts to be considered and their activity. It is to be categorically emphasized that the theory of homogeneous reactions, particularly the laws of both rate and equilibrium, do not necessarily apply to heterogeneous reactions.

A heterogeneous reaction involves two or more phases. Reaction may be between the phases as in the carbon-steam system. If such a reaction is to be described in terms of homogeneous theory, it requires that an activity or fugacity (or the equivalent) be assigned to the carbon — a procedure sometimes frought with ambiguity.

In heterogeneous catalysis, it is considered that the reactions between components of a bulk continuous phase occur at the surface of a second phase which is the catalyst. A description of the process involves diffusion or mass transfer of the reactants to the surface (adsorption), reaction at the surface, and desorption of the products. It is feasible to assign activities to the components at all times, and in this way mathematically describe the reaction in terms of mass transfer rate and equilibrium constants, and surface reaction rate and equilibrium constants. The overall equilibrium will be made up of these several steps. It is possible to describe non-catalyzed heterogeneous reaction in a similar way, e.g., the steam-carbon reaction.

The point to be made is that the concepts of homogeneous reaction rate and equilibrium no longer hold for heterogeneous reactions. There is no correspondence between the two. By the concepts of homogeneous reactions, the role of a catalyst is as an intermediate product which cancels out overall. Hence, the statement that the catalyst has no effect on equilibrium.

On the other hand, in heterogeneous reactions, the overall equilibrium is affected by the coefficients of equilibrium of mass transfer and of the reaction equilibrium coefficient of the surface reaction, a physical situation entirely dissimilar to the homogeneous state. Accordingly, we should suspect that data obtained for homogeneous reactions is inapplicable, or at least suspect, when applied to heterogeneous reactions, particularly catalyzed ones.

Thus, while alkali carbonates enhance the reactivity of the steam-carbon reaction to yield $CO-H_2$ mixtures at lower temperatures (or CO_2-H_2 mixtures via the CO-shift), the further reaction of CO and H_2 is selective to catalysis and may exceed homogeneous equilibrium. This is discussed in later chapters.

1.2 REACTING SYSTEMS

Intimate to the chemical conversion of coal is the type and configuration of the reacting system used. Thus there are batch reactors and flow reactors, and the latter may be further categorized as continuous or semi-continuous. As a further distinction, continuous operations are sometimes on a cyclic basis.

The chemical conversions of coal variously involve reaction of a solid phase (coal) with gases and/or liquids — and may involve a secondary solid phase serving as a catalyst.

Batch operations are for the most part confined to experimental studies. Larger-scale operations, pilot plant or commercial, utilize continuous systems in one form or another. This continuity can be defined in terms of the solid fuel, the gas or liquid phases, or the catalyst, or in terms of all.

As a further distinction, the motion or action of the solid phase or phases is used to denote the type of reaction system or reactor.

For the purposes here the reacting systems are classified as fixed or slowly-moving bed, fluidized bed, and entrained systems. (Fast-dropping beds, for instance, are a variant which will not be considered separately.)

In the fixed or slowly-moving bed, the sized coal (in pieces or "lumps") exists in place and has only a slow lateral or vertical movement as fresh coal is introduced and spent coal (ash or char) withdrawn via a fixed or moving grate. Air or other reactants pass through the grate and through the interstices of the bed. The bed proper may be stirred or otherwise mixed or moved mechanically.

In a fluidized bed, sized particles of the requisite dimensions are suspended in more or less random motion by the upward motion of the fluidizing reactants which are gases (or liquids). Some internal circulation may occur within the bed. A recycle stream classifies the spent coal (ash). Some solids carryover will occur in the gases exiting from the top of the reactor.

In entrained systems, the solid particles are suspended in the gas phase and move at the same bulk linear velocity in the same direction.

The foregoing also applies to catalyst solids as well as solid fuels. Thus a catalyst may be fixed, fluidized or entrained. A special distinction is given to fluidized systems whereby the catalyst is of uniform size (and shape). Such are referred to as "ebullating" beds, as distinguished from "fluidized" beds where a size distribution occurs among the particles. The latter will capture and retain smaller-sized particles that are entrained in the gas phase whereas in the former classification occurs.

FIXED VERSUS SUSPENDED BEDS

Fluidized beds and entrained systems are each a special case of suspended beds: beds in which there is random motion of the particles with respect to one another, in varying degree.

In a fixed bed there is no random motion of the solid particles, only stationary or linear motion. The particles remain in contact with no relative motion to each other. The particles may be without any motion whatsoever as the name implies, or introduced at a point in the reactor and withdrawn at another with linear or one-dimensional movement between points. Relative motion in the same direction is called concurrent, and relative motion in opposite directions is called counter current. Fixed beds are sometimes referred to alternately as slowly-moving beds.

In a suspended bed, there is a random motion of the particles, which are not in continuous contact but may undergo collisions. Suspension is maintained by an interaction of rate, buoyancy, gravity, and elevation. The action is akin to that in classification or elutriation. There may be back-mixing or other effects.

The relative motion of the two phases may be countercurrent or concurrent. Each phase will have its own retention time within the reactor. The respective phases may be introduced at any point, but usually the bottom and/or top are specified.

A fully suspended or entrained system is one in which the retention times of both phases are equal in a concurrent flow arrangement. Flow hypothetically may be either upward or downward, but is usually upward. The ideal is a limiting condition which cannot be completely attained unless the particle densities are the same as the medium. Fully suspended systems are sometimes called transport reactors.

In a fluidized bed the retention time of the solid particles is much greater than that of the gaseous medium. For this situation to exist it is necessary that the gas be introduced at the bottom of the reactor. The solid particles may be introduced at the bottom or top or intermediately, and withdrawn at any specified location. In the limit, the solid particles as charged are not withdrawn. Such is sometimes categorized as a fixed fluidized bed.

Multiphase systems may also occur. For instance, solid, liquid and gaseous phases may co-exist, with varying degrees and kinds of action.

In all cases the solid reactant particles may be used up, converted, or otherwise modified, since this interaction ordinarily is the purpose of the system, unless the particles are a catalyst.

Introduction

1.3 STATE OF THE TECHNOLOGY

Direct combustion remains the most efficient use for coal from an energy standpoint, though not necessarily from a convenience standpoint — the motive for partial combustion and gasification or liquefaction.

With regard to combustion, the fluidized bed is the notable innovation. Capable of high heat transfer rates with smaller system sizes, the characteristic low-temperature operation under pressure offers the potential of low nitrogen oxide emissions and the ready capture of particulates and sulfur. A simplified schematic is indicated in Figure 1.3 with the tubes drawn perpendicular to the plane. Other versions are shown in Chapter 3, along with other types of combustion systems.

Also noteworthy are attempts to combust high-moisture coals and even coal-water slurries.

Other efforts in combustion are more peripheral — e.g., control of stack-gas emissions with add-on equipment. There are also efforts toward pollution control — that is, control and removal of ash particulates and sulfur — during combustion, particularly at the conditions for producing a low-Btu or producer gas.

The many routes for the chemical conversion of coal are diagrammed schematically in Figure 1.4. A part represents known technology; other routes are still under development.

So-called "first generation" technology for producing liquid and gaseous fuels from coal refers to past technologies: simple pyrolysis, the production of producer gas or town gas, fixed-bed hydrogenation under extreme conditions, the low- or medium-pressure production of synthesis gas (CO and H_2) followed by a Fischer-Tropsch conversion to hydrocarbons and/or oxygenated compounds.

In the past the emphasis was on liquid fuels, but methane has now become the hydrocarbon of paramount interest. The principal component of natural gas, methane is also the principal component of synthetic or substitute natural gas (SNG). It is also a feedstock in the petrochemical industry.

The processes of low-, medium- or high-temperature carbonization produce liquid and gaseous products from pyrolysis, plus char or coke solids. The liquids, often referred to as tars and/or oils, are further refinable to motor-grade fuels. Indeed, this was a major source of liquid fuels in Germany during World War II. The coke produced may be burned as a solid fuel, or even gasified with steam. If possessing the requisite properties, the coke also serves as a reducing agent (source of carbon monoxide via partial combustion) in metallurgical practices.

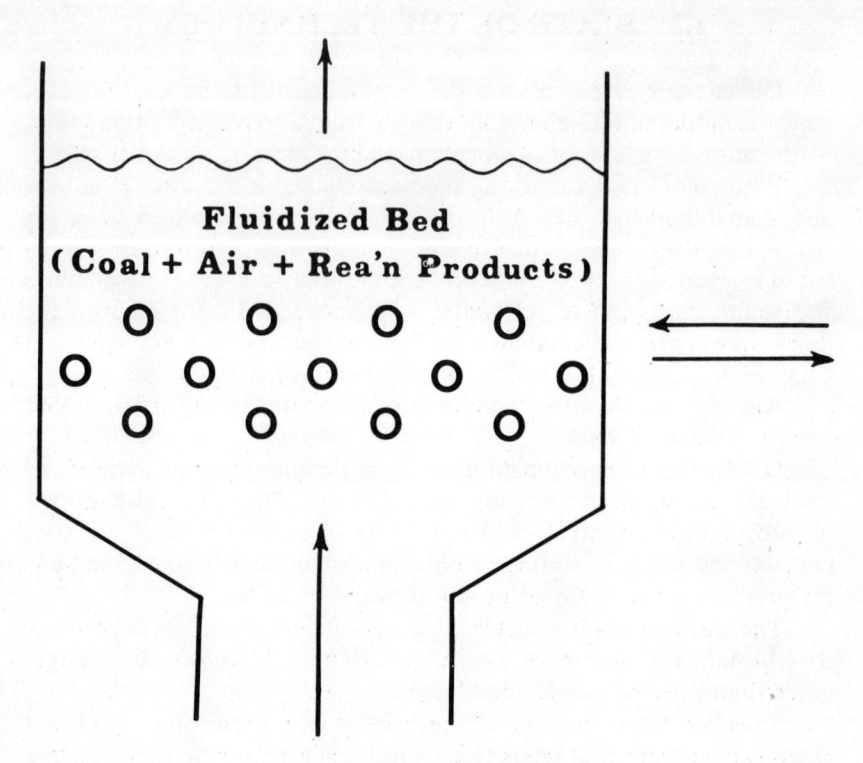

Fig. 1.3. Fluidized combustion.

There is other first generation technology in current use — e.g., many so-called town gas operations yield a producer gas from various types of gasifiers, traditionally fixed or slowly-moving beds operating at atmospheric pressure. The producer gas may be fortified with petroleum distillates to increase the Btu-rating.

A more sophisticated unit is the Lurgi gasifier which uses air or oxygen in a slowly moving (dropping) bed. With oxygen under pressure, a medium-Btu gas is produced which may be used for synthesis. Still other versions of gasifiers consist of fluidized beds or entrained systems (usually at atmospheric pressure) and bear such names as Winkler, Koppers-Totzek, Wellman-Galusha and other company-inspired designators for the gasifier. A few are shown in Chapter 9. With steam and oxygen, these latter gasifiers alternately produce a synthesis gas of H_2 and CO which can be further converted to methane or other hydrocarbons.

Introduction

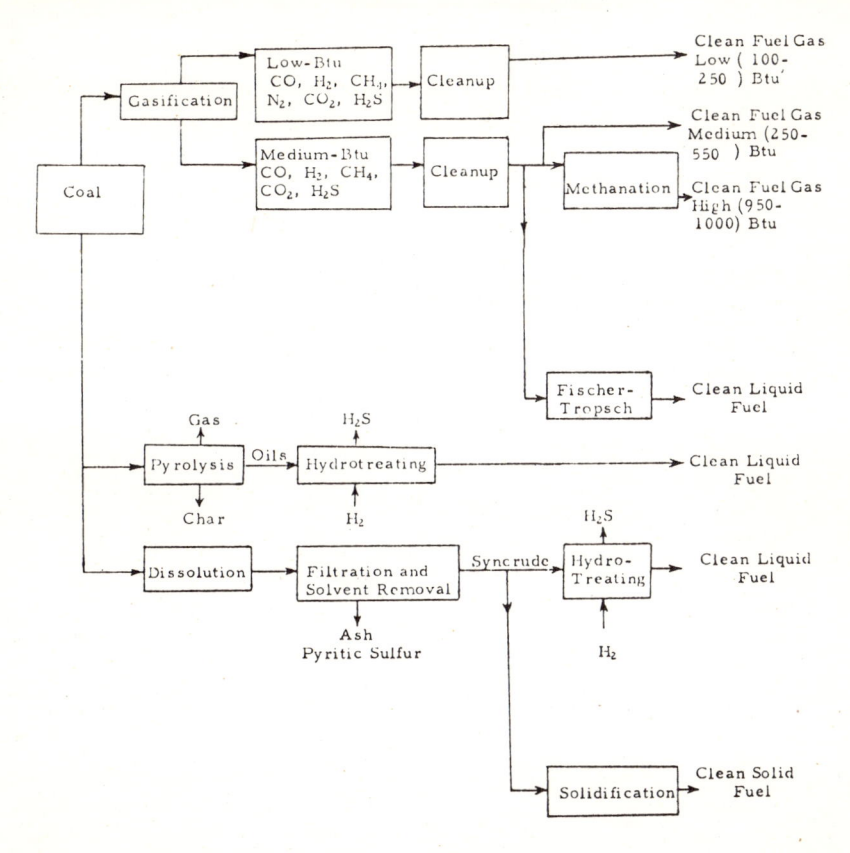

Fig. 1.4. Clean fuel from coal (from W.W. Bodle and K.C. Vyas, "Clean Fuels from Coal — Introduction to Modern Processing," Clean Fuels from Coal, Symposium II, Institute of Gas Technology, Chicago, June 1, 1975.

One of a kind in existence, the SASOL plant in South Africa produces a synthesis gas via Lurgi gasifiers operating with oxygen under pressure. This synthesis gas is further converted to both gasoline fractions and to chemicals by means of Fischer-Tropsch conversions. The former uses a fluidized iron catalyst, the latter a fixed-bed iron catalyst. The technology was originally developed and utilized in Germany prior to and during World War II as a supplement to pyrolysis.

Second generation work, which for the most part has reached or is in the pilot plant and demonstration stages, has been aimed at technical improvements and modifications based on first generation technology, particularly with regard to reactor types and reactor conditions.

A partial listing of second generation technology for producing liquids or high-Btu gas is as follows:

Pyrolysis
COED
Garrett or Occidental
TOSCOAL
Coalcon (hydrocarbonization)
Lurgi-Ruhrgas (hydrocarbonization)

Solvent Action
Consol
Solvent Refined Coal
Solvent Extraction-UOP

Direct Hydrogenation
Hygas (also includes indirect hydrogenation)
H-Coal
Synthoil
Exxon Liquefaction
Hydrane

Indirect Hydrogenation
Oxygen Gasifier:
    Lurgi
    Winkler
    Koppers-Tetzek
    Bi-Gas
    Synthane
    Atgas

Indirect Heat Transfer
    CO_2-Acceptor
    Kellogg Molten Salt
    COGAS
    Agglomerating Ash
    Exxon Gasification

Beyond the above-mentioned, work continues on more economical means for producing hydrogen, such as application of the steam-iron

Introduction   15

process. There is also work on the electrothermal gasification of coal and steam to produce hydrogen and to produce a synthesis gas.

Work also continues on improvements and new concepts for air-blown systems to produce low-Btu gas.

Third generation efforts for the most part may be regarded as catalytic in nature, or else involve still more novel aspects. Additionally, work is being done toward finding uses and applications of the advances of space age technology such as magnetohydrodynamics (MHD) and plasma physics toward coal conversion and utilization.

The enumeration grows almost too unwieldy to keep track of, but is made easier by common features and principles characteristic to each and all. As indicated in the Preface, descriptions of research and development are carried by ERDA (now the Department of Energy) reports, and process descriptions are provided in the handbook of Cameron Engineers and in the book by Howard-Smith and Werner, not to mention the numerous symposia, proceedings, meetings and papers appearing almost continuously.

Moreover, the work is highly partisan and sometimes charged with innuendo, possibly as the partial result of overzealousness. A difficult enough undertaking in the first place, the politics of research and development do not help, and the implementation still remains to be seen.

Rather than focusing on the myriad differences in detail, we will be more concerned with the principles and fundamentals behind the processes. For the record, however, a Supplement of schematic diagrams is appended at the end of this chapter displaying a number of the processes which are under development.

1.4   TRENDS IN RESEARCH AND DEVELOPMENT

An encapsulation of some of the research and development underway is as follows.

The ideal complete gasification process is a single-stage continuous operation. In autothermic operation, O_2 or air supplies heat by oxidation. The process fluid may be used to supply heat (or remove heat as in hot gas recycle). Or externally or indirectly heated vessels can be used.

Interest remains in supplying heat for gasification indirectly through the reactor walls and in other methods to eliminate the use of oxygen. In moving-burden processes, heated particles are transferred to reaction vessels to provide the exothermic heat. Among the media considered are sand, ceramics, and even metal balls containing calcium and sodium chlorides. The molten salts supply the heat of fusion.

The use of nuclear heat to supply the endothermic heat of gasification is under study. Helium is a proposed medium of transfer.

The operation of gasifiers in the slagging region lowers the steam requirement for temperature control. The direct production of CO-rich gases is favored at the higher temperatures of slagging conditions.

Molten slag itself has been used as a heat transfer medium to supply the endothermic heat of gasification, though hot ash may prove a more satisfactory means.

In the area of coal hydrogenation, objectives are cheaper H_2, better catalysts, less severe operating conditions, and simpler operation. Some of these requirements have been met by catalyst impregnation, and reaction for longer residence times at lower pressure with higher catalyst concentrations.

The use of a moving bed of iron oxide to generate H_2 continues a possibility. The iron oxide would be reduced by producer gas in one vessel and oxidized by steam in another to yield H_2. In another version, pulverized coal and iron ore are heated at 1550-1650°F to yield pure H_2.

In hydrogenation in the liquid phase, catalyst-coal-vehicle oil slurries may be preferable over fixed bed catalysts. A different embodiment is in the H-Oil or H-Coal process, which uses a suspended ebullating bed of catalyst.

An objective in high-temperature coal hydrogenation has been the development of a one-step process for the direct conversion of coal to gas or liquids. This would combine liquid- and vapor-phase hydrogenation at temperatures over 950°F. Direct hydrogasification of coal in fluidized beds will largely accomplish this objective for conversion to gas, while the H-Coal ebullating bed is to accomplish this objective for conversion to gasoline.

There is an emphasis toward dry coal hydrogenation in fluidized beds. In laboratory and process development studies, dry hydrogenation in the autoclave has been demonstrated feasible without the use of a carrier or vehicle oil. Embodiments include the IGT Hygas process and the Bureau of Mines Hydrane process.

Hydrocarbonization applies fluidized techniques to hydrogenate dry coal so as to produce oil or gas and a dry char, which may be further used to produce synthesis gas, H_2, or power. It becomes similar to hydrogasification if a high-Btu gas is produced. Agglomeration or coking occurs unless noncoking coals are used.

Introduction

In gas synthesis reactions, improvement over conventional fixed beds lies in such systems as fluidized beds, oil circulation, catalyst slurries, and hot-gas recycle.

Gasification of coal under pressure increases the methane content of the product gases. Thus the oxygen version of the Lurgi gasifier operates at ∼ 350 psi, the Hygas gasifier at ∼ 1000 psi or higher, the Bigas gasifier at ∼ 1000 psi, and the Synthane gasifier at ∼ 1000 psi. This has the additional advantage in that the final product (methane) is generally desired at pipeline transmission pressures (∼ 1000 psi). The disadvantage lies in the complexity of processing and in the reactor vessel requirements.

Gasification with steam-hydrogen mixtures is synergistic, since the presence of steam generates more hydrogen and provides temperature control — permitting perhaps up to a 70% reduction in hydrogen requirements for hydrogasification. The IGT Hygas process uses this approach, plus further gasification with steam-oxygen mixtures.

The gasification of coal to produce a synthesis gas of carbon monoxide and hydrogen ultimately requires a methanation step to produce methane as the final product, though some methane is formed during pressure operations as previously noted.

The methanation step is highly exothermic and employs a nickel catalyst. Even with temperature control of the gaseous phase, the catalyst surface may reach inordinately high temperatures, damaging the catalyst. Thus there is the problem of maintaining temperature control for the catalyst itself as well as the gaseous reacting system. Additionally, the nickel catalyst is sensitive to sulfur poisoning during methanation, and pre-removal of the sulfur (which will exist as H_2S in the synthesis gas) is necessary.

Gas recycle with cooling is one such means of temperature control. (For liquids production via hydrogenation, an oil recycle can be used.) Alternately, the reactor may be constructed as a shell-and-tube heat exchanger, with heat transfer preferably to a boiling liquid. The catalyst pellets are contained in the tubes. The Bureau of Mines has worked with Raney nickel catalyst sprayed on the inside tube surfaces.

A cooperative investigation conducted at a Lurgi gasifier installation in Scotland has demonstrated techniques for temperature control during gas-phase methanation, apparently successfully.

An alternate possibility is liquid-phase methanation in which the catalyst is suspended in a recycle oil. This is the basis for a process under investigation by Chem Systems Inc.

Trends in catalyst research have been directed toward finding more active and economical catalysts, in particular catalysts of nickel and iron.

With regard to the latter are included such forms as treated steel turnings and magnetite. Better and cheaper fixed-bed catalyst systems have been provided by the use of lathe turnings and catalyst sprayed on walls or plates. Other materials, tried chiefly for fluidized beds, are listed elsewhere.

The use of fluidized beds for gasification reactions has been studied considerably, including for methanation. Studies have been undertaken in England and Australia as well as in the United States. The Winkler process, however, remains the only commercial process for the fluidized gasification of coal, though others are under development and testing. The Winkler gasifier emphasizes atmospheric air or oxygen, while other versions under development (Bi-gas and Synthane) will use oxygen under pressure in order to enhance methane formation.

Agglomeration problems in fluidized beds of coal can evidently be partially solved, at least. Studies have shown that, by heating coal rapidly through its plastic range, for instance, agglomeration can be reduced. Preliminary oxidation also destroys coking properties.

Falling particle techniques have been tried for the reaction of coal and steam. Here, coal particles and steam flow downward through the reactor in concurrent fashion.

Fluidized feeders have been developed for feeding powdered coal to a fluidized reactor. An inert gas keeps the coal fluidized in the feeder and the gas-coal suspension is fed by pressure differential to the reactor. Feeding solids into pressurized systems still remains a problem, however, and lock-hopper arrangements remain in evidence.

Steam distillation has been advanced for the *in situ* recovery of coal. In addition to inducing porosity and permeability, a liquid product along with volatiles and hydrogen would be produced. High-temperature natural gas injection is also a proposed method for *in situ* recovery, both of coal and of oil shale.

In experiments using coal in high-temperature plasma research, coal was heated in an inert plasma, yielding as products a very fine residue, CO, C, acetylene, hydrogen, and nitrogen. The products varied with conditions and rate of quenching. An electric arc process produces low-, medium-, or high-Btu gas, depending on reaction and quenching conditions.

ENERGY CONVERSION

Inasmuch as coal conversion as defined is only a special case of energy conversion, the developing new technologies in this general area are also worthy of mention. For the most part, direct conversion methods

Introduction

are of chief interest, offering the potential at least of increased efficiency in the generation of electrical power.

Some of the more promising means of energy conversion, as distinguished from coal conversion, are as follows:

Thermoelectric:

Thermoelectric generators convert heat into electricity, or the reverse. The phenomenon is based upon the thermoelectric or Seebeck effect of the thermocouple. The use of semiconductors has revived an interest.

This method appears destined for small scale application to heating or cooling, rather than power generation. Its application would thus be one of power consumption.

Thermionic:

In thermionic conversion a heated cathode gives off electrons, thus providing direct current from heat. This method is adaptable to bulk power generation, and could be used with a nuclear reactor.

ELECTRIC FIELD

DIRECTION of FLOW
(Ionized Combustion Gases)

MAGNETIC FIELD

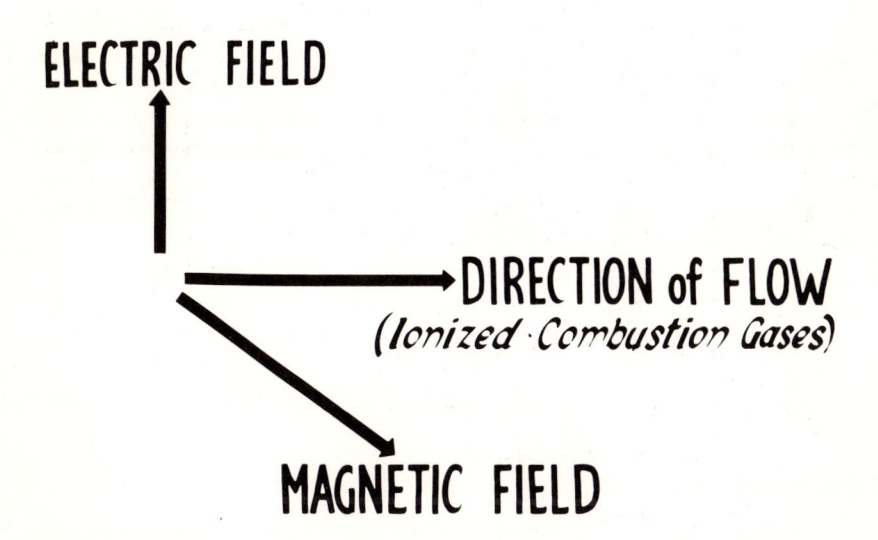

Fig. 1.5. Principle of MHD. The directions of flow and force are orthogonal (the Lorentz effect).

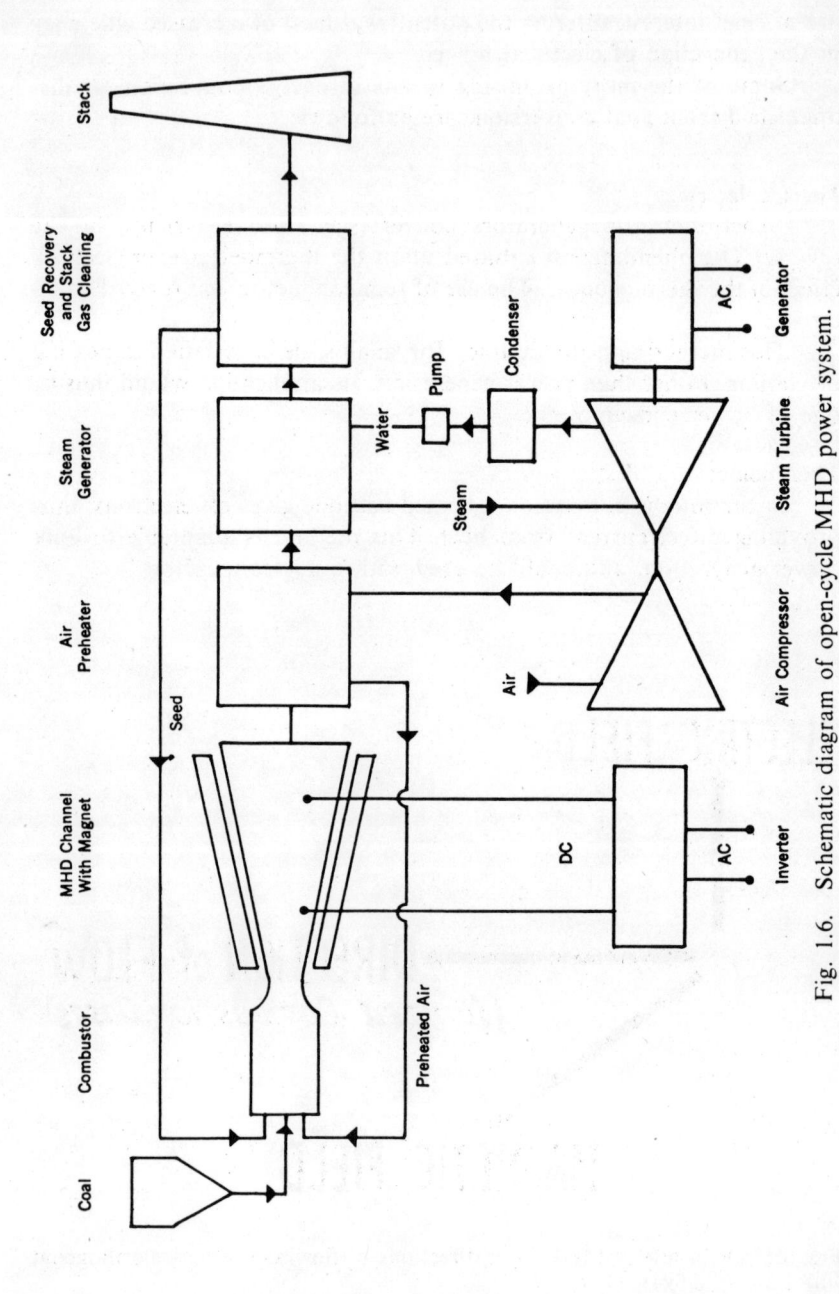

Fig. 1.6. Schematic diagram of open-cycle MHD power system.

Introduction

MHD:

Magnetohydrodynamic generators pass a conducting fluid, usually an ionized gas (plasma), at right angles to a magnetic field. A current is produced normal to the plane and chemical changes take place in the fluid. The principle and applications are diagrammed in Figures 1.5 and 1.6. Seed materials (alkalies) enhance conductivity.

There are differences in opinion as to future application to power generation. The recovery of compounds generated in the plasma may be an important use. The recovery of seed materials is a necessity.

For maximizing efficiency, it is required that the MHD generator serve as a "topping" unit followed by a conventional steam-power cycle. This is the common case for other high-temperature devices.

Fuel Cells:

In the presence of an electrolyte a chemical reaction may be made to proceed at anodic and cathodic surfaces, generating an electrical current in the process. Hydrogen-oxygen cells are the most advanced, due to efforts in aerospace. Cells are classed by operating temperatures, the electrolytes varying from dilute caustics and acids at 300°F, through concentrated acids and caustic pastes, and molten carbonates, to ceramic oxides at 1500°F.

Regenerative fuel cells would be of interest for off-peak power. For central power, the fuel cell evidently could only approach the efficiency of conventional methods.

The use of fuels such as methane first requires reforming to hydrogen, though there is evidence that some fuels — e.g., methanol — can be used directly at the electrode surface.

Electrodes are another problem, platinum being functionally the most satisfactory. Zirconia and other semiconductors are of interest, however — and are much less costly.

A cell is diagrammed in Figure 1.7. The subject, incidentally, is related to catalysis (2, 3).

D.C. to A.C. Conversion:

Advanced methods of energy conversion are mainly d.c. generators. Moreover, the efficiency of extra-high voltage d.c. transmission renders the form attractive and is under development; it must follow that d.c.-a.c. conversion will not be a problem when the time comes.

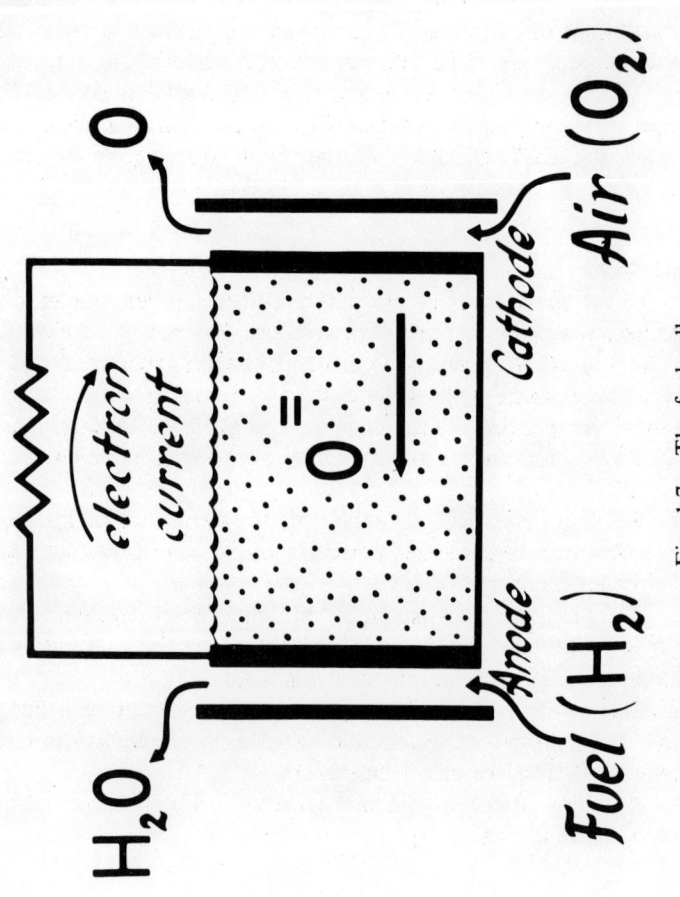

Fig. 1.7. The fuel cell.

REFERENCES

1. Hubbert, M. K., "Energy Resources," in *Resources and Man,* National Academy of Sciences - National Research Council, W. H. Freeman, San Francisco (1969).
2. Hoffman, E. J., *The Concept of Energy: An Inquiry into Origins and Applications,* Ann Arbor Science Publishers, Ann Arbor, Michigan (1977).
3. Vol'Kenshtein, F. F., *The Electronic Theory of Catalysis on Semiconductors,* Pergamon, New York (1963).

Supplement to Chapter 1
MISCELLANEOUS COAL LIQUEFACTION AND GASIFICATION PROCESSES

In a number of cases the schematic flow diagrams are composites and represent a stage in the development of the process.

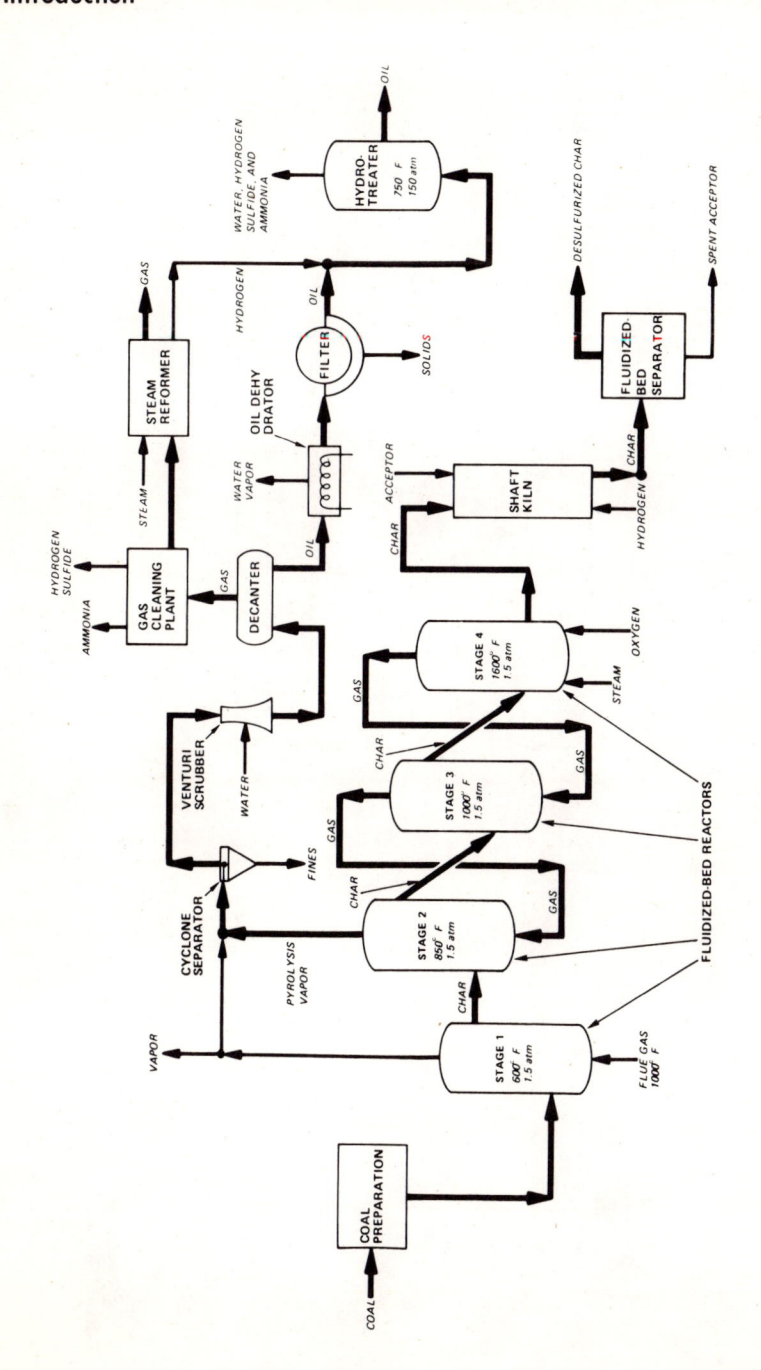

Fig. 1.8. COED process (ERDA).

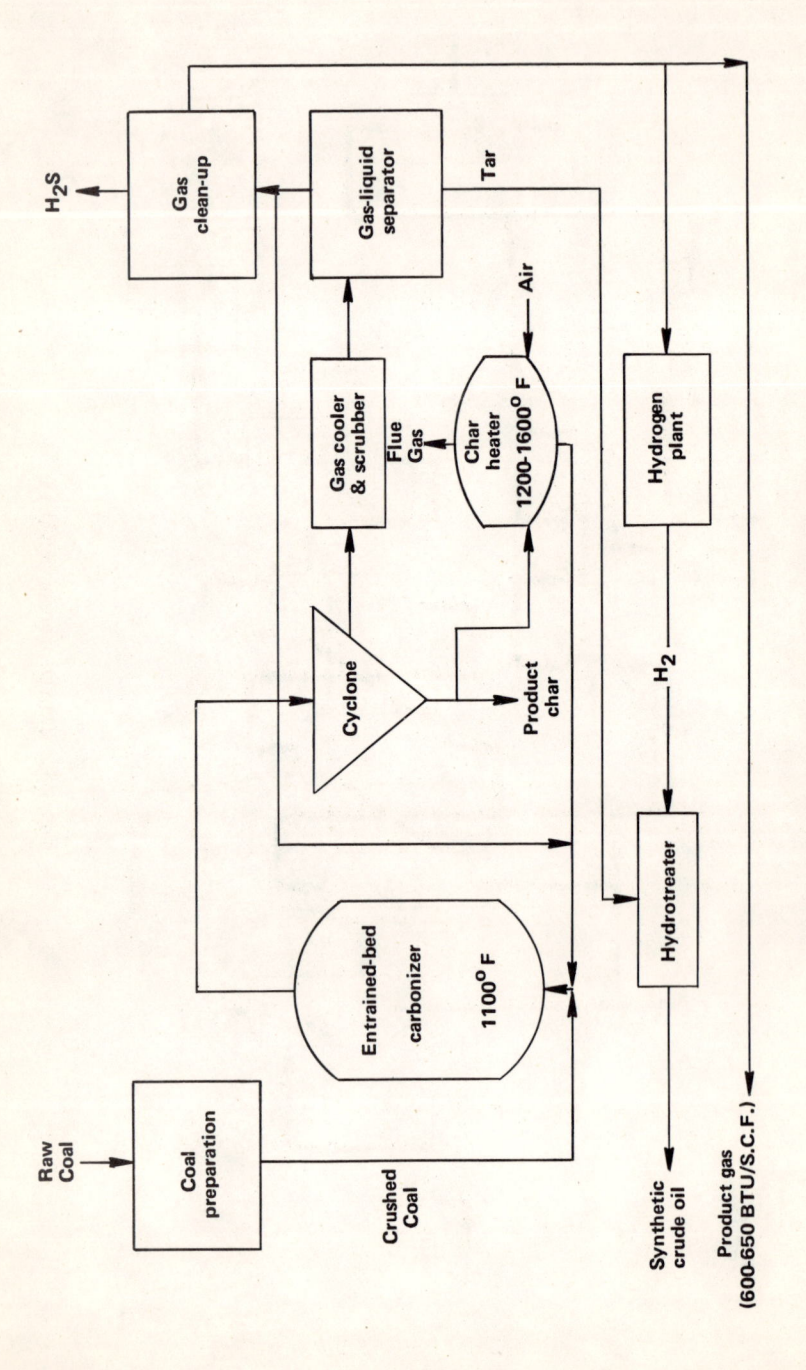

Fig. 1.9. Garrett or Occidental process.

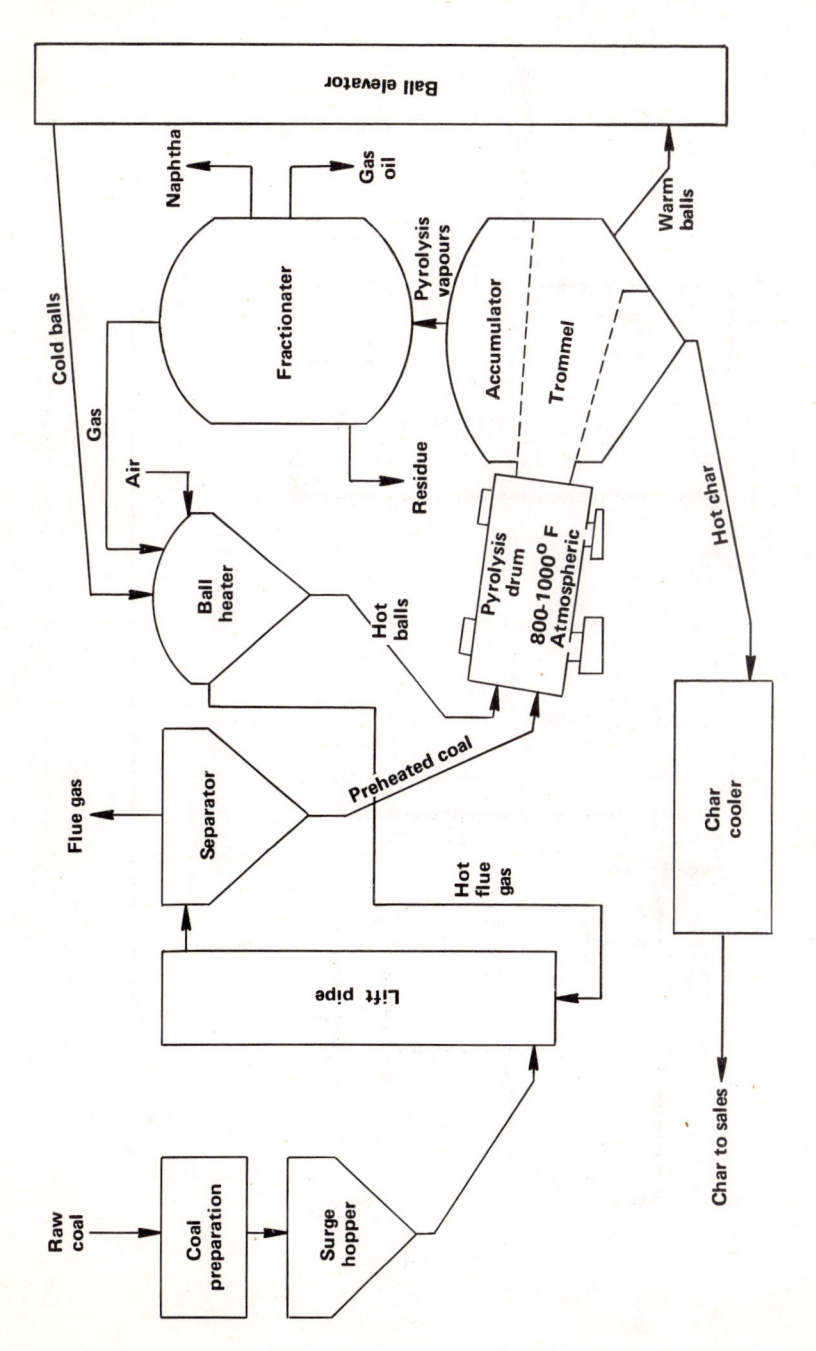

Fig. 1.10. TOSCOAL process.

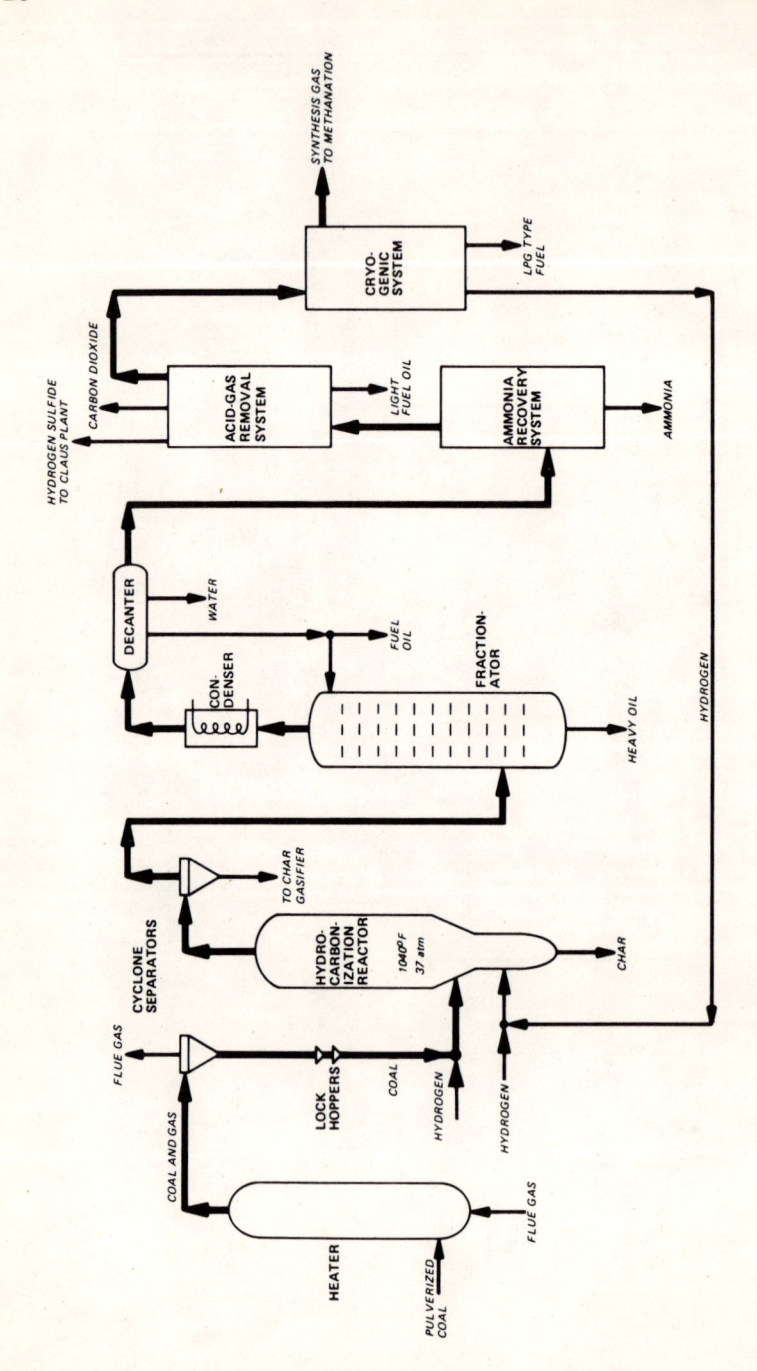

Fig. 1.11. Coalcon Hydrocarbonization process (ERDA).

Introduction

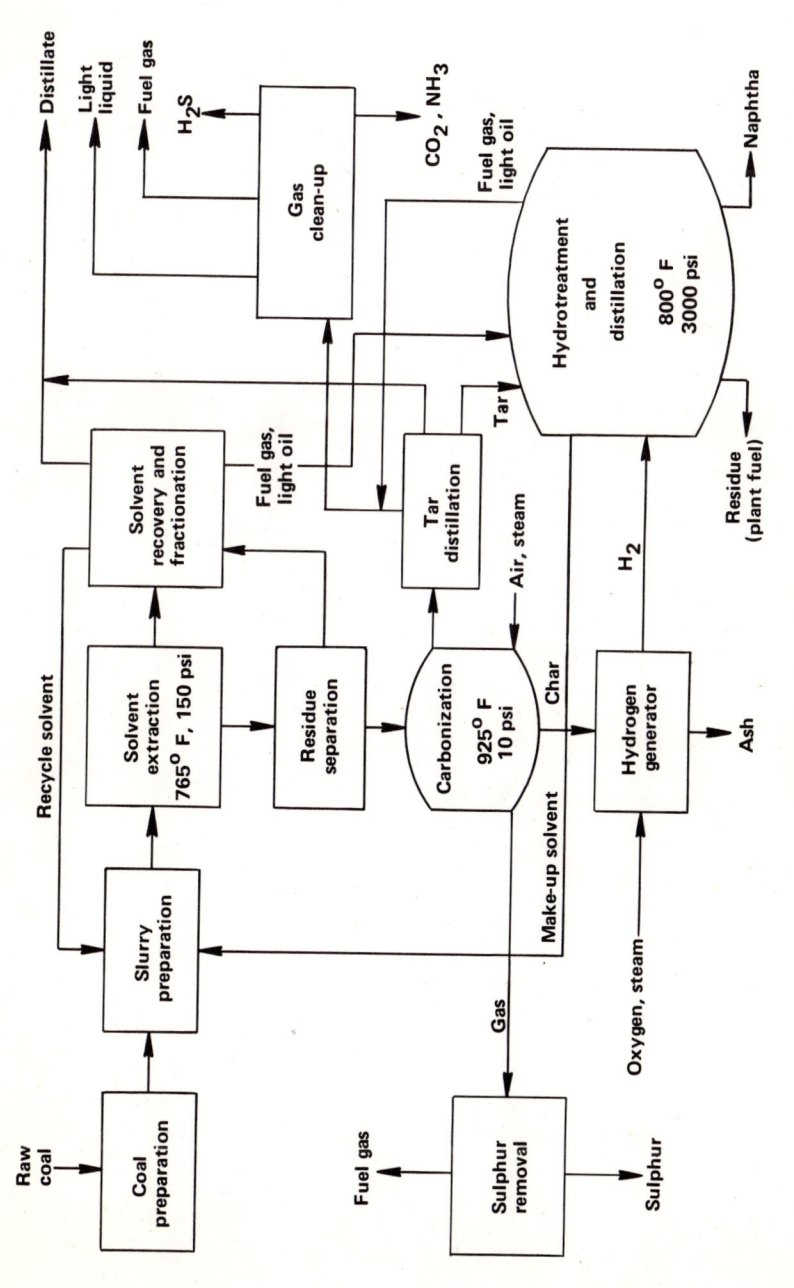

Fig. 1.12. Consol process.

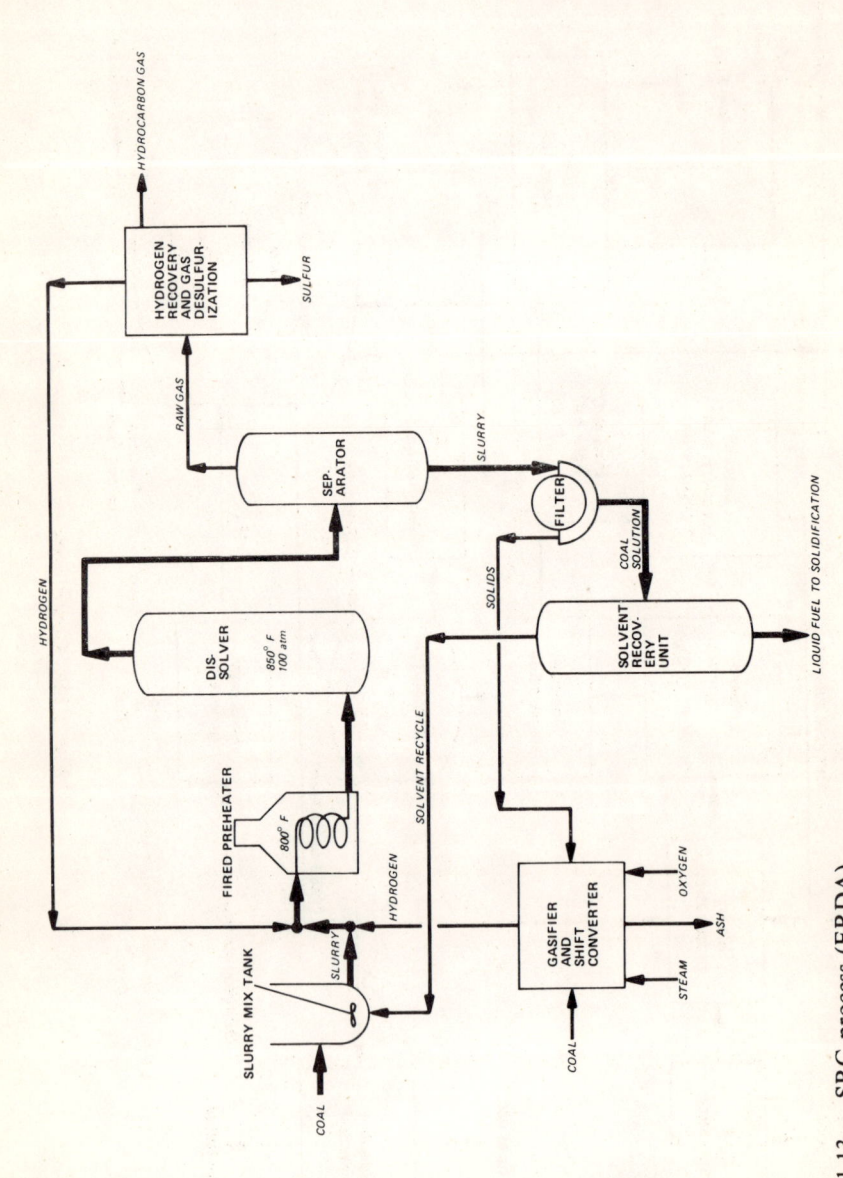

Fig. 1.13. SRC process (ERDA).

Introduction

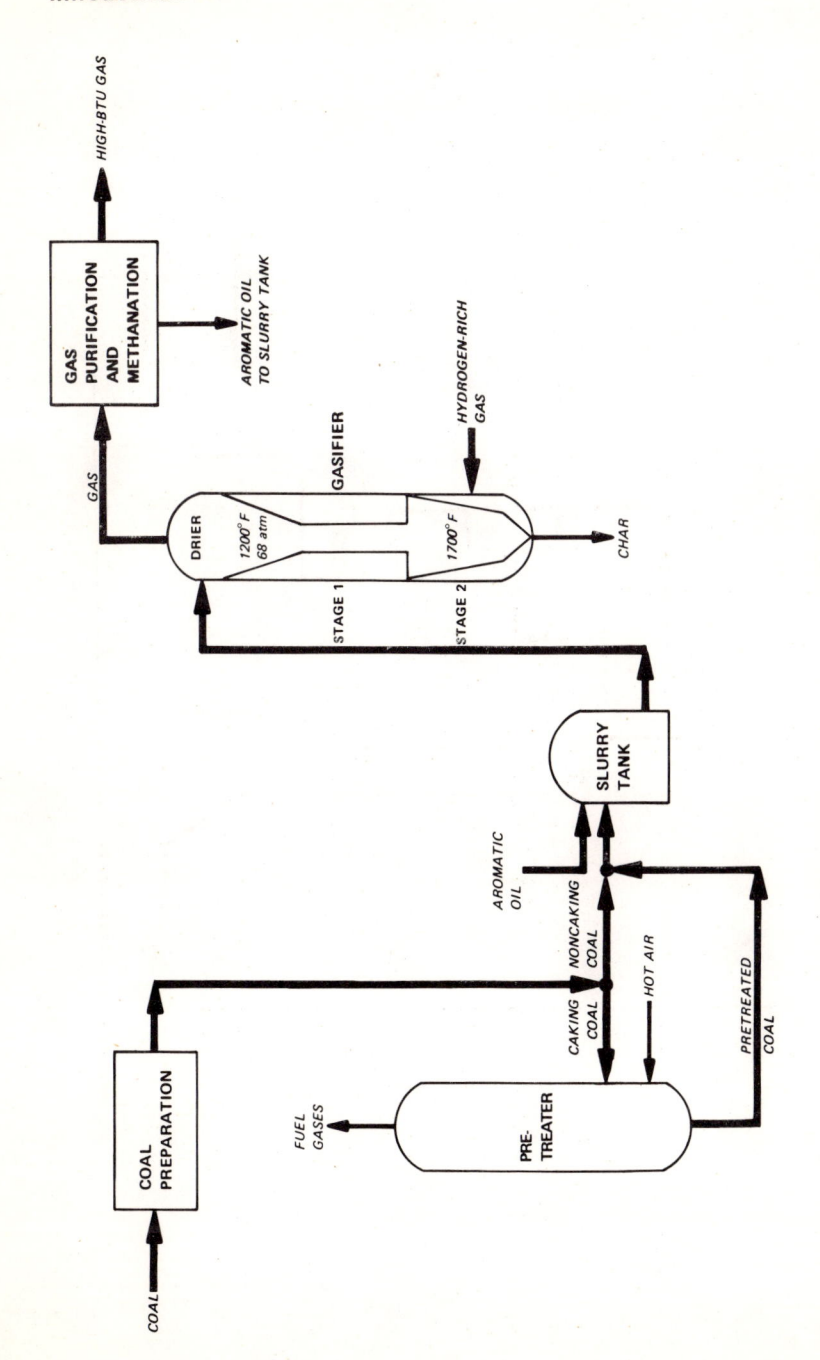

Fig. 1.14. Hygas process (ERDA).

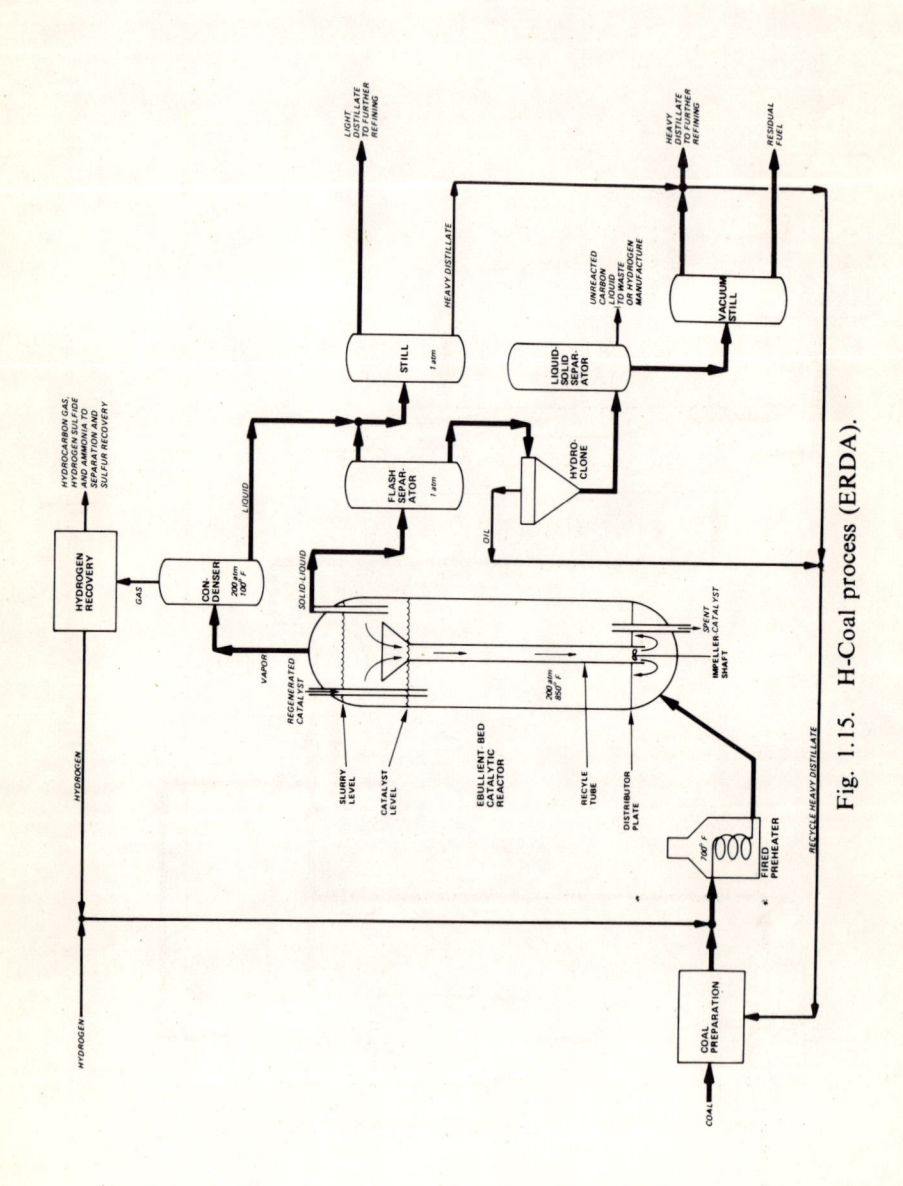

Fig. 1.15. H-Coal process (ERDA).

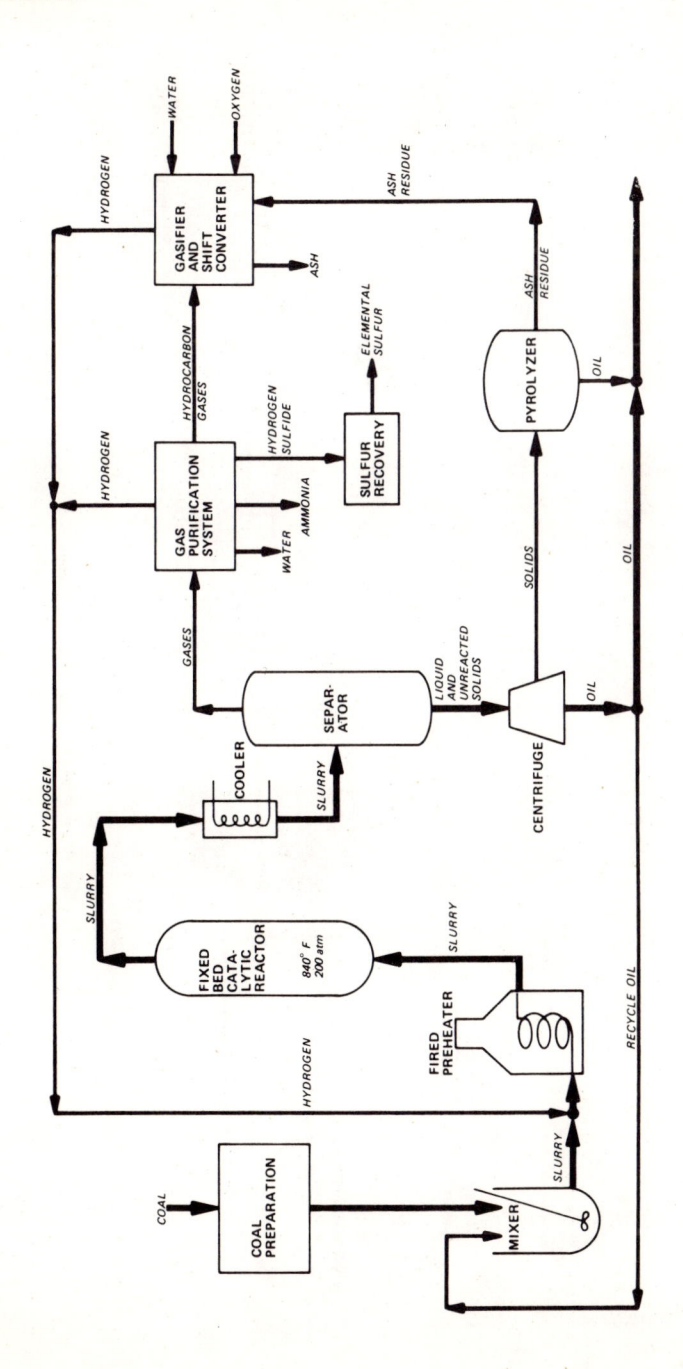

Fig. 1.16. Synthoil process (ERDA).

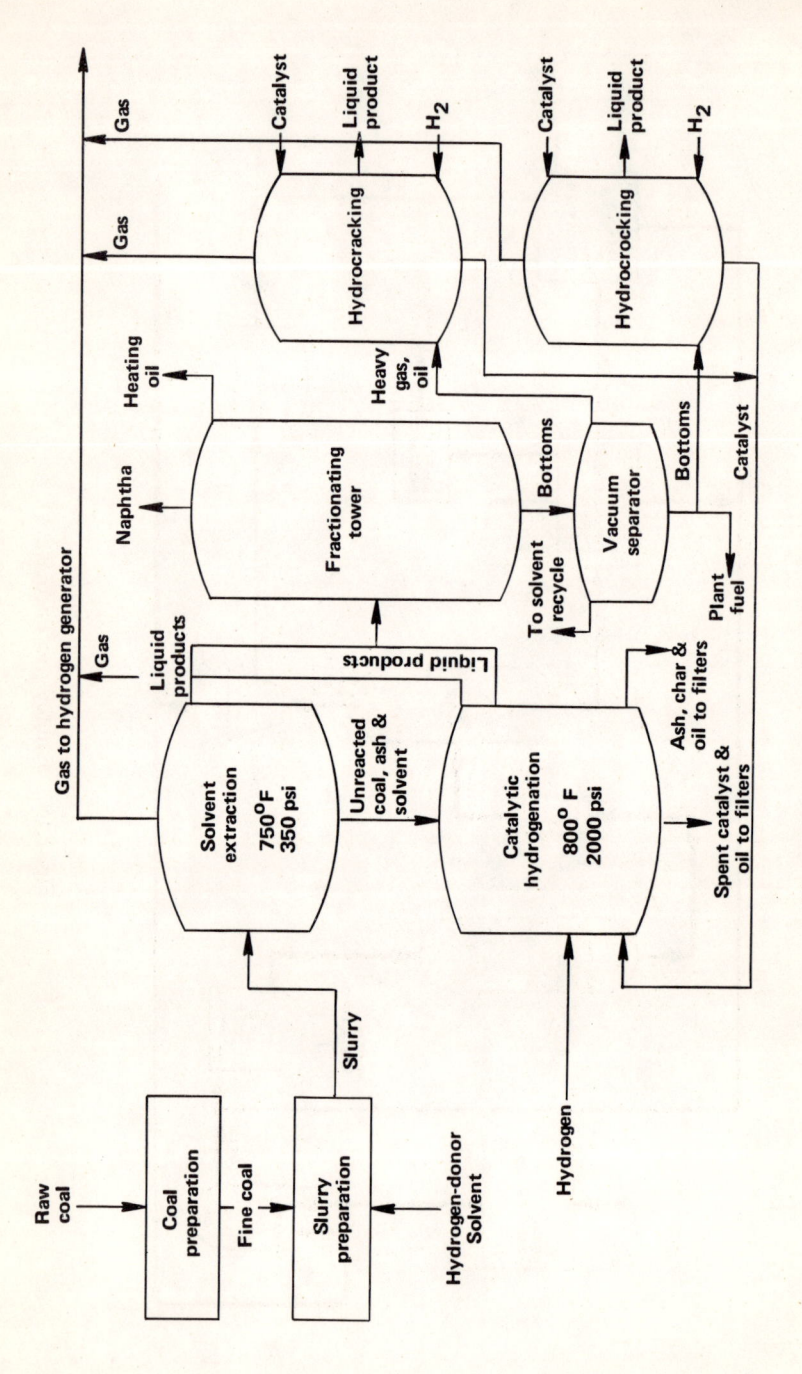

Fig. 1.17. Exxon Liquefaction process.

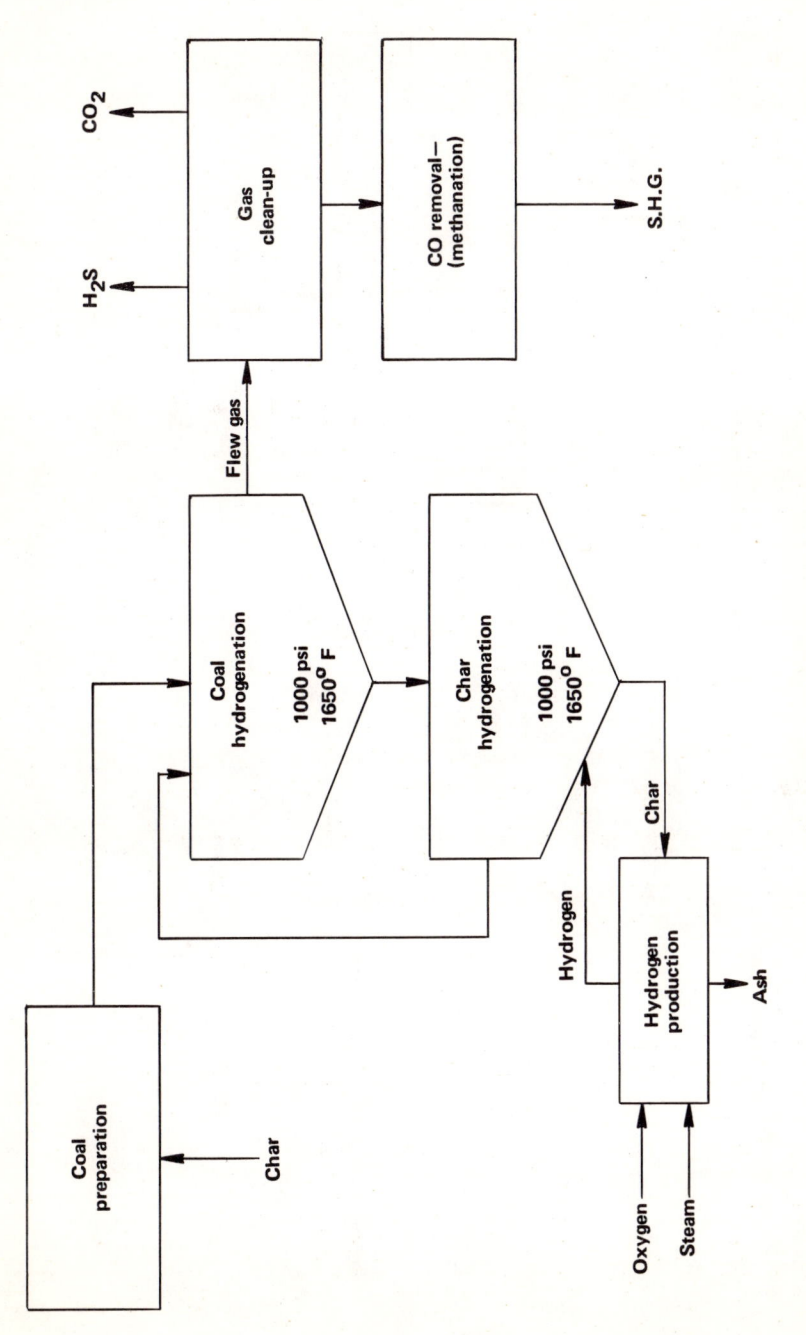

Fig. 1.18. Hydrane process.

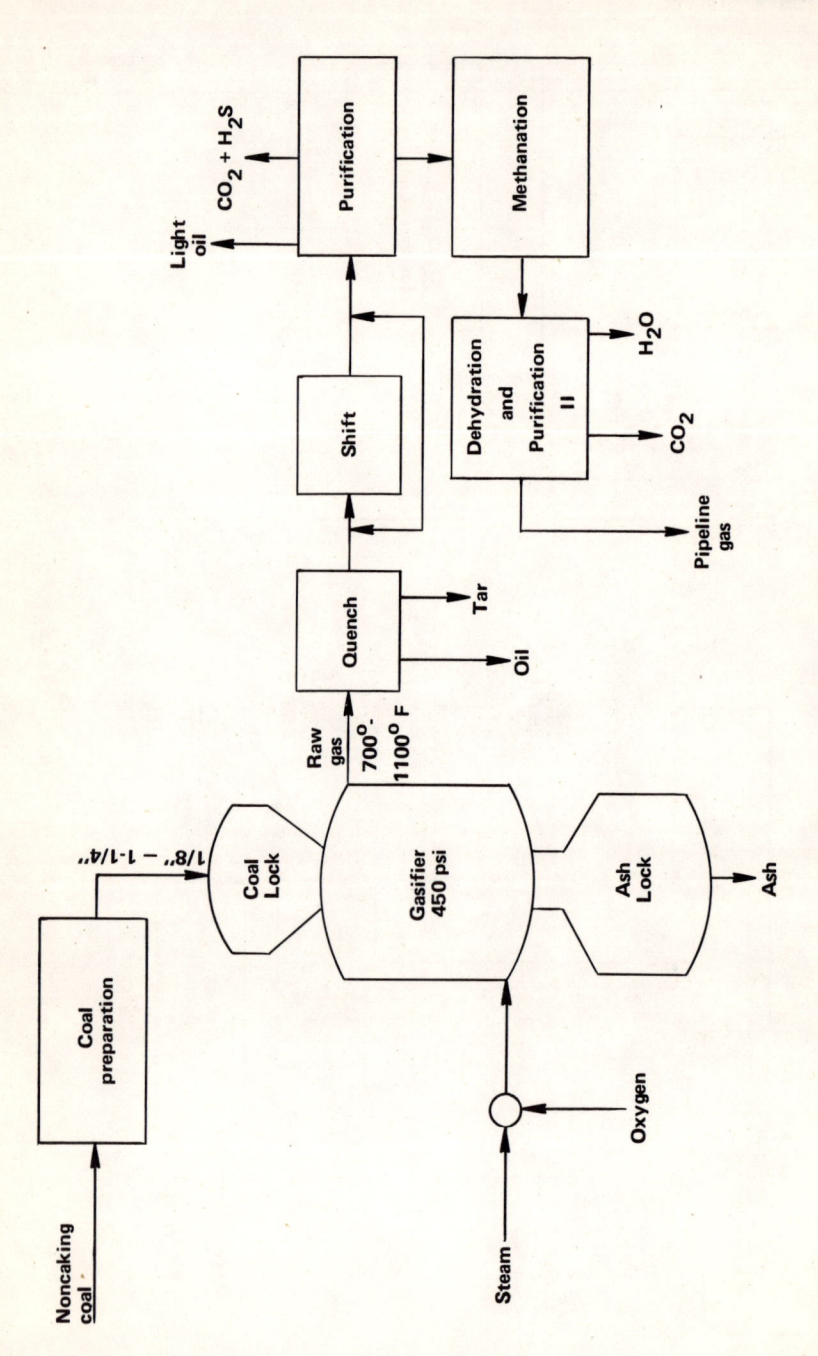

Fig. 1.19. Lurgi process.

Introduction

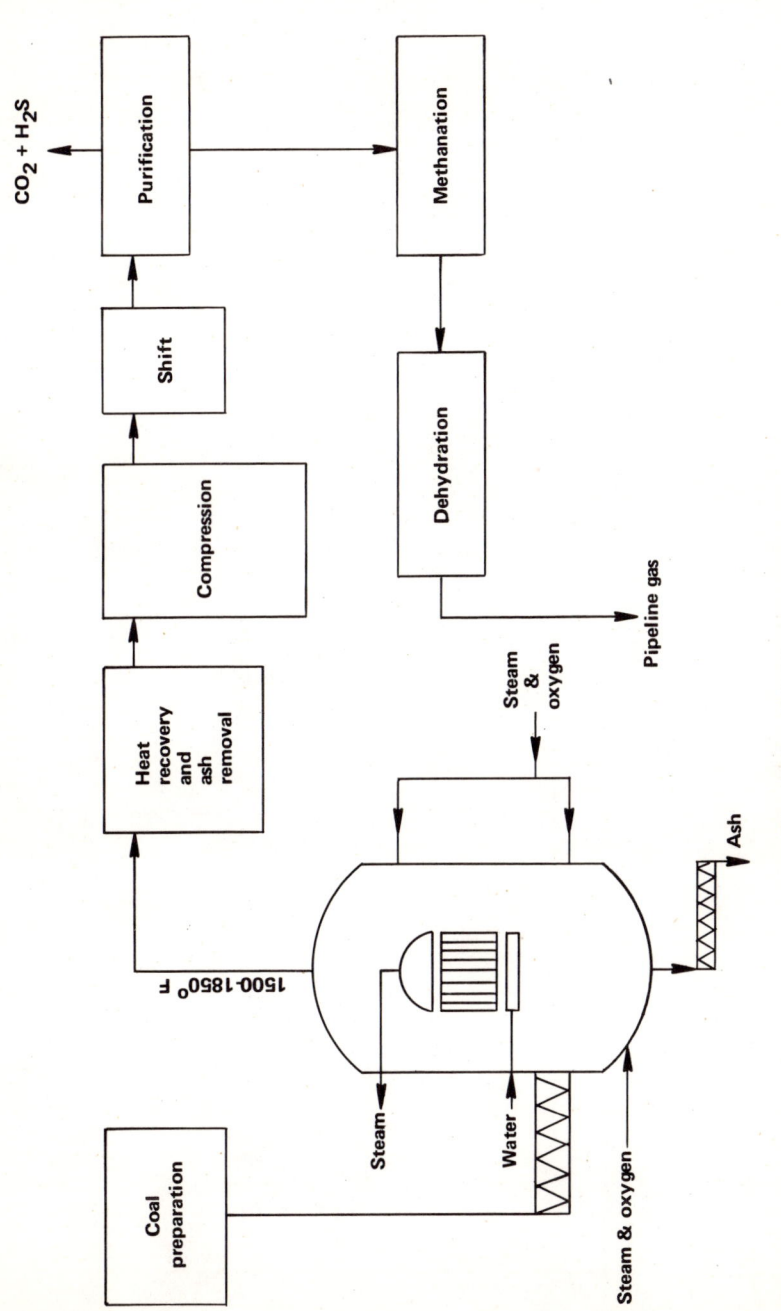

Fig. 1.20. Winkler process.

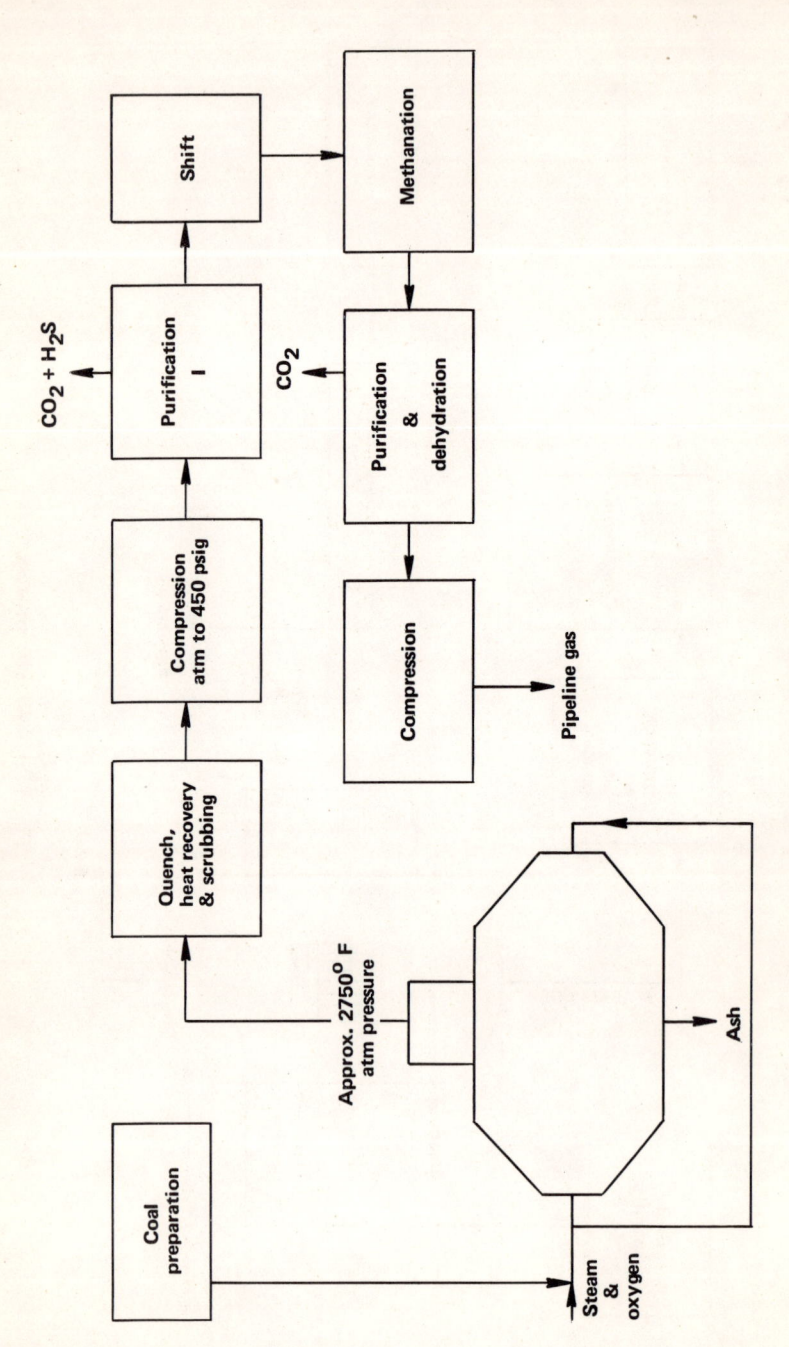

Fig. 1.21. Koppers-Totzek process.

Introduction 39

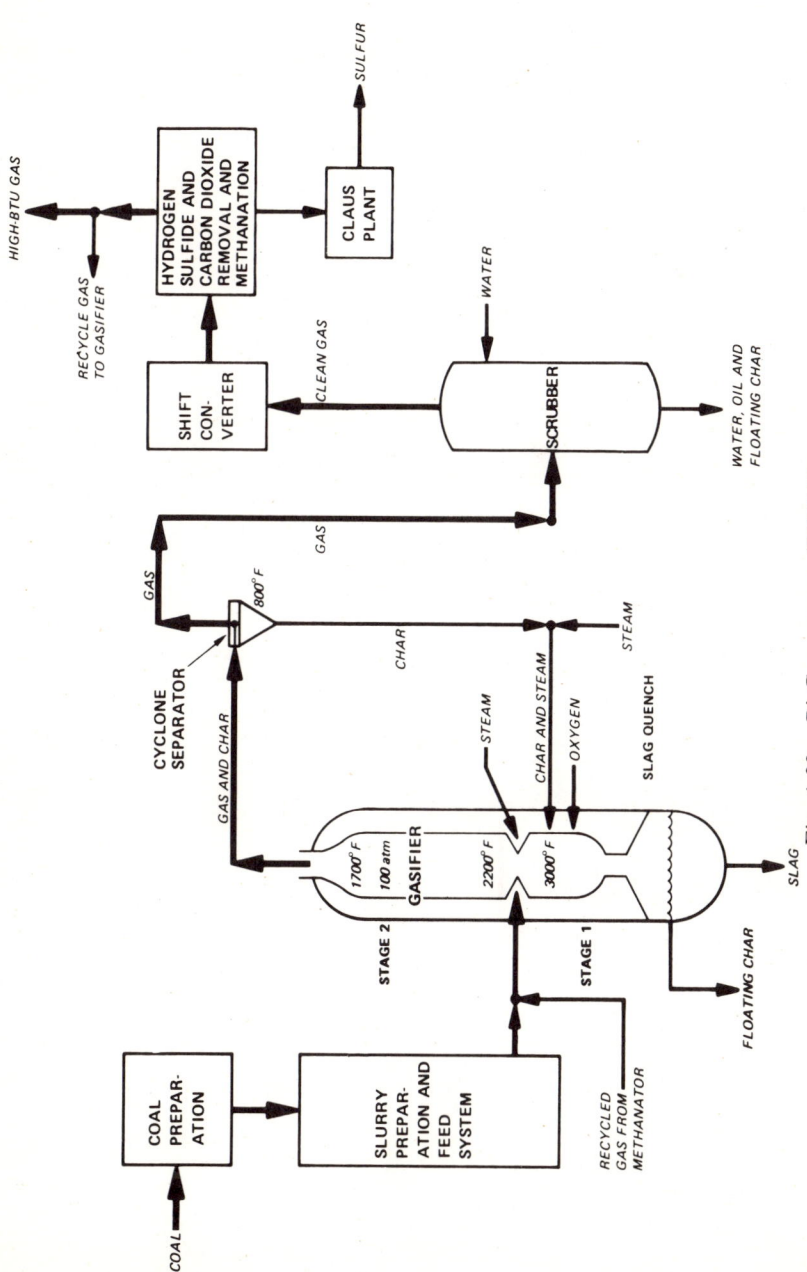

Fig. 1.22. Bi-Gas process (ERDA).

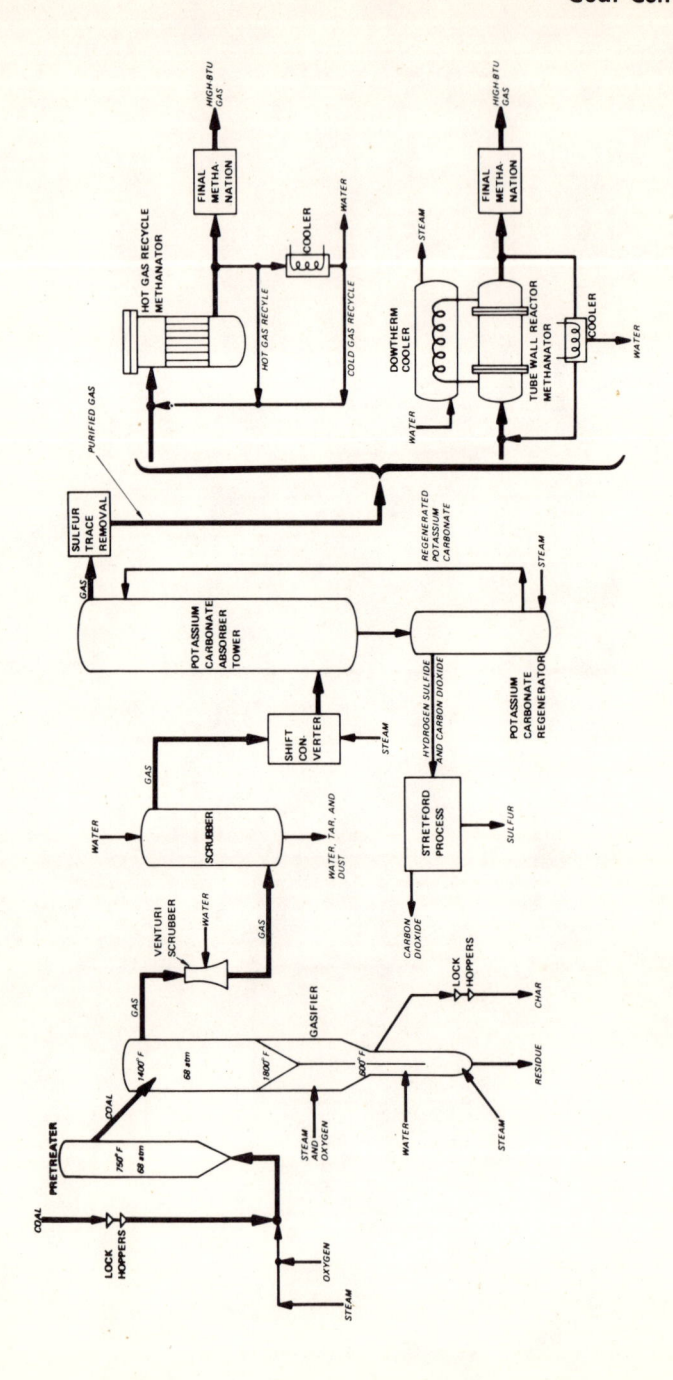

Fig. 1.23. Synthane process (ERDA).

Introduction

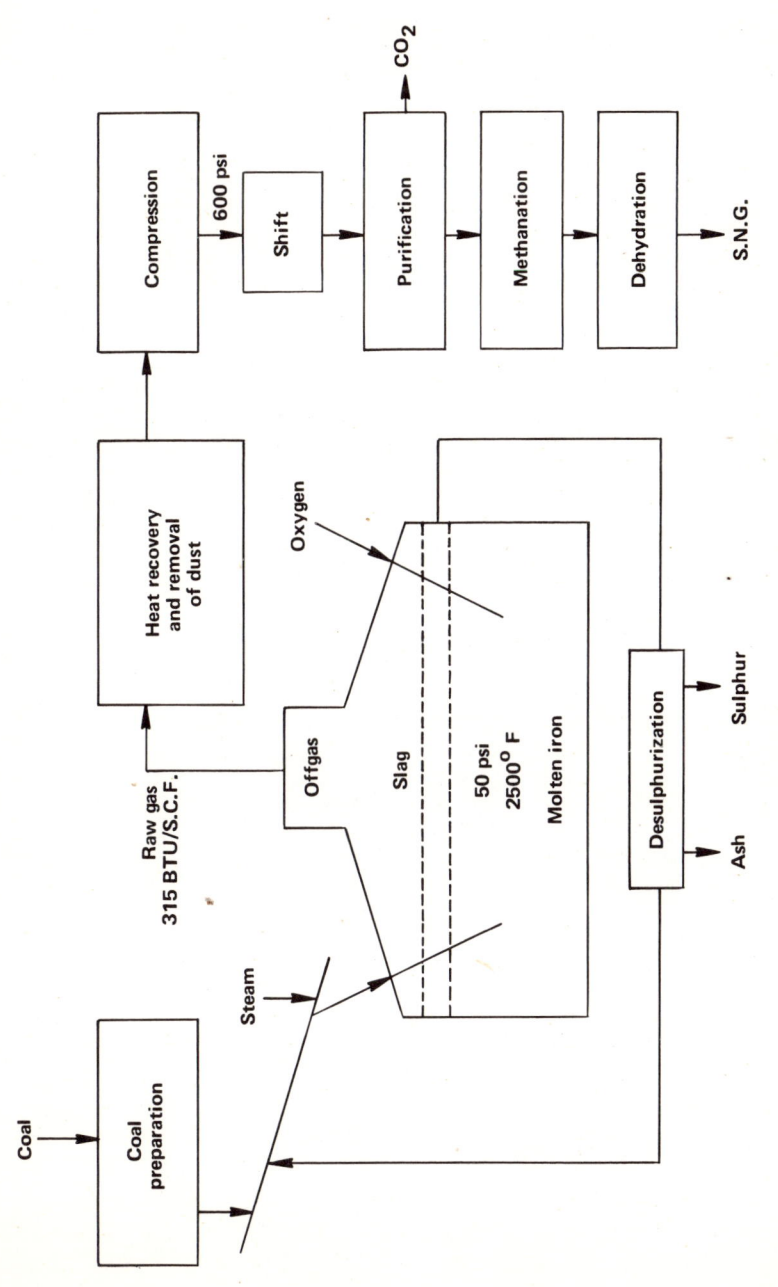

Fig. 1.24. Atgas process.

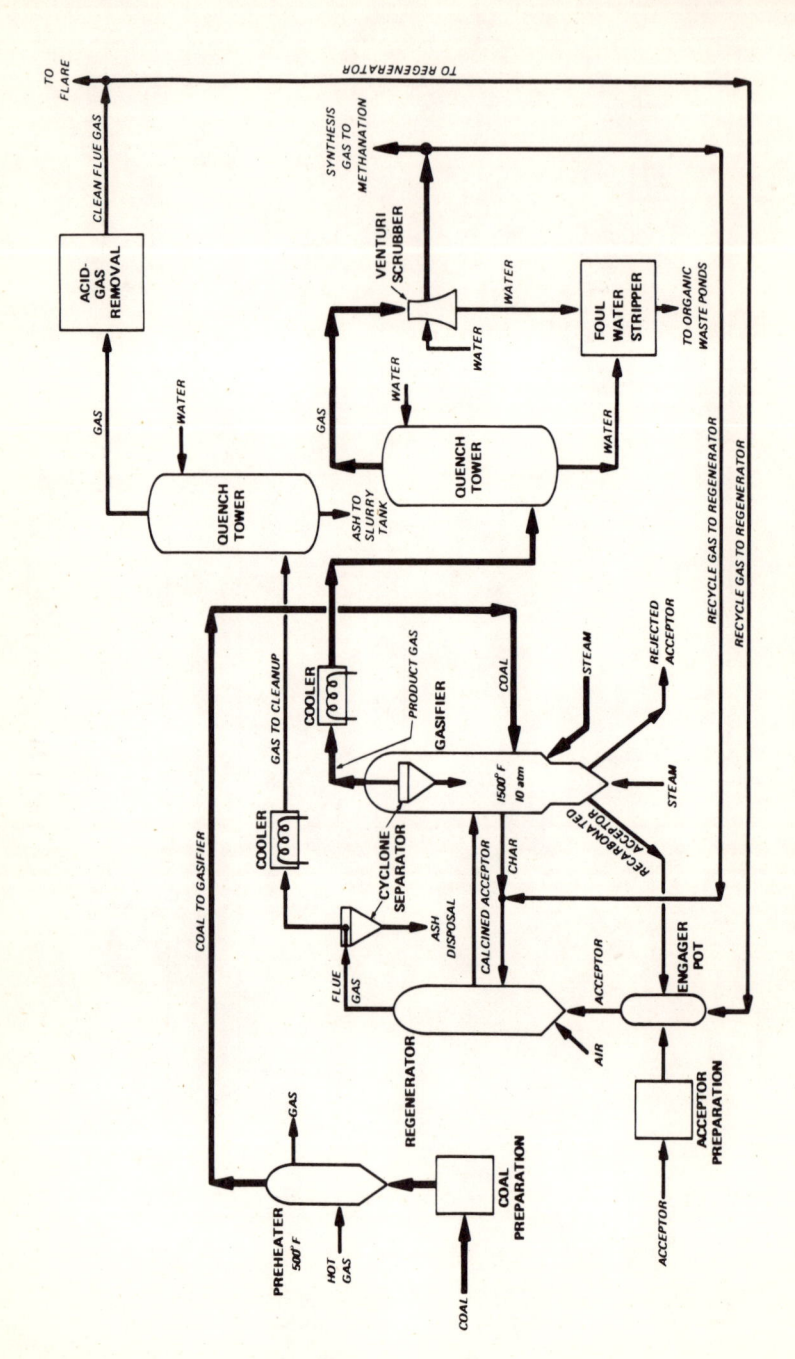

Fig. 1.25. CO_2-Acceptor process (ERDA).

Introduction

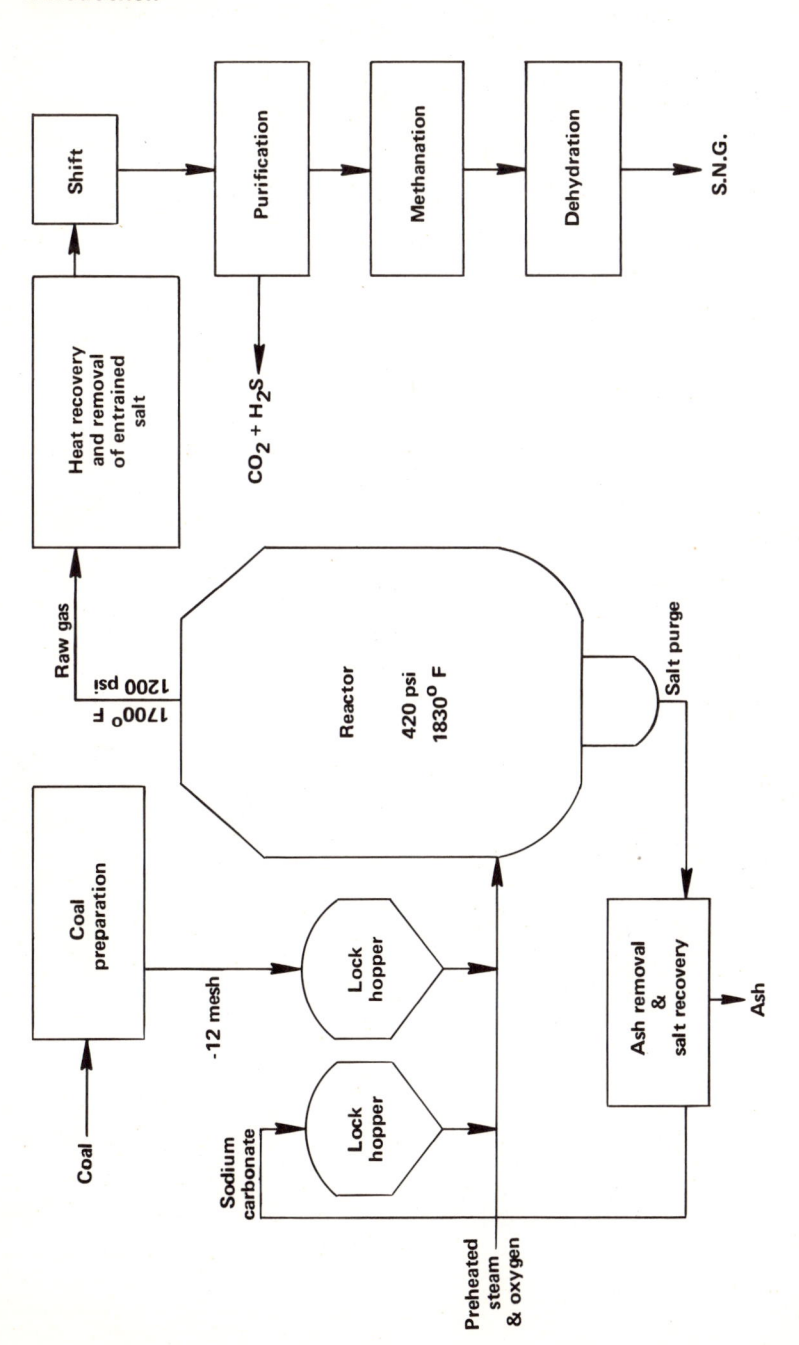

Fig. 1.26. Kellogg Molten-Salt process.

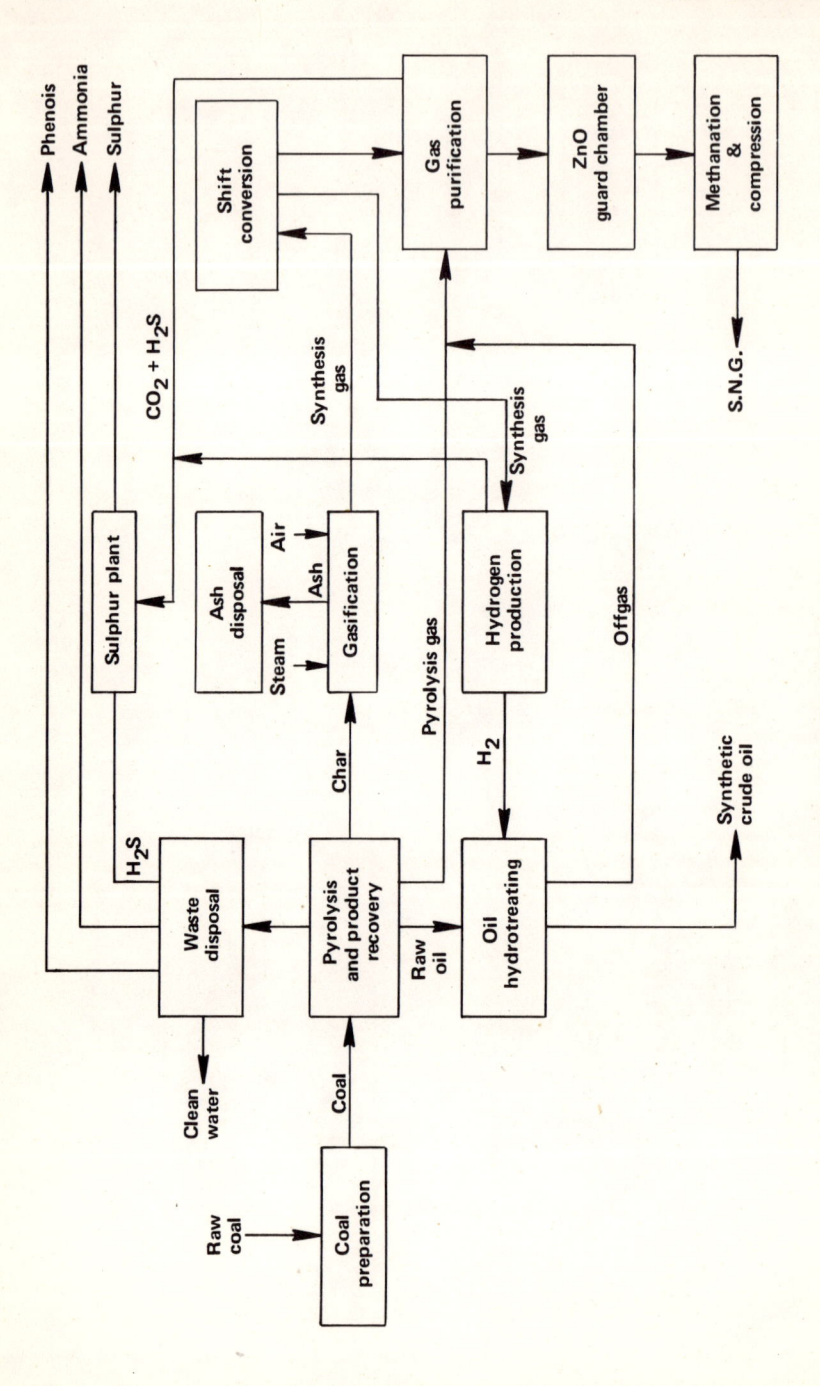

Fig. 1.27. COGAS process.

Introduction

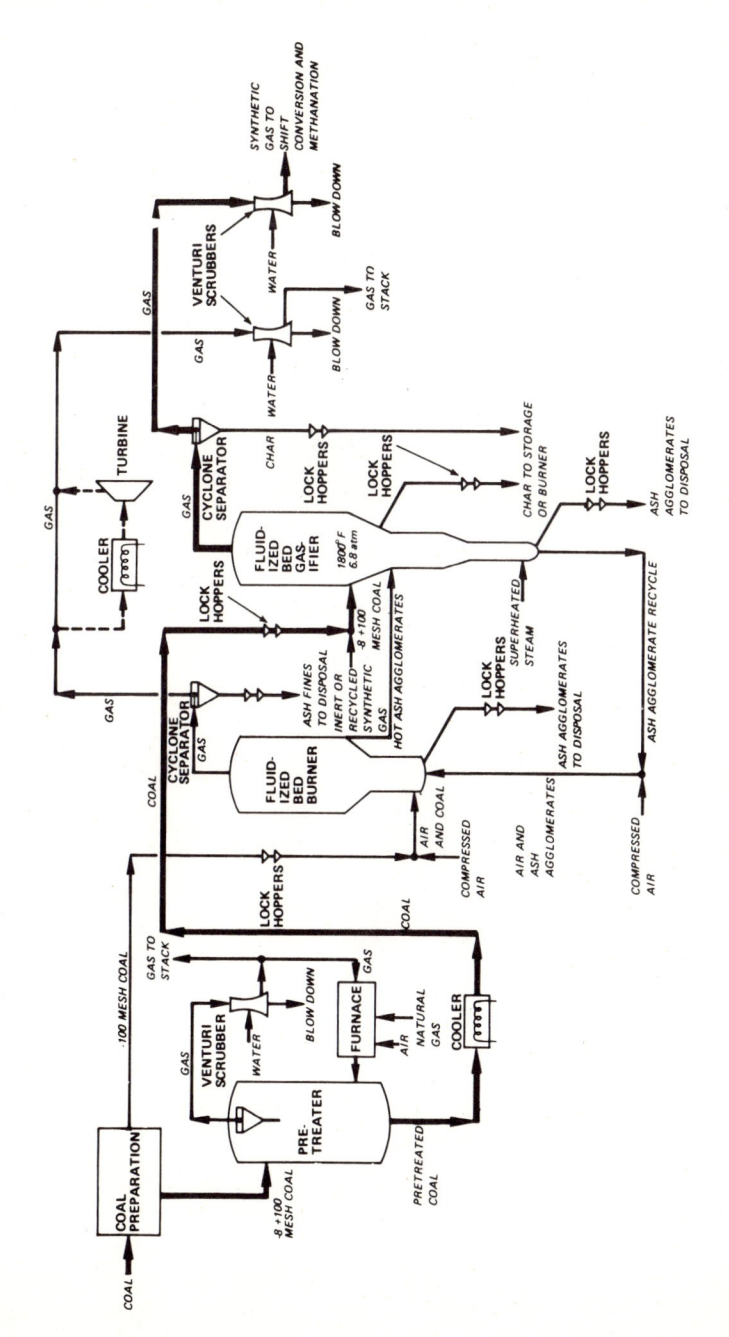

Fig. 1.28. Agglomerating-Ash or Agglomerating Burner process (ERDA).

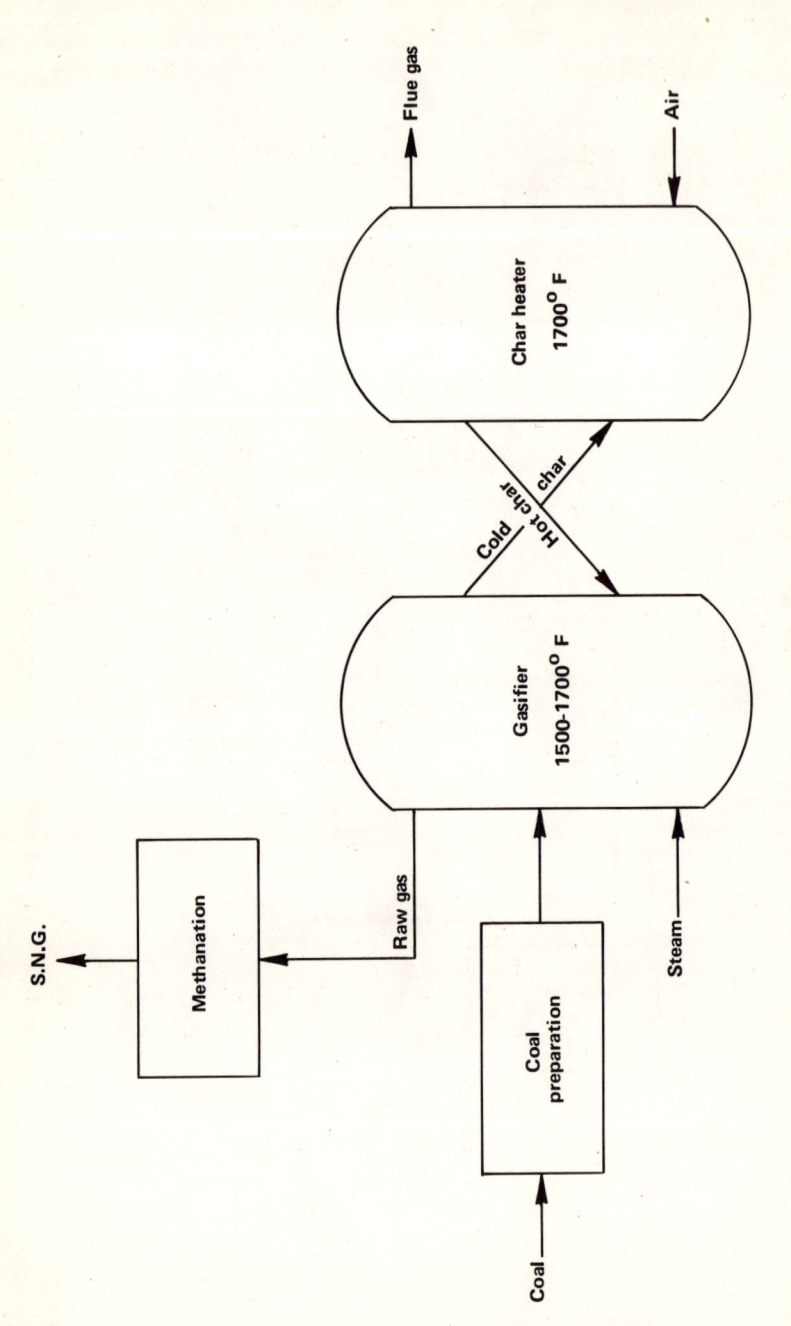

Fig. 1.29. Exxon Gasification process.

Chapter 2

PHYSICAL AND CHEMICAL CHANGES IN COALS

As befits a substance of such bewildering complexity, the characterization of coal properties and behavior has necessarily been by empirical means not presaged by theory. Though coals are subjected to increasingly sophisticated testing, the transient nature of the substance may yield to change merely by the test itself.

That coal is no simple carbon structure is given measure by Figure 2.1, which is one version of the way coal is put together. Increase this detail randomly a thousand-fold and some idea of the possible permutations is provided.

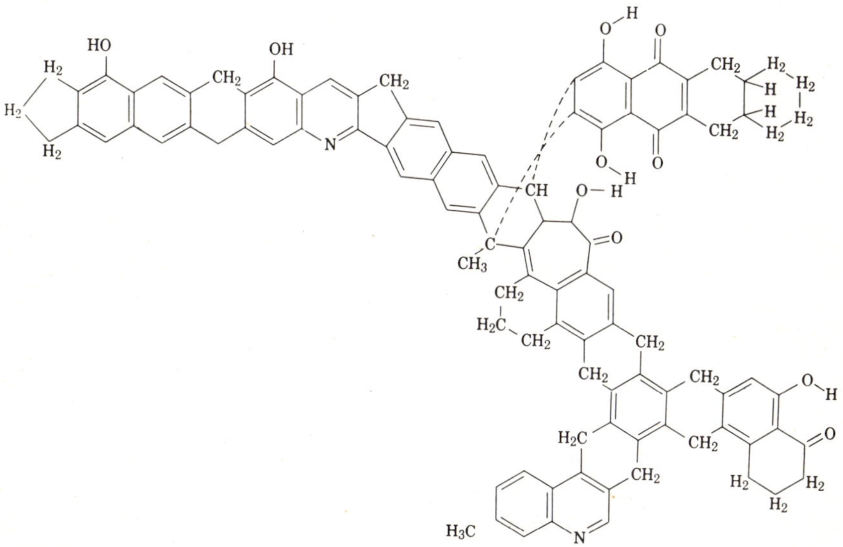

Fig. 2.1. Molecular structure for coal (82% C) proposed by Given. [From P. H. Given, *Fuel, 39,* 147-53 (1960), by permission of the publishers, IPC Business Press, Ltd. Consult *Chemistry of Coal Utilization,* Supplementary Volme, Wiley, New York (1963), p. 293. Also see G. R. Hill and L. B. Lyon, *Ind. Eng. Chem., 54* (6), 36 (1962) and W. Weiser, ACS Div. of Fuel Chem., *Preprints, 20* (2), 122 (1975).]

Nevertheless, macroscopic differences in properties and behavior are detectable and measurable, and the coal structure can be changed or broken down and converted to less complex substances by a number of means. Testing precedes the application and identifies the use.

2.1 THE CHARACTERISTICS OF COAL

Coal has been defined by Francis (1) as a compact stratified mass of mummified plant material which has been modified chemically in varying degree and is interspersed with smaller amounts of inorganic matter. At times there may also be present the degradation products of microscopic animal life. Additionally, free moisture is present in greater or lesser degree. The history of the formation is the precursor of the final product. Coalification is a highly complicated process sequence, dependent upon both circumstances and path.

While Francis and others — notably Peter Given at Pennsylvania State University — have extensively explored the entire route in the formation, composition and makeup of coals, we shall be concerned here, in introductory fashion, with only the main distinguishing features of analysis, rank and petrography — all of which have bearing on the reactivity.

ANALYSIS

Coals are analyzed by well-defined procedures (e.g., ASTM Standard D-271) to yield either or both proximate and ultimate analyses. The comparison is as follows, as frequently reported:

Proximate	**Ultimate**
% Moisture	% Moisture
% Ash	% Carbon (C)
% Volatile Matter (VM) or Volatiles	% Hydrogen (H)
% **Fixed Carbon**	% Oxygen (O)
100% (total)	% Nitrogen (N)
	% Sulfur (S)
Btu	% **Ash**
% Sulfur	100% (total)

Physical and Chemical Changes in Coals

The percentages above the line add up to 100% and are on the basis of weight. In the ultimate analysis, the moisture content may alternately be entered into the hydrogen and oxygen content.

The data is customarily reported on an "as received" basis, or else on a moisture-free or dry basis. Alternately, it may be reported on a moisture and ash-free basis, or "maf" basis.

The heat of combustion is preferably determined experimentally as a part of the proximate analysis, but if necessary can be estimated from Dulong's formula using the ultimate analysis:

$$\text{Btu/lb} = 14{,}544\, C + 62{,}028\, (H - \tfrac{O}{8}) + 4{,}050\, S$$

where the compositions C, H, O, S are in weight fraction. There is some error in applying the formula to western coals. Of interest is the fact that the oxygen present is considered already combined with hydrogen. The above yields a gross heating value — that is, considers the combustion product water to be condensed. (See Appendix.)

Along with the principal components of the ash, such as silicates and aluminates and other majors, trace elements are sometimes determined. For the composition of the ash is becoming of increasing interest from the standpoint of trace elements, which are either of economic significance or pose an environmental threat. The U.S. Geological Survey is in fact conducting analyses of representative coals for trace elements. Mass spec. analyses of ash residues are becoming of routine acceptance as a means of monitoring pollution hazards from the combustion of coal.

Another characteristic of coals sometimes provided by testing is ash fusibility. Temperature of fusion is a variable depending upon degree, a consequence of time-dependent tests. Such designators are used as initial deformation temperature, softness temperature, hemispherical temperature, and fluid temperature to describe the stages within the test. Still other designators may be used.

Typically, softening points may range from 2200-2800°F, though sometimes higher or lower. The presence of alkalies lowers the temperature. Below ~ 2200°F, the ash may be characterized as "slagging". This is of consequence in burner and gasifier operations.

RANK

The proximate analysis is used to classify coals by rank. Four general classifications are used in this country: anthracite, bituminous, subbituminous, and lignite. Peat is outside the domain of the properties used. The breakdown is presented in Table 2.1 and diagrammed in Figure 2.2, according to the American Society for Testing Materials.

Legend: F.C. = fixed carbon; V.M. = Volatile matter; Btu = British thermal units.

Class	Group	Limits of Fixed Carbon or Btu Mineral-Matter-Free Basis	Requisite Physical Properties
I. Anthracitic	1. Meta-anthracite	Dry F.C., 98 percent or more (dry V.M., 2 percent or less)	Nonagglomerating †
	2. Anthracite	Dry F.C., 92 percent or more and less than 98 percent (dry V.M., 8 percent or less and more than 2 percent)	
	3. Semianthracite	Dry F.C., 86 percent or more and less than 92 percent (dry V.M., 14 percent or less and more than 8 percent)	
II. Bituminous §	1. Low-volatile bituminous coal	Dry F.C., 78 percent or more and less than 86 percent (dry V.M., 22 percent or less and more than 14 percent)	Either agglomerating or nonweathering *
	2. Medium-volatile bituminous coal	Dry F.C., 69 percent or more and less than 78 percent (dry V.M., 31 percent or less and more than 22 percent)	
	3. High-volatile A bituminous coal	Dry F.C., less than 69 percent (dry V.M., more than 31 percent); and moist ‡ Btu, 14,000 ‖ or more	
	4. High-volatile B bituminous coal	Moist ‡ Btu, 13,000 or more and less than 14,000 ‖	
	5. High-volatile C bituminous coal	Moist Btu, 11,000 or more and less than 13,000 ‖	
III. Subbituminous	1. Subbituminous A coal	Moist Btu, 11,000 or more and less than 13,000 ‖	Both weathering and nonagglomerating
	2. Subbituminous B coal	Moist Btu, 9,500 or more and less than 11,000 ‖	
	3. Subbituminous C coal	Moist Btu, 8,300 or more and less than 9,500 ‖	
IV. Lignitic	1. Lignite	Moist Btu, less than 8,300	Consolidated
	2. Brown coal	Moist Btu, less than 8,300	Unconsolidated

* This classification does not include a few coals which have unusual physical and chemical properties and which come within the limits of fixed carbon or Btu of the high-volatile bituminous and subbituminous ranks. All these coals either contain less than 48 percent dry, mineral-matter-free fixed carbon or have more than 15,500 moist, mineral-matter-free Btu.

† If agglomerating, classify in low-volatile group of the bituminous class.

‡ Moist Btu refers to coal containing its natural bed moisture but not including visible water on the surface of the coal.

§ It is recognized that there may be noncaking varieties in each group of the bituminous class.

‖ Coals having 69 percent or more fixed carbon on the dry, mineral-matter-free basis shall be classified according to fixed carbon, regardless of Btu.

¶ There are three varieties of coal in the high-volatile C bituminous coal group, namely, Variety 1, agglomerating and nonweathering; Variety 2, agglomerating and weathering; Variety 3, nonagglomerating and nonweathering.

TABLE 2.1. ASTM Classification of Coal by Rank.

Physical and Chemical Changes in Coal

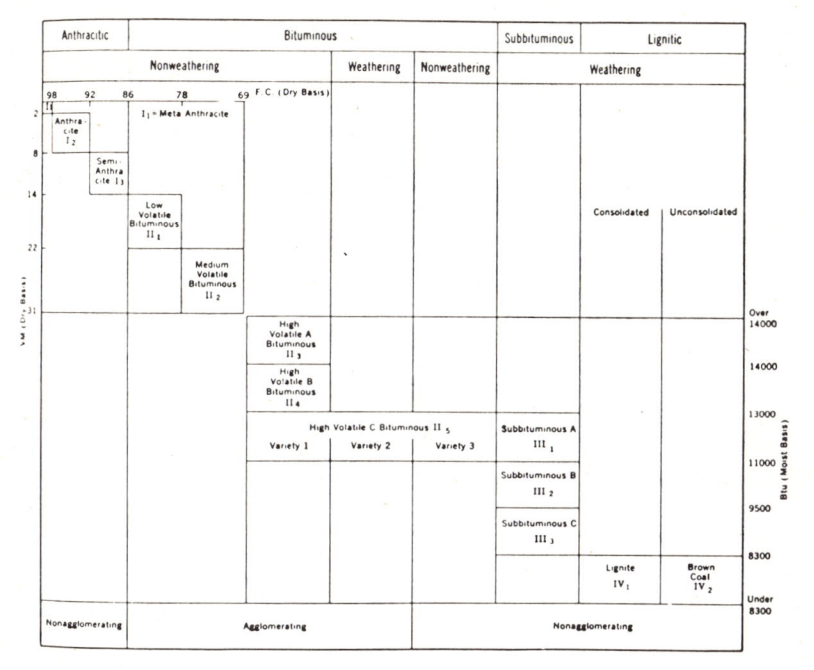

Fig. 2.2. Diagram of A.S.T.M. classification of coals by rank. [From B. C. Freeman, *Econ. Geol.*, 33, 570-1 (1938). Consult *Chemistry of Coal Utilization*, Vol. I, Wiley, New York (1945), p. 59.]

Within a given formation there may be variations in rank, due to pressure or other effects (2). Thus folds or inclines in the formation may cause a lowering of the volatiles and moisture, producing an increase in rank. Variation of coal rank with the ultimate C, H, O analysis is shown in Figure 2.3.

There are other systems of classification than that of the American Society for Testing Materials (ASTM). In England the National Coal Board has a ranking based on moisture and ash-free volatiles and on coking properties. There is an international classification system for coals which is in limited use on the continent. The principal breakdown is based on dry, ash-free volatile matter and the moisture, ash-free calorific value.

Still other descriptive appellations are used to describe coals. Thus in Europe, there are "brown" coals, which somewhat resemble lignite. The divisions are all steps in the coalification process, and give rise to thought of providing a more inclusive and scientific basis for grading coals. The action of coals during heating is one such possibility.

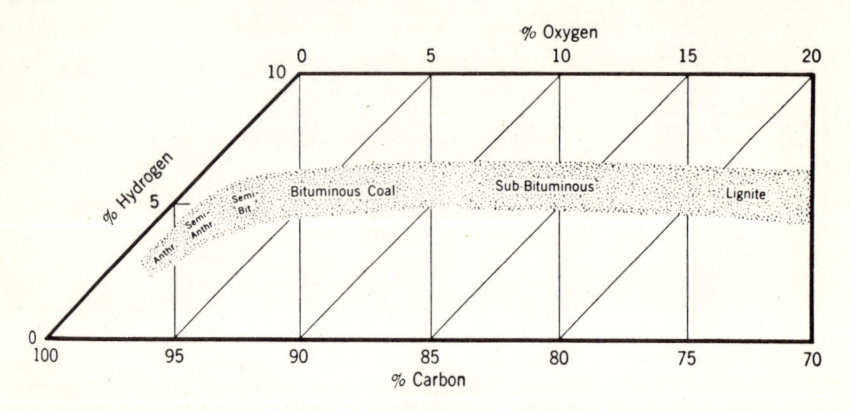

Fig. 2.3. Triaxial diagram of coal ultimate analyses. [From H. J. Rose in *Chemistry of Coal Utilization,* Vol. I, Wiley, New York (1945), p. 47.]

The volatiles given off during the heating of coal consist of
    Combustible gases
    CO_2
    H_2O vapor

The combustible gases are principally CO, H_2, and hydrocarbons, with some oxygenated materials. The residue consists of a char. For coals with higher proximate volatiles, the char goes down. The calorific value of the volatiles is higher than for the coal. The true value in combustion is not known, however, for the makeup of the volatiles changes during heating, etc. Accordingly, the value obtained in analysis will be different than that occurring in combustion. Furthermore, heat may be added (or removed) at varying rates during combustion.

Essenhigh (3) has proposed that the behavior of coal can be explained by considering two homogeneous components:

    Volatiles V_e
    Volatile residue V_r

where $V_e + V_r = V_t$, the total volatiles. In addition, there will exist the fixed carbon, denoted at C_f. The sequence of events, on heating, is that a major transition will occur at about 650°C, yielding the volatiles V_e and a solid residue. Further heating of the solid residue yields the volatile residue V_r and the fixed carbon C_f.

It is next proposed by Essenhigh that the coal be characterized by an "index of rank" which is the ratio

Physical and Chemical Changes in Coals

$$\frac{V_e}{V_r + C_f}$$

It is advanced that better correlation with behavior is obtained using this ratio, than with using V_t or C_f. A continuous evaluation is provided instead of dividing coals into groups and subgroups according to rank.

From the standpoint of combustion, one way to regard all complex fuels is to consider them as mixtures of three elementary fuels (4):

1) Solid carbon
2) Gaseous hydrocarbons
3) CO and H_2

This is usually done in varying degree, though it may be a gross oversimplification.

PETROGRAPHY

While coals have been and are distinguished by such terms as bright, splint, semisplint, cannel and boghead, and by banded vs. nonbanded, etc., a system of using mineral or rock terminology has come about. This system refers to the petrographic description (1, 2, 5).

Thus, for example, on the macroscopic scale of description there are vitrain, durain, clarain and fusain, considering coal as a rock. On a microscopic scale the corresponding "macerals" — analogous to the mineral components of a rock — are vitrinite, durinite, clarinite and fusinite, each of which has its own origins. This is not the only grouping of types, as still others are advanced along with various subcategories. The terms vitrite, durite, clarite, fusite, etc. are also adopted for rock types based upon maceral makeup.

Additionally, there are arguments over lignitic versus cellulosic origin (1), not to mention the processes of formation via humic acids or ulmins, and bacterial vs. inorganic theories. Suffice to say, the final structure is a complicated conglomerate as suggested schematically only in part by Figure 2.1.

A large question mark is the appearance of organic sulfur — that is, sulfur present in the coal molecule or structure. While inorganic sulfur is expected to occur naturally as a mineral component of ash or non-organic portions of the coal (e.g., as sulfate or sulfide), sulfur is of only minor concentration in plant-life — though a vital one (along with nitrogen) as a building block for proteins. The conclusions may be inferred that inorganic sulfur is converted in part to H_2S by the processes of coalification — a phenomenon observed in deep-sea hydrothermal vents. Moreover, certain forms of marine life metabolize H_2S.

Microbiology also gets into the act, perhaps the most common occurrence being the anaerobic conversion of organic material to methane and carbon dioxide. It all starts to come around to the origins of both coal and oil and natural gas, and whether there is an element of commonality.

PHYSICAL AND THERMODYNAMIC PROPERTIES

Coal has been viewed not only as an organic "rock" but as a solid colloid (5). It has porosity, giving rise to the measurement of pore volume, pore surface and permeability. This physical structure is both macro-pore and micro-pore, and somewhat resembles the structure of zeolites.

Moreover, the behavior of the heat of wetting with the properties of the wetting fluid shows a similarity with molecular sieves. Thus, from another point of view, coal can be regarded even as a molecular sieve.

The specific gravity of coals varies over a wide range. Limiting figures supplied (2) are 0.5-1.30 for lignites, 1.15-1.50 for bituminous coals, and 1.40-1.70 for anthracite. A representative figure for subbituminous and bituminous coals, for engineering calculations, is 1.30 or about 80 lb/cu ft. Ash components undeniably play a significant role as does the presence of free water.

The hardness of coals varies from point to point, though anthracite is more uniform: 2.75-3.00 Mohs for anthracite and 2.50 to 2.75 Mohs for semianthracite. An effective or average overall hardness is of practical consideration since it is related to grindability and size reduction of the bulk coal. In this regard, empirical testing is used. An example is Hardgrove grindability, an ASTM method which provides an empirical measure of relative energy requirements for size reduction, or the degree of size reduction for a given energy input. A correlation versus rank is shown in Figure 2.4.

Of greater significance are direct tests on the energy requirements for size reduction. Though a great deal of theory has been generated to relate size reduction to the variables and parameters involved, the final judgment is obtained from experimental determinations. Representative data are provided in Table 2.2.

Specific heats or heat capacities for coal become speculative since the coal decomposes (pyrolyzes) when heated. Furthermore, the decomposition reactions are alternately endothermic and exothermic, as evidenced by differential thermal analysis. Representative thermograms

Physical and Chemical Changes in Coals 55

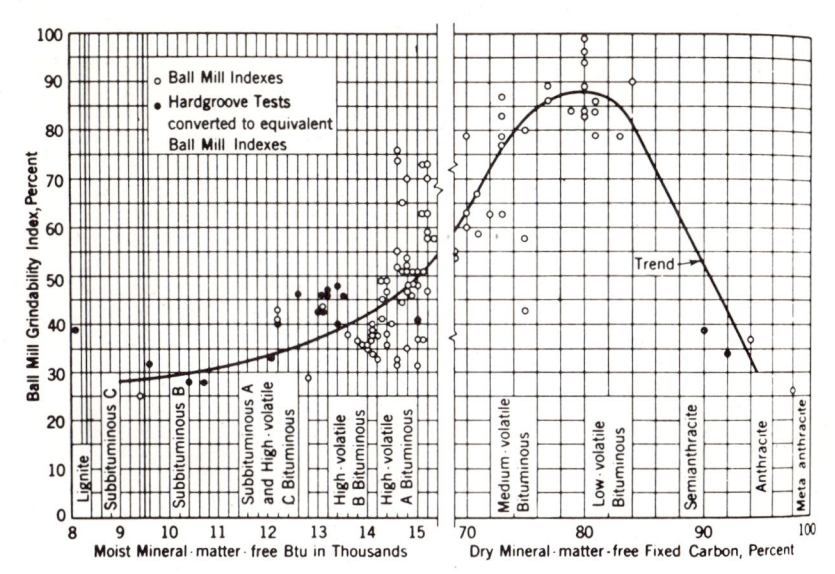

Fig. 2.4. Grindability vs. rank. [From W. A. Selvig and F. H. Gibson in *Chemistry of Coal Utilization,* Vol. II, Wiley, New York (1945), p. 156.]

are shown subsequently in Figure 2.5. Moreover, coal solids show "second order" transitions in going through what is sometimes called the plastic phase.

Therefore, the enthalpic behavior involved in heating a coal becomes a very complicated function indeed, involving what may be termed solid, liquid and gaseous phases. For engineering purposes, however, a mean specific heat of 0.25 can be used for the coal solids. This value may be slightly higher at lower temperatures.

Other properties or measurements of sometime interest include electrical conductivity, dielectric constant and magnetic susceptibility. These and other properties are discussed by Tschamler and de Rutter (6). Mechanical properties are covered by Brown and Hiorns (7).

The physical and also the chemical properties of coal relate to its subsequent preparation and use. Preparation — which includes screening, breaking and crushing, washing, cleaning and de-watering — has been covered in the AIME publication *Coal Preparation* (8), and in *Chemistry of Coal Utilization.*

Comminution, the fragmentation or breaking up of coals, is used to effect a degree of separation into higher- and lower-grade materials by mechanical, hydraulic, magnetic or chemical means. Aside from the elimination of other ash components, there are also attempts to eliminate

TABLE 2.2. Size Reduction Data.*

Material Ground	Purpose	Mill Used	Capacity per Hour	Feed Size Inches	Product Mesh	Total Power H. p.	Power per Ton H. p.	Remarks
Limestone	Treatment in cement mill	10-ft. × 66-in. conical ball mill with air classifier	24 tons	−1/2	99% of −100; 90% of −200	475	19.8	Life of lining, over 12 years; ball consumption, 0.132 lb. per ton
	Agricultural limestone	7-ft. × 36-in. conical ball mill in closed circuit with vibrating screen	13 tons	−1/2	100% of −20; 60% of −100; 40% of −200	130	10	
	Grinding and drying mixt. of limestone and coke for special cement	10-ft. × 60-in. conical ball mill with reverse-current air classifier	20 tons	−2	80% of −200	395	19.75	Feed contains 6% moisture; product less than 1%; temp. of preheated air, 500° F.
	Wet grinding for filler	8-ft. × 30-in. conical pebble mill in closed circuit with wet classifier	1500 lb.	1 1/2	99.9% of −350	40	.....	Grinding media, lumps of limestone
Cement clinker	Manufacture of Portland cement	7-ft. × 39-ft. 3-compartment cylindrical ball mill in open circuit	80 bbl.	−3/4	92% of −200	600	7.5 (per bbl.)	
	Manufacture of Portland cement	10-ft. × 66-in. conical ball mill with reverse-current air-classifying system	74.5 bbl.	−1/4	93.1% of −200	522	6.95 (per bbl.)	Total h. p. includes power to elevate product 50 ft.; liners last over 10 years; ball consumption, 0.097 lb. per bbl.
Feldspar	Manufacture of pottery	8-ft. × 60-in. conical pebble mill with air classifier	1.5 tons	−1/2	99% of −200	89	59.4	
Silica sand	Pottery and glass trade	7-ft. × 22-ft. cylindrical pebble-tube mill in closed circuit with 14-ft. diam. centrifugal air separator	1.9 tons	−20 (mesh)	99% of −325	185	97.5	.....
	Pottery filler	8-ft. × 48-in. conical pebble mill with air classifier	1 ton	−20 (mesh)	98% of −325	77	...	
	Engine sand	4-ft. × 8-ft. rod mill in closed circuit with vibrating screen; wet grinding	22 tons	3/16	100% of −10	45	2	Feed, sandstone tailings
Coke	Preparation of material for sintering	5-ft. × 10-ft. rod mill in open circuit; dry grinding	11 tons	−1 1/2	−4 (10% of −100)	55	5	Coke contains 13% moisture but does not pack in peripheral-discharge rod mill
Iron oxide	Wet grinding for paint trade	4.5-ft. × 16-in. conical ball mill in closed circuit with hydraulic classifier	1 ton	50 (mesh)	99.99% of −200	20	...	.....
Lithopone	Wet grinding for paint trade	4.5-ft. × 15-ft. cylindrical pebble-tube mill in closed circuit with hydraulic classifier	1.1 tons	1/4 (and finer)	99.9% of −300	40	...	
Chromite	Manufacture of chromium salts	7-ft. × 48-in. conical ball mill with classifier	4.25 tons	−3/4	97.6% of −200	161	38	Hardness of chromite, 5.5
Ill. bituminous coal	Firing boiler	7-ft. × 6-ft. × 6-ft. tricone mill with air classifier	5 tons	−3/4	95% of −200	137	27.4	Coal contains 16.5% moisture; temp. of preheated air, 450° F.
Eastern bituminous coal	Pulverized fuel for boilers	10-ft. × 72-in. conical ball mill with air classifier	26 tons	−3/4	70% of −200	486	18.7	Coal contains 4% moisture; temp. of preheated air, 400° F.
Manganese	.....	6-ft. × 36-in. conical ball mill with air classifier	1.5 tons	−3/4	99% through 200	70	46.6	Feed contains 5% moisture; product, less than 1%; preheated air is circulated for drying in the mill
Ilmenite	Titanium pigment	8-ft. × 72-in. conical ball mill with air classifier	6.1 tons	−40 (mesh)	96.8% of −200	320	52.5	Feed consists of concentrates; hardness, 5.5; ball wear, 1/4 lb. per ton

*W.H. Withington, "Ball, Rod, and Tube Mills," *Ind. Eng. Chem.*, 30 (8), 897–904 (Aug., 1944).

Physical and Chemical Changes in Coals

sulfur (e.g. sulfides) in the process. A newly formed venture in chemical comminution is that of the Syracuse Research Corporation and Catalytic, Inc. Another effort, involving the use of iron carbonyl to activate sulfides for magnetic separation, originated at Hazen Research. However, as has been demonstrated at MIT, magnetic separation itself can be used to effect extraction of pyrites. A listing of physical and chemical coal beneficiation processes is given in Table 2.3.

TABLE 2.3. Coal Beneficiation Processes*

Process	Developer
Physical processes	
CFRI oil-agglomeration process	Central Fuel Research Institute, Dhanbad, India
Multi-stream coal cleaning system	General Public Utilities Corp.
Two-stage froth flotation	U.S. Bureau of Mines
Electrophoretic separation	U.S. Bureau of Mines
Magnetic separation	Aquafine Corp., Brunswick, Ga. (two projects)
Magnetic separation	Hazen Research, Golden, Colo. Inex Resources, Denver, Colo.
Ilok process	Ilok Powder Co., Inc., Washington, D.C.
Chemical processes	
Battelle hydrothermal process	Battelle Memorial Institute, Columbus, Ohio
Meyers process	TRW, Inc., Redondo Beach, Calif.
Ledgemont oxygen leaching	Ledgemont Laboratory (Kennecott Copper), Lexington, Mass.
"Sulfur oxidation"	KVB, Inc., Tustin, Calif.
Not disclosed	Atlantic Richfield
Chemical comminution	Syracuse University Research Corp., Syracuse, N.Y.
Chemical cleaning	Otisca Industries Ltd.

*From J. M. Evans, "Alternatives to Stack Gas Scrubbing," *Coal Processing Technology*, Vol. 3, American Institute of Chemical Engineers, New York, 1977. See Also R. A. Meyers, *Coal Desulfurization,* Dekker, New York (1977) and *Sulfur in Fossil Fuels,* Short Course, Chem. Eng. Dept., Univ. of Houston (Jan., 1978).

2.2 MOISTURE IN COALS

Water or moisture in coals affects both the physical properties and chemical behavior, including the further reactivity. Accordingly, a brief discussion of this topic follows, with cited literature references provided for additional reading. A noteworthy discussion on the condition of water in coals is by Gauger (9). Another on the effects of water is by Whittingham (10).

The water present in coal may be "free" or combined as "bound" water. Interestingly, in this connection, is that in calculating the heat of combustion by Dulong's formula — as was previously mentioned — all of the oxygen is assumed already combined as water.

While "free" water can be defined by experimental test, the exact condition or state of the "bound" water remains elusive. As far as determining the heat of combustion is concerned, this is of no consequence since only the overall performance is measured. This performance, however, must be referenced to the exact moisture content of the coal as used in the measurement. Greater or lesser amounts of moisture in the coal as subsequently used will affect the heat of combustion and the heat delivered by the fuel.

Free moisture in the coal is by definition in the liquid form. During combustion, evaporation and superheating occurs in reaching flue gas temperatures. This amounts to a heat loss. In the measurement of the heat of combustion, however, the total water product is condensed. This contributes a heat gain and gives a so-called "gross" value for the heat of combustion. To obtain the "net" value, the latent heat of the water product is subtracted. In actual furnace performance, the superheat involved must also be subtracted, including that of the other combustion products as well as the water. All of which can be confusing unless a strict accounting is made in the heat balance.

The following effects on coal preparation properties have been related to moisture content (9):

1) Grindability decreases as moisture increases.
2) Sieving capacity decreases as moisture increases.
3) Density increases as moisture content is decreased down to $\sim$ 5%.
4) Bulk density decreases as moisture increases to $\sim$ 5%, then levels off.

The most important technique for the study of the behavior of water in coal is by means of vapor pressure (9). The vapor pressure will vary

Physical and Chemical Changes in Coals

with retained water content in a hysteresis-type curve. The retained moisture of a given coal will be a function of the capillary radius: the larger the radius, the less retained. In coalification, as a coal proceeds to a higher rank, the capillaries decrease in size — and the retained water also decreases.

It is stated that water was present during coalification (9), and that the affinity of coal surfaces to water or air will depend upon which was present when the surface was produced.

Moisture is not only important for combustion, but may also be critical in briquetting, for instance. For lignite, for example, an optimum briquette strength is obtained at 16% moisture content. On the other hand, combustion uses may allow a higher water content. Thus in the firing of lignite, acceptable ranges of properties are (10):

Moisture	32- 43%
Ash	4- 12
Btu	6000-7500
Hardgrove grindability	30- 60

A high ash fusion temperature is also desired.

ANALYSIS

It is customary to consider the water content of coals as being only that which can be driven off at 105°C (9), but cutoff may also be determined at 104-110°C or 219-230°F. Representative moisture contents for various coals are as follows:

East Appal. Bit.	1- 4%
West Appal. Bit.	2- 8
Ind., Ill., Iowa, Mo. Bit.	8-17
Subbit.	12-30
Lignites	20-45

Alternate means of establishing a moisture content of coals is by boiling with alcohols and with xylol.

The combined water is represented by the oxygen content of the coal, plus the equivalent amount of hydrogen necessary to react to form H_2O (4). The remaining hydrogen is termed free or available.

DRYING

Prior intensive drying of coal produces the following effects during combustion (11):

1) Accelerated ignition rate
2) Increased luminosity of the volatiles flame
3) Increased swelling and coking

During pulverizer operation, the air temperature may drop several degrees due to the latent heat of evaporation of the moisture (12). Up to half of this effect may be offset by dissipative heat from the pulverizing. It is also noted that if the coal temperature gets above 300°F during pulverization, the coal particles may stick together.

Saturated steam has been used for the partial dehydration of lignite at high pressure (9). The system was heated at 13 + atm., and then blown with hot air. The water content is reduced to above 15%.

During the ignition stage in fuel beds, water present in the coal is driven off (11). The main coal source of water vapor in the final waste gas will be from the volatile flame, along with small amounts of methane and hydrogen released during burning-out. By this stage, combustion is essentially at "dry" conditions.

Moisture, conversely, may also produce beneficial effects in combustion, as will be examined in detail in Chapter 3.

2.3 THE ACTION OF SOLVENTS

The action of solvents upon coal is the mildest form of chemical conversion. In addition to the purely solvent action there are also molecular rearrangements which are not reversible. The operation may be described as one of depolymerization followed by repolymerization. The processes are accented by temperature. Hydrogen added to the mixture terminates the pattern of rearrangement. Good solvents produce a swelling action on the coal, with the solvent action related to the solvent surface tension, solvent internal pressure and capillarity (13).

The solvent action of various solvents on coal may be correlated against internal pressure (14), defined as

$$\text{Internal pressure} = \frac{(\Delta H_v - RT)\rho}{M}$$

Physical and Chemical Changes in Coals   61

where

ΔH_v = latent heat of vaporization

R = gas constant, in consistent units

T = temperature, absolute

M = molecular weight

ρ = density

The higher the internal pressure, the greater the solvent action on coal, demonstrable from tabulated values for various solvents. Notable among the solvents appearing in the literature, in increasing effectiveness are naphthalene, tetralin, pyridine, aniline, m-cresol, and phenol. Coal tar fractions are effective solvents while paraffinic hydrocarbons are not.

Tetralin is somewhat unusual in that it is a hydrogen transfer solvent, yielding naphthalene, and providing hydrogenation action in solvation (15). At temperatures above 400°C it will also decompose to naphthalene, a temperature which is above the decomposition point of most coals (14).

The utilization of fractions similar to coal tar-like fractions as a solvent is a feature of commercial hydrogenation recycle (16). Partially reacted liquid is recycled to form a paste with the ground coal fuel. The coal is partially dissolved and reacted while the recycle oil is also reacted to a further stage of conversion.

Temperature and pressure have a solution effect. It has been stated that temperature and hydrogen pressure alone tend to convert some coals to a liquid state (17). The higher the rank, the more difficult the operation. In fact, in general the lower the rank or the higher the hydrogen content of a coal, the more easily it is dissolved in a solvent (14). For example, pyridine is a very efficient solvent for bituminous coals but has no effect on anthracite. Unfortunately, pyridine cannot be recovered completely due to retention in, or reaction with, the coal system.

Solvents which have pronounced activity at temperatures of less than 100°C are termed specific solvents (15). Preheating the coal will increase solubility up to eight-fold. At elevated temperatures, coal decomposition takes place in the presence of non-specific solvents. With a specific solvent in the presence of hydrogen under pressure, a depolymerization takes place rather than decomposition.

A particular solvent will show a maximum effect with coal rank (13). Solvents have a swelling effect on coal substances, and the solution effect evidently does not reach an equilibrium, only a reduced rate.

For a given coal there is an optimum temperature for solvent extraction (14). This optimum temperature is at the decomposition temperature of the coal.

The effect of particle size apparently has little effect on solvent action, with or without accompanying hydrogenation (14). This may be interpreted with respect to equilibrium rather than rate.

The coal solution process or solvent extraction process has been observed to go through several stages including colloidal dispersion and finally true solution (14). This sequence involves depolymerization and repolymerization and will be affected by whether hydrogen is added to the system to cause chain termination. Others have observed the extract phase to be a colloidal dispersion. Thus the extract phase may be classed, depending upon the solvent, as a true solution or as a colloidal dispersion.

Removal of the solvent from the extract phase leaves most products of the same type as the original but smaller in molecular size, due to a mild pyrolyzing effect of the solvent (14). This provides one means for estimating the structure of coal. The extracts may have a molecular weight from a few hundred to (usually) several thousand (15).

The soluble products are sometimes called extracts or bitumens (14), a source of some confusion. A part of these soluble products appears as oils, fats, acids, alcohols, esters, resins, hydrocarbons, and waxes, which may be occluded within the solid structure of the coal system. A part of the soluble product appears to result from thermal depolymerization of the coal and another part from the solvent.

The bitumen extracts are also separable into solid and oily materials. The extract has been characterized as an organosol, consisting of an oily phase and a dispersed phase (micelle phase) containing protective and humic substances. Extracts from low rank bituminous coals contain all solid material with no oily material (13). In the Pott-Broche extraction process, the addition of hydrogen to the extract was noted to form solid deposits.

The extract yield from plants and wood is classified into waxes and resins (14). The yield from peat is divided in the same way. The extract is amenable to conventional tests such as saponification number, acid no., ester value, iodine no., solubility, etc.

A number of solvents may be used to obtain an extract from brown coals, including alcohol, benzene, and toluene. The resins may be separated from the waxes by the use of acetone, which selectively dissolves the resins. A waxy product, called Montan wax, with a certain range of specifications, is obtained from the solvent extraction of coals, and is of industrial importance.

Physical and Chemical Changes in Coals 63

The asphaltenes are defined as that portion soluble in benzene, CS_2, or the equivalent, but insoluble in n-pentane, n-hexane, or the equivalent (18, 19). The oils and resins are soluble in both.

The structure of petroleum asphalt is regarded similar to that of coal but its chemical action is very different (18). For instance the $KMnO_4$ oxidation rate of coal is many times that of asphalt.

Heating the extracted bitumen causes an increase in acidic, basic, phenolic, and ether-soluble materials up to about 425°C, after which a decrease occurs (14). Coal oxidation is said to reduce the amount of soluble material.

In some cases the bitumen from extraction will have poorer coking qualities than the original coal, depending upon the solvent. If the bitumen extract is recombined with the residue and coked, the solid portion of the extract promotes swelling and the liquid portion of the extract promotes cementing. It has been noted that the extract from a poor coking coal added to the insoluble residue from a good coking coal produces a good coke, while the extract from a good coking coal added to the insoluble residue from a poor coking coal produces a poor coke. These and other studies have been aimed at the mechanism of coking, and are as yet inconclusive.

The insoluble residue from the benzene pressure extraction of coal always has a lower H/C ratio than the original coal and suggests a benzenoid structure containing less hydrogen than benzene (which is 6/6) and corresponds to about 4/6. The ability to absorb oxygen has been attributed to the residue only. The residue will have a higher decomposition temperature than the coal.

Bituminous coals will have a marked difference in composition compared to lower-rank coals. The role of waxes and resins is no longer clear. The extracts from bituminous coals will yield aliphatic hydrocarbons rather than acids, alcohols, esters, carbohydrates, etc. It has been claimed that hydrocarbons are produced from the hydrolysis of the waxes followed by splitting. It has also been advanced that bituminous coals consist of two forms of polymerized material: one from the resins and waxes, and the other from cellulose. This is at variance with the benzenoid theory of structure. Bituminous coals contain no appreciable amounts of alcoholic hydroxyl, carboxyl, or methoxyl groups, but will contain hydroxyl and carbonyl groups (13).

Oxidation of both the extract and the residue indicates a benzenoid structure for coal (14). The exact nature of chemical substances from extraction studies of lower-rank coals has been identified, but not for bituminous coal. Soluble resins have been reported, but the exact chemical nature is unknown. Occluded materials such as hydrocarbons have been identified.

The asphaltic, tar-like higher molecular weight compounds obtainable from solvent extraction and/or hydrogenation are considered to be structurally little different from rubber, vinyl compounds, or linoleum (18). All are colloidal dispersions of an aggregate imbedded in a matrix. Asphaltenes are regarded the aggregate, and resins and oils the matrix. The relative proportions determine the nature of the product.

2.4 PYROLYSIS

The pyrolysis of coal involves the physical distillation of components, and the pyrolytic decomposition (and recombination) of components. With the exception of moisture, it is an academic question whether all products evolved are decomposition products. For the coal itself is in but one state of a decomposition process.

The pyrolysis process *per se* is generally understood to occur in the absence of added air or other added substances. Thermal decomposition, however, is also a feature of combustion, partial combustion, hydrogenation and reaction with steam or still other materials wherein conversion takes place as the temperature is elevated. The conversions occur to different degrees and in different directions, however, and are covered in subsequent sections and chapters.

Combustible gases along with carbon dioxide and water are evolved during pyrolysis. Hydrogen sulfide and ammonia, as well as other sulfur and nitrogen compounds appear in varying degree depending upon the coal composition and the variables of pyrolysis. Condensable oils and tars are distilled, which contain oxygenated compounds as well as hydrocarbons. Aromaticity is a feature, as is the presence of branched structures. In common with petroleum-derived oils, the makeup may be spoken of as saturated (straight- and branch-chain), unsaturated, aromatic and naphthenic, with the accent on the later two categories.

Varying amounts of sulfur are liberated principally as H_2S. The remaining sulfur is bound with the carbonaceous char and as part of the ash constituents (e.g., as sulfide and/or sulfate). It becomes in part a matter of temperature and time. A representative ball-park figure for estimation purposes is that half the total sulfur content is driven off during pyrolysis.

Physical and Chemical Changes in Coals

A part of the sulfur driven off appears in the condensed aqueous phase in the form of various sulfur compounds in combination with ammonia or other compounds also driven off, and is subject to the myriad phase and chemical equilibria involved.

The rate of removal of pyrolysis products has a direct effect upon the course of decomposition and the nature of the products, as well as do the other reaction conditions (15).

Each particular coal will have a definite decomposition temperature, marked by the evolution of gases. Treatment with organic liquids reduces the decomposition temperature and raises the yield of volatile matter.

At a given elevated temperature, coal will first show an increase in the fluidity property with time, followed by a decrease. This maximum is possibly explained by the sequence of events: coal → fluid → coke.

While the rate of gas evolution will show a steady rise with temperature, the rate of evolution of the light hydrocarbon gases goes through a maximum.

Pretreatment by oxidation increases the yield of the oxides of carbon, and of water. Coking properties are also reduced. On the other hand, pretreatment with hydrogen improves the coking properties.

As already mentioned, preheating of a coal will enhance the action of a solvent on the coal. Preheating will also increase the yield of low molecular weight material in hydrogenation reactions.

Heating coals in the presence of water or steam has a decided effect. Lignite undergoes significant changes in composition in the presence of saturated steam at around 600°F. Peat and brown coals heated with water under pressure at 660°F showed evidence of coalification. It is concluded that heating in the presence of water increases the rank of coal. No satisfactory explanation or mechanism is yet available.

Certain compounds and other materials have been observed to have an effect upon the pyrolysis of coal. These include $ZnCl_2$, $AlCl_3$, NaOH, H_3PO_4, active clay, ferric oxide, and active carbon. There was a significant decrease in tar, an increase in coke formation, and, save for the H_3PO_4 and carbon, an increase in gas production. For the $ZnCl_2$, the H_2/CH_4 ratio of the gas produced increased from 0.3 to 2.0.

The compounds NaOH, Na_2CO_3, and NaCl all have an effect in decreasing order as enumerated. All serve to lower tar formation, increase gases and low molecular weight hydrocarbons, decrease higher molecular weight hydrocarbons, increase the fraction of unsaturates, and increase hydrogen formation over methane formation.

Pyrolysis has been performed in melts of metallic compounds, ranging in severity from the relatively low-melting zinc chloride through sodium carbonate to sodium hydroxide. Melts of the latter are particularly destructive to the coal molecule.

Of the several theories of coal structure, one interpretation assumes coals to consist of condensed cyclic-hydrocarbon units of the order of 26 carbon atoms linked together by oxygen bridges. Pyrolysis causes a rupture of these units with accompanying fusion and decomposition. Solvents are assumed also to effect rupture but yield a different variety of products.

Thermal behavior in the heating of coal below the decomposition temperature may be analyzed in a relative fashion by means of "differential thermal analysis". The differential change in temperature produced by heating a coal at a given rate is taken in ratio to the change in temperature produced in a reference substance. The derivative, plotted against temperature, is referred to as a thermogram. A positive slope for the derivative indicates an exothermic effect; a negative slope, an endothermic effect.

Peat, for instance, gives evidence of exothermic reactions, reaching a peak at about 700°F. With higher rank there is a less pronounced peak, but at increasingly higher temperatures. An endothermic effect occurring between 250-300°F is attributed to the latent heat of evaporation of water.

Representative thermograms for several coals are presented in Figure 2.5.

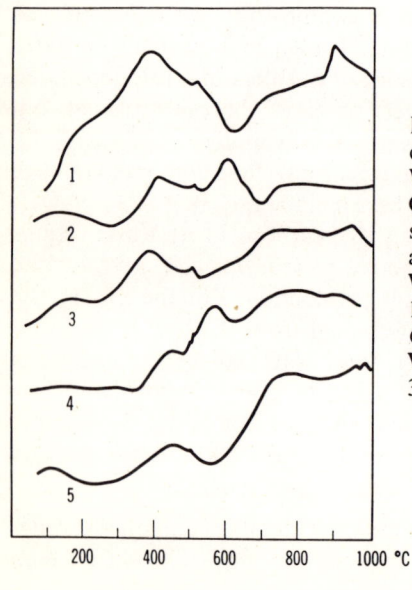

Fig. 2.5. Thermograms of American coals. 1. North Dakota lignite; 2. Wyoming high-volatile C bituminous; 3. Colorado bituminous coal; 4. Pennsylvania low-volatile coal; 5. Semi-anthracite from Alaska. [From W. L. Whitehead and I. A. Breger, *Science*, 111, 279-81 (1950). Consult *Chemistry of Coal Utilization*, Supplementary Volume, Wiley, New York (1963), p. 368.]

Physical and Chemical Changes in Coals

Another manner of determining the behavior of coals by heating is by means of the thermogravimetric balance. The weight percent residue remaining is plotted against temperature. This integral or cumulative curve may be differentiated to yield the customary bell-shaped curve of the Gaussian or normal distribution. The cumulative curve is very similar in shape to the true-boiling-point or ASTM curves (volume percent distilled over versus temperature) used to analyze crude oils or petroleum fractions. Thermogravimetric curves also serve to characterize coals. Examples are shown in Figure 2.6.

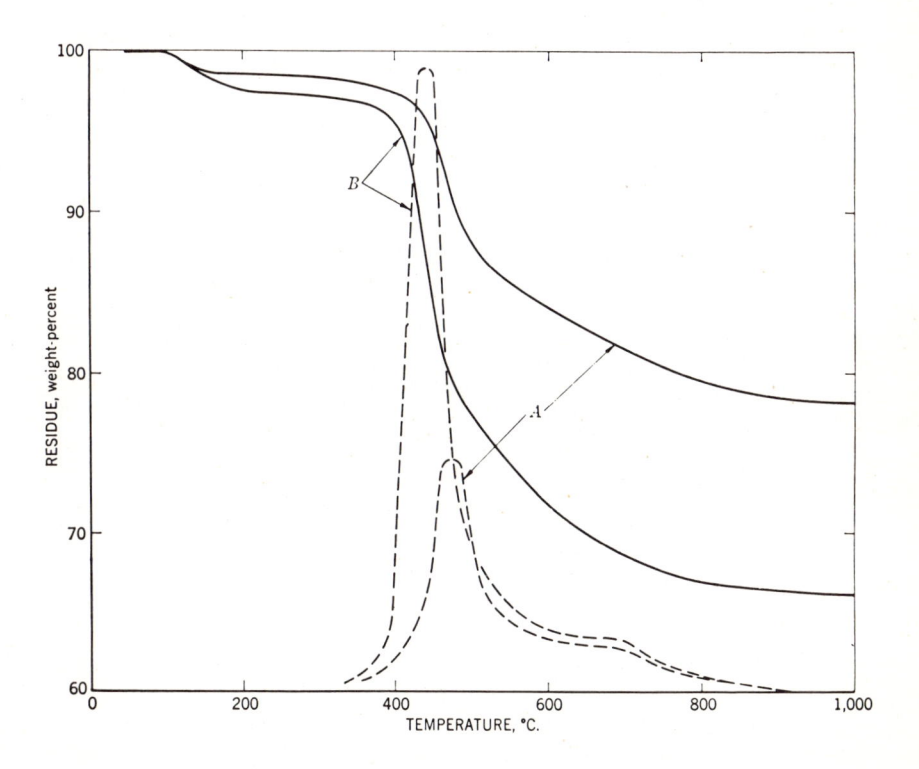

Fig. 2.6. Thermogravimetric curves for two coals — integral and derived. [From H. C. Howard in *Chemistry of Coal Utilization*, Supplementary Volume, Wiley, New York (1963), p. 363. Adapted from A. F. Boyer, *Compt. rend. congr. ind. gaz - 69th Congr.*, 1952, pp. 653-58 and 668-71.]

Carbonization:

Commercial pyrolysis, referred to as carbonization, may be carried to very high temperatures (~ 1000°C or 1800°F) to yield coke as the predominating product of interest. Decomposition is severe with an overhead high in gaseous products. Heating is by indirect means.

Coke is used chiefly for blast furnace operations in the steel industry and also for other pyrometallurgical reductions. In addition to having a high calorific value as fuel and providing the source for carbon monoxide (and hydrogen) as the reducing agent, cokes must have a high mechanical strength and exhibit high porosity and permeability. These properties are to be sustained as the mixture moves down through the bed. Metallurgical-grade coals which produce the required coking properties are the exception.

Low-temperature carbonization, though of less commercial importance, is of more interest to this study. The temperatures involved are on the order of 500-700°C or 900-1300°F. Direct methods of heating have been used as well as indirect, and a wider variety of products are recoverable.

Agroskin [20] divides the carbonization of coal into three temperature ranges, low temperature carbonization to 550°C, intermediate carbonization at approximately 700°C and high temperature carbonization from 900-1050°C.

Upon heating, coal undergoes three major endothermic reactions [21]. Water is removed at about 150°C, primary degasification or chemical decomposition occurs between 350 and 500°C, and secondary degasification takes place between 600-700°C. Behavior depends on the coal and the method of heating [22]. Lower rank coals evolve large quantities of gas and produce char. Coking coals evolve gas and become plastic. During the secondary degasification period the composition of the gas product is temperature dependent and varies greatly with different coals.

Gas liberation begins at about 250°C with tar vapors and considerable amounts of water being formed at about 300°C. The average gas yield during the medium range is 60 to 80 m^3/ton, while in the high temperature range the yield is from 300 to 350 m^3/ton. Gas composition during the mid-range is predominantly methane and hydrogen. At high temperatures the gas products exhibit a decrease in CO_2 and an increase in CO and H_2.

Powell [23] states that at about 350°C the coal becomes plastic, and water and small amounts of gas are liberated. Low temperature carbonization provides low gas yields of 3000 to 6000 ft^3/ton of coal. The

Physical and Chemical Changes in Coals

gas yield in the 800°C and higher range is 8000 to 12,000 ft³/ton. The volatilization process is quite complex and the gas products depend on several factors.

Four main products are obtained by low-temperature carbonization processes: the final semi-coke or char, a tar consisting mostly of complex liquid hydrocarbons, fixed gases, and aqueous liquor (24). The proportions are determined in part by the rate and time of heating and the condensation levels.

Both fixed bed and fluidized systems have been used. The fluidization process requires direct contact with another heating medium, while fixed bed systems may be heated directly or indirectly.

Among the carriers for direct heating have been steam, air, recycle gases, sand, and metal balls containing a salt to take advantage of the latent heat of fusion. Superheated steam at 1000-1200°F has worked but was found too expensive for carbonization purposes. No comment was made about any carbon-steam reaction.

An example of a fluidized bed pyrolysis pilot plant operation utilized nitrogen or steam (25). A solids residence time of 40 minutes was used at reactor temperatures of from 800-2000°F, with the fluidizing gases preheated to 1200°F maximum and the solids preheated to 300°F. No catalyst was employed. The 3-inch diameter reactor was also heated externally. About 50 percent char was left at 2000°F and 75 percent char at 800°F. The tar yield decreased and the gas yield increased with temperature: about 50 percent gas at 2000°F but nil at 800°F. No comment was made regarding any carbon-steam reaction.

The COED process developed by the FMC Corporation involves a multistage pyrolysis to yield an oil which could be mixed with a refinery oil feedstock or otherwise be processed by conventional petroleum refining methods.

The commerical embodiment most often takes place in ovens or kilns which are heated externally by a fuel gas, usually byproduct gas from the coke ovens themselves. There are a great many setups that have been used, one of which is shown in Figure 2.7.

The more modern trend is toward fluidized pyrolysis, as illustrated by the proposed schematic of Figure 2.8. If char or coke is desired as a product, then a hot fluidizing gas is used, composed of the inert products of combustion.

For a given coal, as the carbonization temperature is increased the distribution of carbon, hydrogen and oxygen in the coke residue falls off.

Furthermore the yield of gases increases with temperature while the yield of char or coke falls off. The yields of tar and oil varies, but remains somewhat uniform.

Some representative yields from industrial retorts per ton of charge, are as follows:

Char	~	1500 lb.
Tar	~	20 gal.
Oil	~	3 gal.
Aqueous liquor	~	30 gal.
Ammonia	~	2 lb.
Gas	~	4000 cu.ft.
		(~ 700 Btu/SCF)

The actual yield of course varies with the process, the carbonization temperature, and the coal used.

The liquids from pyrolysis or carbonization are in the main unsatisfactory as fuels without further upgrading. While the aromatic and cyclic nature indicates good anti-knock properties as motor fuel, the presence of oxygenated materials (including phenolic compounds) requires further treating, including hydrogenation. Hence the accent on hydrogenation during pyrolysis, a technology discussed in Chapter 5.

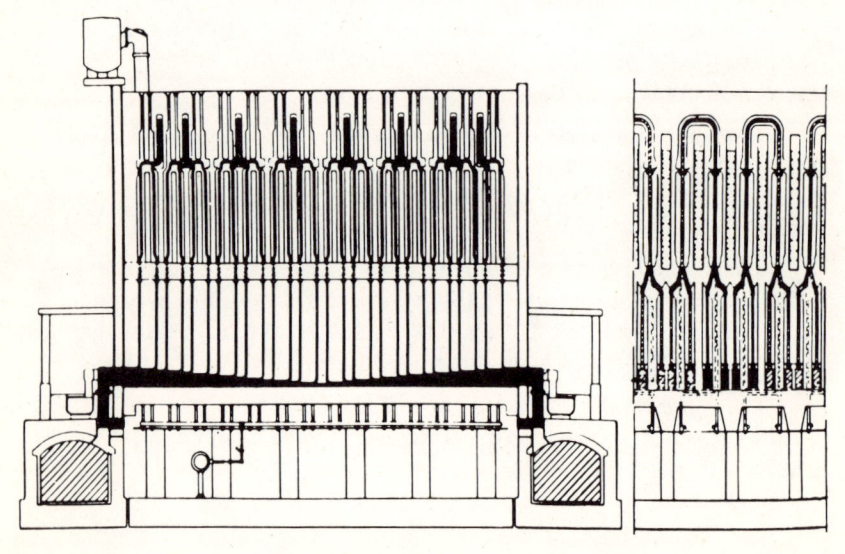

Fig. 2.7. Koppers-Becker underjet oven.

Physical and Chemical Changes in Coals

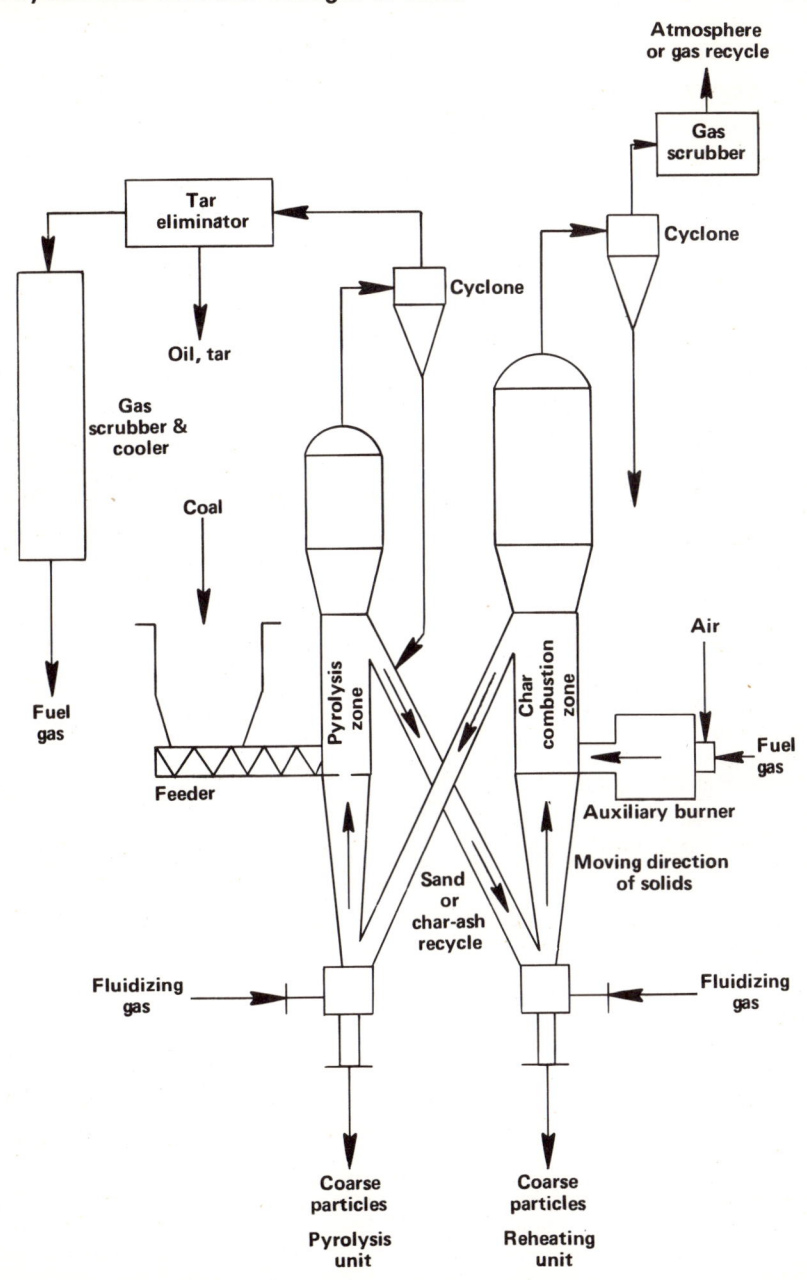

Fig. 2.8. Two-bed circulating fluidized-bed pyrolysis system (Kansas State University).

Effects During Firing:

Most literature on pyrolysis is concerned with slow decomposition and has little relevance to the firing of pulverized fuel (p.f.) (26). Industrial carbonization is such a slow decomposition process. A process which could be considered to be intermediate in rate of decomposition would be the firing of fuel in a packed bed, whereas p.f. firing in an intrained system involves rapid decomposition.

As an example of the differences which occur between slow heating and rapid heating, with cellulose a char is produced upon slow heating but little or no char is produced upon rapid heating.

The yield of volatiles for example is dependent not only upon the temperature-time heating relation but upon particle size. Moreover, it can be expected that the composition of the volatiles will depend upon the heating history.

The combustion characteristics of a coal are affected by the time-temperature rate of heating. With rapid heating, more volatiles are formed. In the combustion of p.f. a high rate of heating is involved: $10^4 °C/sec$. The total period of decomposition is less than one second and involves a rapid transfer of volatiles from the surface. Such rapid decomposition occurs with particle sizes less than 100 microns.

Perhaps most significantly, the gasification of coals is noted to increase with moisture content (27). This is attributed to the secondary reaction of steam and/or CO_2 with carbonaceous matter.

When heated, some coals pass through a plastic stage wherein the particles fuse together. These "caking" or agglomerating coals produce a semi-coke which must be broken up for adequate combustion. The term "caking" is not necessarily synonymous to "coking," since some caking coals do not produce coke, while some free-burning coals do. Exposure to air reduces this property.

Effect of Moisture:

Moisture slows down the rate of carbonization (9). For the carbonization of brown coals, the tar yield is greater for higher moisture content. The comment was made that the yield may be more affected by the nature of heating. In further comment, the yield of tar may be increased due to a steam stripping effect.

Observation of the effects of moisture during pyrolysis is enhanced if the pyrolysis is conducted under vacuum. This is similar to the effect of moisture (steam) in increasing the volatility of heavy crude oil fractions during vacuum distillation.

Physical and Chemical Changes in Coals

During the vacuum distillation of coal there is no significant yield until thermal decomposition takes place, except for minor amounts of water vapor, CO and CO_2, and low molecular weight hydrocarbons (28). In fact, for dried coal, CO and CO_2 are the principal gases evolved under vacuum at room temperature. But for newly-mined undried coal, hydrocarbons will predominate. A representative sequence of products for the heating of a coal under vacuum is approximately as follows:

200°C	Copius evolution of water, CO, and CO_2
200-300°C	Sulfur containing gases and olefins
310°C	Liquids other than water
350°C	Rapid gas evolution and much viscous oil
750°C +	Vigorous hydrogen evolution due to secondary decomposition

In comment, in addition to distillation occurring more readily at lower pressures, there may be an azeotropic effect due to the presence of water (steam distillation), even at the higher temperatures. This azeotropic effect increases the effective volatility of the higher-boiling materials. And perhaps most importantly, the types and quantities of materials evolved will be different in the presence of significant amounts of moisture.

2.5 ADDITION AND SUBSTITUTION REACTIONS

Coal undergoes reaction with the halogens and with oxidizing and reducing agents. Coal undergoes hydrolytic reactions with bases, but does not react significantly with acids except whereby the acid acts as an oxidizing agent. Decomposition accompanies reaction to a lesser or greater extent depending upon the severity of conditions.

The chlorination of coals yields both gaseous and solid products (13). Reaction with chlorine vapor at 435°F for 24 hours yielded a product of about 50 percent chlorine. Among the gaseous products obtained were CCl_4 and hexachlorobenzene. A chlorinated residue was observed to give off HCl gas above 570°F and chlorine gas above 1100°F.

Chlorination of suspensions of coal in various solvents has produced a variety of chlorine-containing unidentified products. Some were of a resinous nature.

Both iodine and bromine react with coal. It is suggested that hydrogen is replaced mainly in the aromatic structure. Aliphatic substitutions evidently do not occur at the same conditions.

Chlorine trifluoride has been reacted with coal at 400-750°F. Solids, liquids, and gases were produced. The liquids were composed of chlorofluorocarbon oils, and were about 100% by weight of the coal charge. A typical oil product contained about 60 percent of the halogen reacted, and the oil about 20 percent. The solid residue amounted to only about five percent by weight of the coal charged.

The air-oxidation of coals below the ignition temperature appears to progress through three stages with an increasing yield of gases accompanied by changes in the solid residue. In the second stage alkali soluble products appeared and increased in concentration. These include humic acids. The proportion of oxygen in the gaseous products increased over that in the solid, and only aromatic hydrogen was left in the solid. In the third stage, the water-soluble material increased.

Preliminary oxidation of coals lowers the coking properties (24).

The nature of the oxidizing reaction and products depends upon the agent used. Ozone converts the carbon directly to water-soluble products (13). The Dow Chemical Company has oxidized bituminous coal suspensions in aqueous NaOH with oxygen under pressure to yield humic acids and water-soluble aromatic acids, and Na_2CO_3 (29). The acids could be converted to resins, with commercial possibilities.

The oxidation of coal with alkaline potassium permanganate solution will first yield humic acids (13). Further oxidation of the humic acids yields water-soluble subhumic acids. Hydrogen peroxide can also be used to oxidize the humic acids. Among the volatile products obtained are acetic and formic acids. Nitrates are also produced from the nitrogen of the coal. Boiling alkaline $KMnO_4$ solution evolved CO_2 and yielded oxalic and other nonvolatile acids.

Hot concentrated nitric acid reacts to give black nitric-acid-insoluble humic acids, then brown nitric-acid-soluble humic acids, and finally water-soluble acids. Vapor-phase reaction with nitrogen oxides has been studied toward the effect on subsequent pyrolysis, with no significant conclusions (15). Acetic acid is produced by reaction with nitric acid-potassium dichromate mixtures (13).

Oxidation in air of younger or lower-rank coals will lower the yield of volatile matter and tar; similar oxidation of older or higher-rank coals will increase the yield of volatile matter. The carbon is apparently converted in the older coals. The middle ranked bituminous coals have the highest yield of humic acids. Low-temperature absorption of oxygen on coal forms peroxides at the coal surface.

Physical and Chemical Changes in Coals

Humic acids, intermediates in the progressive oxidation of coal, have been deduced to be derivatives from aromatic ring systems. In fact, low-temperature oxidation of coals is one of the means used in the attempt to deduce the structure of coal.

The reaction of coal and steam can be considered an oxidizing reaction with H_2O the oxidizing agent, in the water gas reaction

$$C + H_2O \rightarrow CO + H_2$$

However, at different conditions, under a different context, coal and water undergo hydrolytic-type reactions.

Results have been reported for the hydrolysis of a coal with 5 N sodium hydroxide at 660°F for 24 hours. The residue contained less oxygen than the coal. Analyses of the products in weight percent show 2.8 percent gases, 3 percent liquid phenols, 5 percent solid phenols, 13 percent fatty acids, 15.6 percent liquid hydrocarbons, 22 percent carbonates as CO_2, and 28.3 percent insoluble residue.

Compounds identified included phenol, and acetic, propionic, butyric and succinic acids. The overall reaction was as if water were added to coal. The acids might have derived from alcohols, ketones or aldehydes by oxidation or reduction. Thus apparently oxidation and reduction by hydrogenation is accompanied hydrolysis. A reaction sequence is possibly as follows:

$$R'OR + H_2O \rightarrow ROH + R'OH$$
$$RCH_2 CH = CHR' + H_2O \rightarrow RCH_2 + R'CH_2CHO$$
$$C + 2H_2O \rightarrow CO_2 + H_2$$

The reduction reaction of coal which is of pervading interest is hydrogenation. While other reducing agents have been tried to a very limited extent, it is the reaction with hydrogen that remains of paramount importance. At the conditions for sufficient reactivity to occur, decomposition takes place along with hydrogenation. As previously mentioned, hydrogenation is also used in the presence of solvents, though there is an increasing interest toward the reaction of the dry coal and hydrogen, particularly as applied to fluidized gasification reactions.

Hydrolytic reactions of coal are described in the next section. The reaction of coal with hydrogen and with steam are both of an importance to merit study unto themselves, and will be described in later chapters.

HYDROLYSIS

The hydrolysis of coal with alkali gives small yields of alkali-soluble degradation products which are phenolic or acidic in nature. Hydrolysis of coal occurs with NaOH in dilute concentrations and all the way to 100% alkali (30). Elevated temperatures are required.

Lower rank coals are more reactive to hydrolysis. As the age (or rank) of a coal increases, the solubility in alkali decreases (14). This solubility behavior may be used as a reactivity index after oxidation. The steam distillation of coal can also be used as a measure of composition and reactivity.

For coals an optimum temperature for extraction will exist just below the decomposition temperature.

The catalytic depolymerization of coal has been suggested as one means for making coal more reactive (31). The object would be to find a catalyst which would depolymerize the coal under mild conditions so as to recover monomeric units.

Alkalies:

Three types of hydrolytic reactions are recognized (32):
1) Hydrolysis.
2) Oxidation by water to give CO, H_2, and CO_2.
3) Hydrogenation.

With coal, no hydrolytic breakdown occurs with acids. Reaction occurs with aqueous alkalis at higher temperature and evidently involves hydrolysis; there is also reason to suspect a water-gas reaction yielding CO_2 and H_2. With charcoal or carbon, at 350°C, the hydrolysis proceeds as

$$C + 2NaOH + H_2O = Na_2CO_3 + 2H_2$$

In comment, this could explain in part the catalytic effect of alkali compounds on the water-gas reaction. At the same time this offers a partial explanation of the effect of moisture in promoting combustion.

There is evidence that liquid water itself may react with coal as the particle temperature is raised during combustion. For instance, it is probable that condensation reactions involving the elimination of water were involved in coalification. Reversal by hydrolytic reaction would then produce degradation. This has been partly confirmed with lower-rank bituminous coals, and also with higher-rank coals.

With fused alkalis at 650°C, carbon reacts to produce hydrogen. Brown or bituminous coals are converted to a brown, alkali-soluble product. The lower-rank coals are more reactive (down to 400°C), and even more dilute alkali solutions affect the coal. For example, 5N NaOH at 250°C will tend to make coal plastic such that the particles cake

Physical and Chemical Changes in Coals

together and form coke. In fact, the development of coal plasticity may be related to incipient hydrogenation reactions at the surface.

As another example, 10N KOH at 280°C converted 75% of a brown coal to an alkali-soluble product. Of the insolubles, half were water-soluble. It was mentioned that oxygenated compounds such as ethers and alcohols were produced. In comment, it is barely possible that the catalytic effect of alkalis on the water-gas or coal-steam reactions involves the presence of a condensed aqueous phase (which acts as a flux).

Experiments on the extraction of peat with water for 8 hrs. at 250°C gave the following yields (14):

Residue	62.9%
Extract	13.7
Unidentified	23.4

The unidentified fraction probably consists of water, CO_2, volatile acids, aldehydes, etc.

The long heating of peat or brown coal bitumens (benzene soluble) converts the bitumen to benzene-insoluble materials.

It is noted that an alkali with an organic solvent is even better in breaking down coal (32). One example uses a mixture of 10% KOH plus the monomethyl ether of ethylene glycol. This is an area which calls for further investigation. As another example, the extraction of peat or brown coal with water-benzene mixtures enhances the recovery of bitumens (14).

Experiments on the extraction of coal with 10% NaOH at the boiling point yielded 97.5% residue after two hours. Results using KOH-ethylene glycol monomethyl ether mixtures were comparable. The residues from these extractions are very active and will ignite in air at 150°C.

ORGANIC SOLVENTS

The action of organic solvents has a mild pyrolizing effect on coal (14). The extract product from the extraction of coals may be colloidal, consisting of an oily phase and a dispersed phase, as found in asphalts.

Solvent extraction has provided one means for studying the coking properties of coal.

The absorption of pyridine vapor provides a measure of reactivity. Young coals absorb it rapidly, old coals very little. There has been note made of the pyridine extract yield of coal as a correlating parameter. As the yield of pyridine extract increases, the ignition temperature decreases and the tendency toward spontaneous combustion increases. It was found that the ability to absorb oxygen lies almost wholly in the residue left after pyridine extraction.

Waxes obtained by the action of solvents on coal are oxidizable in air at higher temperatures, in the presence of alkali.

At times, coal itself may act as an oxidizing agent. For instance, in the extraction of coal by benzene at elevated temperatures, the benzene may be converted to phenol and acetaldehyde. This is contrary to the usual role of coal as a reducing agent.

OTHER EFFECTS

Lignite, notably, is a potential source of uranium and other minerals (33), and cols have adsorptive properties for the recovery of metals (34, 35). Leonardite, an alkali-soluble oxidized coal, may serve as a soil conditioner, surfactant for oil production, and leach for ionic metal capture (36).

REFERENCES

1. Francis, W., *Coal: Its Formation and Composition,* Arnold, London (1954).
2. Moore, E. S., *Coal: Its Properties, Analysis, Classification, Geology, Extraction, Uses and Distribution,* Wiley, New York (1922, 1940).
3. Essenhigh, R. H., "Influence of Coal Rank on the Burning Times of Simple Captive Particles," *J. Eng. Power,* 85 (3), 183-190 (1963).
4. Haslam, R. T., and R. P. Russell, *Fuels and Their Combustion,* McGraw-Hill, New York (1926).
5. VanKrevelen, D. W., *Coal: Typology - Chemistry - Physics - Constitution,* Elsevier, Amsterdam (1961).
6. Tschmler, H. and E. deRutter, "Physical Properties of Coals," in *Chemistry of Coal Utilization,* Supplementary Volume, H.H. Lowry Ed., Wiley, New York (1963).
7. Brown, R. L. and F. J. Hiorns, "Mechanical Properties," in *Chemistry of Coal Utilization,* Supplementary Volume, H. H. Lowry Ed., Wiley, New York (1963).
8. *Coal Preparation,* D. R. Mitchell Ed., American Institute of Mining and Metallurgical Engineers, New York (1943).
9. Gauger, A. W., "Condition of Water in Coals," in *Chemistry of Coal Utilization,* Vol. I, H. H. Lowry Ed., Wiley, New York (1945), pp. 600-626.
10. Markley, G. F., "Progress in Lignite Firing," *Power Engineering,* 71 (11), 55-57 (Nov., 1967).

11. Whittingham, G., "Some Effects of Moisture on the Combustion of Coal," British Coal Utilization Research Association, Bulletin *27* (6), 229-236 (June, 1953).
12. Griswold, J., *Fuels, Combustion and Furnaces,* McGraw-Hill, New York (1946).
13. Dryden, I. G. C., "Chemical Constitution and Reactions of Coal," in *Chemistry of Coal Utilization,* Supplementary Volume, H. H. Lowry Ed., Wiley, New York (1963), pp. 232-295.
14. Kiebler, M. W., "The Action of Solvents on Coal," in *Chemistry of Coal Utilization,* Vol. I, H. H. Lowry Ed., Wiley, New York (1945), pp. 677-760.
15. Howard, H. C., "Pyrolytic Reactions of Coal," in *Chemistry of Coal Utilization,* Supplementary Volume, H. H. Lowry Ed., Wiley, New York (1963), pp. 340-394.
16. Storch, H. H., "Hydrogenation of Coal and Tar," in *Chemistry of Coal Utilization,* Vol. II, H. H. Lowry Ed., Wiley, New York (1945), pp. 1750-96.
17. Rapoport, I. B., *Chemistry and Technology of Synthetic Liquid Fuels,* Second Edition, Translated from Russian, Published for the National Science Foundation, Washington, D.C. by the Israel Program for Scientific Translations, Jerusalem (1962).
18. Silver, H. F., B. E. Davis, and W. E. Duncan, "Coal Hydrogenation Study," Natural Resources Research Institute, University of Wyoming, June, 1965.
19. Hawk, C. O., and R. W. Hiteshue, "Hydrogenation of Coal in the Batch Autoclave," U.S. Bureau of Mines Bulletin 622 (1965).
20. Agroskin, A. A., *Chemistry and Technology of Coal,* U.S. Dept. of Int. and Nat. Sci. Foundation, Washington, D.C. (1966), p. 234. U.S. Dept. of Commerce, Clearing House for Federal Scientific and Tech. Information, Springfield, Va.
21. Glass, H. D., "Differential Thermal Analysis of Coking Coals," *Fuel, 34,* 253 (1955).
22. Van Loon, W., and H. H. Smeets, "Combustion of Carbon," *Fuel, 19,* 119-21 (1950).
23. Powell, A. R., "Gas From Coal Carbonization - Preparation and Properties," in *Chemistry of Coal Utilization,* Vol. II, H. H. Lowry Ed., Wiley, New York (1945), p. 922.

24. Wilson, P. J., and J. D. Clendenin, "Low-Temperature Carbonization," in *Chemistry of Coal Utilization,* Supplementary Volume, H. H. Lowry Ed., Wiley, New York (1963), pp. 395-460.
25. Jones, J. F., M. R. Schmid, and R. T. Eddinger, "Fluidized Bed Pyrolysis of Coal," *Chem. Eng. Prog., 60* (6), 69-73 (1964).
26. Field, M. A., D. W. Gill, B. B. Morgan, and P. G. W. Hawksley, *Combustion of Pulverized Coal,* British Coal Utilization Research Association (BCURA), Leatherhead, Surrey (1967).
27. Morinaga, K. and Y. Jomoto, "Coal Slurry Injection into the Blast Furnace," *Iron Steel, 40* 92 (March, 1967).
28. Howard, H. C., "Vacuum Distillation of Coal," in *Chemistry of Coal Utilization,* Vol. I, H. H. Lowry Ed., Wiley, New York (1945), pp. 761-773.
29. Montgomery, R. S., "Coal Chemicals from Coal Oxidation Products," presented at 143rd National Meeting, American Chemical Society, Division of Fuel Chemistry, Cincinnati, Ohio, January, 1963 (Vol. 7, No. 1).
30. Wilson, P. J., and J. H. Wells, *Coal, Coke, and Coal Chemicals,* McGraw-Hill, New York (1950).
31. Ouchi, K., Imuta, K., and Y. Yamashita, "Catalytic Depolymerization of Coals. 1. Depolymerization of Yubari Coal by p-Toluene Sulphonic Acid as Catalyst," *Fuel, 44,* 29-38 (Jan., 1965).
32. Howard, H. C., "Chemical Constitution of Coal: As Determined by Hydrolytic Reactions," in *Chemistry of Coal Utilization,* Vol. I, H. H. Lowry Ed., Wiley, New York (1945), pp. 418-424.
33. Breger, I. A., M. Deul and S. Rubinstein, "Geochemistry and Mineralogy of a Uraniferous Lignite," U.S. Geological Survey, Washington, D.C., Jan. 15, 1955. Presented in part at ACS Div. of Fuel Chemistry, 126th Meeting, New York City, Sept., 1954.
34. Moore, G. W., "Extraction of Uranium from Aqueous Solution by Coal and Some Other Minerals," U.S. Geological Survey, Denver, Colo., Feb. 20, 1954.
35. "Australian Breakthrough in Metals Extraction," *Mining Engineering,* p. 68, October, 1961.
36. Swanson, V. and T. G. Ging, "Possible Economic Value of Trona-Leonardite Mixtures," U.S. Geological Survey Prof. Paper 800-D, pages D71-D74, Geological Survey Research 1972.

Chapter 3

REACTIONS WITH AIR OR OXYGEN (Combustion)

Combustion, carried to completion to yield carbon dioxide, is a conversion yielding a product representing the limit of degradation either for further energy or for chemical uses.

The reactions of carbonaceous materials with oxygen or oxygen-bearing mixtures are but instances of oxidation, a concept which has been generalized way beyond the original conceptions, e.g. to embrace electron transfer. The amount of literature on the subject of oxidation is large as befits these phenomena of both such general and specialized interest.

There is also a voluminous literature for the more special case of the combustion reactions. Any compound which will react with oxygen is in a sense potentially a combustible material, the matter being one of degree, depending upon what temperature levels, reaction rates and heats of reaction are involved. If the reaction occurs at levels whereby it maintains itself, then the reaction can be described as combustion. It may be further characterized by incandescence of the reaction system.

In the reaction with coal, the carbon-oxygen reaction is the more important, the hydrogen-oxygen reaction being secondary. Two independent equations may be written, in simple terms:

$$C + \tfrac{1}{2}O_2 \rightarrow CO \qquad (1)$$
$$CO + \tfrac{1}{2}O_2 \rightarrow CO_2 \qquad (2)$$

From these, other algebraicly dependent equations may be formed, such as

$$C + O_2 \rightarrow CO_2 \qquad (3)$$
$$CO_2 + C \rightarrow 2CO \qquad (4)$$

The heat of reaction for (1) is -26.4 kcal/g-mole at standard conditions, and -67.64 for (2), giving -94.05 for the overall reaction to CO_2. The equilibrium constants for (2) favor lower temperatures, but even at higher temperatures the values are so high as to be of no consequence to combustion. The equilibrium constant for reaction (3) also favors lower temperatures. (See Chapter 4.)

However, at higher temperatures the equiibrium constant for reaction (3) tends to approach that for reaction (1). This indicates that if CO is to be the product, then higher temperatures should be employed.

Though the stoichiometric reaction equations are simple enough to write, there is a confusing variation of hypotheses as to the sequential reaction mechanism. This is caused in considerable part by the fact that the reaction is heterogeneous, involving solid and gaseous phases — a matter also related to catalysis.

For instance, as one hypothesis, surface oxides are considered to form and decompose to CO and CO_2. For another hypothesis, a surface reaction is considered to be $C + CO_2 \rightarrow CO$ accompanied by a film reaction $CO + O_2 \rightarrow CO_2$. Later concepts have returned to the formation of surface oxides (at active centers) followed by decomposition to form the carbon oxides. Etc.

The diffusional surface reaction conception is amenable to treatment by the methods developed for heterogeneous catalysis. In this way, intuitive arguments can be methodically replaced by mathematical models, which can be compared one against another and a most probable mechanism assigned. Thus, for instance, the reforming of CO_2 to CO is analyzed to be diffusion controlling.

Studies and reviews either concerned primarily with the combustion of coal or having bearing on aspects to be emphasized — some of which have been previously cited — include those of Field et. al. (1), Jost (2), Start (3) and Whittingham (4). Others will be referred to as discussion proceeds.

To quote Field et. al. in comment on the state of the art of combustion theory:

"There is no mathematical model against which to match experimental values and the experimental studies themselves have not yielded empirical relations of any degree of generality."

Apropos of combustion, Boudart (5) brings up the following questions:

"Is it possible to obtain valid kinetic information from a direct study of flames?"

and

"Is it permitted to use the classical formulation of chemical kinetics in the treatment of combustion processes?"

Reactions with Air or Oxygen (Combustion)

To the first of these, it is remarked that because of numerical techniques and computer availability, the equations involved can be handled. To the second question, it is commented that classical kinetics is based on equilibrium concepts, whereas flame reactions are of a rapidity to cause severe departure from equilibrium. Moreover, the rates of heat and mass transfer may become the determining factors. (The term "equilibrium" as used above has the special connotation that at any point the equation of state, at the least, will hold.)

While on first examination a chemically reacting system appears to be one phenomenon, further examination may reveal a more complex order of events. Such is the case with combustion. What appears at first to be a simple stoichiometric reaction between a carbonaceous material and oxygen, on further analysis proves better explained by taking into account heterogeneous effects involving heat and mass transfer, and reaction at a surface.

The reaction of a gaseous hydrocarbon, even with air or oxygen, is not ordinarily a truly homogeneous system. Rather it also involves gradients in composition and temperature (and pressure) with respect to surfaces and interfaces.

The situation becomes even more complicated for the combustion of liquids or solids. Here, change of phase is involved as well as heat and mass transfer between phases. This, plus the fact that variation also occurs with respect to time and position, generates a very complicated system for study, indeed. Moreover, the reaction rates encountered are very fast, making the taking of experimental data very difficult.

For these and other reasons, both the quantitative experimental data and the corresponding mathematical models and correlations have not advanced to the degree corresponding to more well-behaved phenomena. Analyses have been made, however, and do provide a means for describing — and to a lesser extent correlating — phenomena occurring during combustion.

Moreover, the processes of chemical reaction have long been considered to be associated in some way with the presence of water or water vapor. The remark is particularly apt for homogeneous reactions in the vapor phase, and is pertinent to combustion reactions.

It is these phenomena which we wish to discuss, and their relation to the practical and controlling variables of combustion systems. The effect of varying amounts of water upon the reacting system is also of outstanding interest, particularly as pertains to the combustion of coal. Additional material related to the formation of carbon monoxide (and hydrogen) is also presented in later chapters, notably Chapter 4.

3.1 INFLUENCE OF COAL QUALITY

The parameters of rank and moisture content are regarded as determining factors in combustibility. "Combustibility," however, pertains to both heating value (heat of reaction) and facility of reaction — items of sometimes opposing effects. Thus lower rank coals may be more "reactive" than higher rank coals, though exhibiting a lower Btu-rating. Hence the statement that there is apparently no influence of rank on coal combustibility (6). At the same time, anthracites are generally more difficult to burn than are bituminous coals, having fewer volatiles (7). It becomes a trade-off between fixed carbon and volatiles (8).

The lower the rank of a coal, the greater the wettability to water; but the higher the rank, the greater the wettability to tar or pitch (9). A high moisture content is associated with a high unit surface area of the coal (10). This is especially so for retained moisture after drying.

Coals become harder to grind as the percentage volatiles decreases, up to a point, as denoted previously by Figure 2.4.

Lignite usually serves as the more extreme example of low-grade fuel of high moisture content. Problems encountered in lignite combustion are to a lesser degree applicable to other systems. Lignitic coals give up their moisture more slowly than harder coals (9), but have a higher volatile content, which helps offset the effect of high moisture (11). In the oxidation of lignite at low temperatures, there occurs a progressive decrease in oxidation rate (12), whereas a high characteristic rate of oxidation is one criterion for the tendency toward spontaneous combustion (10).

For the burning in the pulverized form, it appears essential to dry lignite and brown coals down to 15-20% moisture (9). For the burning of pulverized lignite the lowest possible ash and moisture contents are desired, and the highest grindability and Btu content (11). A high ash-fusion temperature is also desirable.

Significant changes occur in the physical and chemical properties of coal during storage and exposure to air (10). These effects are accelerated by temperature and are considered due primarily to oxygen and moisture content. These effects apparently are somewhat related, for the higher the bed moisture content of coals, the higher the characteristic rate of oxidation. High moisture also usually correlates with high oxygen content of the coal.

There are advantages and disadvantages to drying coal for pulverized systems (13). Among the advantages are:

1) The capacity of the pulverizer is increased (after drying).

Reactions with Air or Oxygen (Combustion)

2) Dry coal doesn't pack readily.
3) A higher flame temperature and lower stack temperature may be obtained.
4) The possibilities of spontaneous combustion are decreased.

Among the disadvantages are:
1) Increased equipment costs.
2) Increased operating costs.
3) Additional floor space.

Uniformity has been mentioned as the single most important property for a fuel (14). As far as burning is concerned, the two outstanding characteristics are coking and free-burning (15). These factors, more than any other, determine the type of equipment to be used for burning. While there are many gradations in between, most eastern coals for example are coking, while interior coals tend to be free-burning.

The term caking is used synonomously with agglomerating. When heated, caking coals pass through a plastic stage wherein the particles fuse together. These masses of semi-coke so formed only allow air and combustion gases to pass through cracks, and the pieces must be broken up for combustion. As previously suggested, however, the term "caking" does not necessarily mean the same as "coking". For instance, there are many caking coals which do not make good coke; on the other hand, some free-burning coals make good coke. The quality of coke — the residue from carbonization — is judged upon its behavior in usage, and is related to such properties as reactivity, true and apparent density, combustibility, adsorptivity, electrical conductivity, and compression strength. Usually caking and coking coals have a lower oxygen content, but when exposed to air, pick up oxygen. At the same time the coking property is reduced. This effect is noted in weathered coals.

The reaction characteristics of cokes are notably dependent upon the preliminary heat treatment (16). Thus a coal heated to 800°C gave faster reaction rates than one heated to 650°C. The rate was said not to depend upon particle size.

3.2 LOW-TEMPERATURE REACTIONS

Chemical and physical changes occurring in coal during storage have been noted as increases and decreases (10):

Increases
    Weight
    Oxygen content
    Hygroscopicity
    Ignition temperature

Decreases
    Carbon content
    Hydrogen content
    Heating value
    Coking power
    Average particle size

These changes are generally attributable to the presence of oxygen and moisture.

Moreover, the combustion properties of coals are changed after atmospheric oxidation. These changes are but ill-defined due to the inherent lack of criteria. It is noted, however, that the ignition temperature may be raised.

These effects are examined in more detail as follows.

ATMOSPHERIC OXIDATION

The statement has been made that all coals, except anthracite, react with oxygen at ordinary temperatures (13). The oxidation is continuous over wide ranges of time and conditions, and the rate increases with temperature and decreases with particle size. Coals which have a high percentage of textural moisture oxidize more easily. This is attributed to the moisture evaporating leaving a large surface area for absorption. However, once weathered, further oxidation occurs less readily.

There is an opinion that weathered coals are inferior in firing qualities. In other words, drying reduces the firing qualities. It is also noted that coking properties are lowered by weathering.

Atmospheric oxidation tends to reduce the amount of soluble material in coal, though some coals will show an increase (17). Oxidized carbon surfaces may have either acidic or basic properties (14), which affects solubility. The exhaustive, low-temperature oxidation of fine, bright coals will yield products which are almost wholly soluble in caustic solutions (10). In fact, this solubility may be used as a measure of the extent of coal oxidation.

In some respects low-temperature oxidation is similar to combustion. The gases given off by low-temperature oxidation are the same as for combustion: CO, CO_2, H_2O. The chief difference is in the lower rate,

Reactions with Air or Oxygen (Combustion)

and the fact that about half of the oxygen consumed remains with the coal — and may even cause a gain in weight.

The oxygen absorbed by coal is retained strongly, and recovered largely as CO and CO_2, and as water vapor (14). Even at combustion temperatures, carbon surfaces have been shown to absorb and retain oxygen.

Oxygen which combines with carbon at low temperatures (200-300°F) goes almost wholly to CO (13). Heating oxidized coal to just below the decomposition temperature drives off CO, CO_2, and H_2O (10).

Low-temperature oxidation has been regarded as forming a coal-oxygen complex as an intermediate, which decomposes to CO, CO_2, and H_2O. Yavorskii (18) states that in low-temperature oxidation the oxygen combines with available free valences and with oxygen compounds in the coal. Under these circumstances, the reactivity of the fuel depends upon the presence of these unstable compounds and *not* on the total amount of volatiles. In comment, the measure of low-temperature reactivity would not necessarily be an indication of combustibility.

The low-temperature oxidation of coal is highly exothermic, with the rate of reaction increasing rapidly with temperature (10). Under appropriate conditions, this sequence can lead to spontaneous combustion. The initial rate of low-temperature oxidation will be the highest, for a given temperature; then the rate falls off. As would be expected, it is higher for fine coals than coarse. In most instances, low-temperature oxidation occurs only at the surface; diffusion is slow. The rate of reaction and heat dissipation will determine local temperatures during low-temperature oxidation.

Some coals, when preheated, may have an increased oxidation rate effect; others may not.

Carbonizing properties change after atmospheric oxidation. Coke strength goes down at times but at other times increases. Yields of tar decrease after oxidation, but the yields of CO and CO_2 increase, as does NH_3. During low-temperature oxidation, a loss in coking power occurs, which may be very rapid. The period may be as little as 48 hrs., up to two weeks, though some coals are not affected.

MOISTURE

Comparisons between dried and wet coal show only a slight effect of moisture on the low-temperature oxidation rate (10). If absolutely the last trace of moisture is removed, the rate is reduced — but the point is academic since water is a reaction product. There appears to be general agreement that if some water is present, then variations in water content have little effect on oxidation rate.

Low-temperature oxidation in dry O_2 has been found less than in aqueous solutions if the sulfate ion product is removed. The sulfate inhibits the oxidation.

When pure carbon is heated with dry oxygen, the gases evolved contain both CO and CO_2 (13). Moreover, the carbon does not glow as with wet oxygen. Moist oxygen which is absorbed by charcoal will be evolved as CO_2 at 212°F. However, with dry oxygen, evolution occurs only very slowly at 842°F (450°C). Above this temperature, the product gases are mainly CO.

PYRITES

There have been arguments as to the effect of pyrites on coal oxidation. Schmidt (10) states that if the pyrites exist in layers or veins there is little effect. If in a finely divided state, the pyrites may play an important role.

Iron disulfide occurs in coal in two crystalline forms: pyrite and marcasite. Both oxidize and will be referred to as pyrites. The exothermic low-temperature oxidation reaction can be considered as follows, in g-moles:

$$2FeS_2 + 7O_2 + 2H_2O = 2H_2SO_4 + 2FeSO_4 \quad - 62{,}000 \text{ calories}$$

The remark is made that storing coal where it would be wet repeatedly (by rains) favors pyrite oxidation. This would not be the case for coal stored completely immersed. It is further asserted that the consensus is that pyrites play only a minor role in changes in the coal, though some hold they serve to pilot spontaneous ignition.

SPONTANEOUS COMBUSTION

Schmidt (10) discusses the merits of characteristic rates of oxidation, and its relation to spontaneous combustion. Walker (19) has indicated that spontaneous combustion of coal can take place by either a dry or wet combustion mechanism. A review of the literature on spontaneous combustion is presented by Walker.

There is a critical temperature below which spontaneous combustion will not start (13). Usually, if coal can be stored 90 days without igniting, then chances are it won't (10). There are of course other variables that enter into the picture, such as rate of heat loss. Adiabatic small-scale apparatus has been used to measure the temperature rise on exposure to air. Also, aqueous solutions of oxidizing agents have been tried, but are difficult to correlate.

Reactions with Air or Oxygen (Combustion)

Features which indicte a tendency toward spontaneous combustion are:

>High characteristic rate of oxidation
>High friability
>Presence of finely divided pyrites

The effects may compensate one another. For example, the high rate of oxidation of bituminous coals is balanced by a lower friability.

Essential features involved in spontaneous combustion have been listed by Haslam and Russell (13):

1) Oxidation of coal is continuous over a wide range of time and conditions.
2) Rate increases with temperature and decreases with particle size.
3) For a given coal, there is a critical temperature below which spontaneous combustion will not occur.
4) Weathered coals have a higher critical temperature.
5) Coals containing a high percentage of textural moisture lose it on exposure to air, thus offering a large surface for oxygen absorption. As a corollary, coals with high textural moisture and high oxygen content have lower critical temperatures.
6) Sources of heat include the effect of moisture; wetting of coal increases the rate of oxygen absorption.
7) Oxidation and spontaneous combustion occur in stages.

In spontaneous ignition, the moisture and oxygen sorption are somehow related (9). It is believed the heat of wetting is significant, and the heat of adsorption of water vapor should be even greater.

Porous and high-sulfur coals give the most trouble with regard to spontaneous combustion, especially when damp (20). Since oxidation occurs at the particle surface, p.f. is very susceptible and will usually ignite if left undisturbed for 2 or 3 days.

It was originally thought that the oxidation of pyrites would generate enough heat for spontaneous combustion by the aforementioned reaction (13):

$$2FeS_2 + 2H_2O + 7O_2 = 2FeSO_4 + 2H_2SO_4$$

However, calculations indicate that 2% pyrites would only raise the coal temperature to 260°F. Consequently it was assumed pyrites are not responsible, though their presence may contribute.

An unusual mechanism for causing spontaneous combustion has been proposed (10). The processes of evaporation and condensation may lead to the movement of water vapor through a coal pile, which can produce an internal temperature rise due to the effects of wetting.

Means for preventing spontaneous combustion include storing the coal in small piles, storing under water, and keeping storage bins and silos airtight, especially at the bottoms and sides (20). Chimney effects in storage piles are eliminated by spreading and rolling in horizontal layers.

3.3 THE ENERGETICS OF COMBUSTION

More generally, the combustion of carbonaceous materials — which contain hydrogen and oxygen as well as carbon — involves a wide variety of reactions among reactants, intermediates, and products. The reactions occur simultaneously and consecutively, in both forward and reverse directions, and may at times approach a condition of equilibrium. Moreover, there is a change in the physical and chemical structure of the fuel particle as it burns, recognized as the "mode of combustion".

Inasmuch as the complexity of coal molecular structures and substructures defies any sort of exact analysis, for manageability discussion is usually confined to the system

C, H or H_2, O or O_2 (and N_2), H_2O, CO, CO_2

Even with this simplification, not only are competing reaction rates involved, but reaction equilibria as well — not to mention phase changes. And in addition to the thermochemistry of the competing reactions, there is the matter of heat and mass transfer with the surroundings — and the latent heat change for water. The system remains complicated even with simplification.

Haslam and Russell (13) present a tabulation of combustion reactions, as do other sourcee. Wilson and Wells (21) list the principle combustion reactions in the following form, as part of the overall conversion:

1) $C(s) + O_2(g) = CO_2(g)$ - 169,290 Btu
2) $2C(s) + O_2(g) = 2CO(g)$ - 95,100
3) $C(s) + CO_2(g) = 2CO(g)$ + 74,200
4) $2CO(g) + O_2(g) = 2CO_2(g)$ - 243,490
5) $2H_2(g) + O_2(g) = 2H_2O(g)$ - 208,070
6) $C(s) + H_2O(g) = CO(g) + H_2(g)$ + 56,490

Reactions with Air or Oxygen (Combustion)

7) $C(s) + 2H_2O(g) = CO_2(g) + 2H_2(g)$ + 38,780
8) $CO(g) + H_2O(g) = CO_2(g) + H_2(g)$ − 17,710

In the convention used above, a negative heat of reaction indicates exothermicity. With the heat of reaction in Btu's, the reactants and products are in lb-moles.

It should be observed that a part of the reactions are dependent. That is, there are only four independent equations in the enumeration.

Reaction (1) is faster than (2) or (3) and such that the latter would be controlling. The relation between composition and distance up through a *fixed coal bed* indicates that equilibrium will finally be attained (21). It is found that CO_2 reaches a maximum and falls off, while CO builds up to equilibrium. This can be explained by the reaction $CO_2 + C = 2CO$.

Reactions which conceivably may occur at the surface of the coal char are (1):

$$C + O_2 = CO_2$$
$$C + \tfrac{1}{2}O_2 = CO$$
$$C + O = CO$$
$$C + CO_2 = 2CO$$
$$C + H_2O = CO + H_2$$

In addition, the gaseous products may react:

$$CO + \tfrac{1}{2}O_2 = CO_2$$
$$H_2 + \tfrac{1}{2}O_2 = H_2O$$
$$CO + H_2O = CO_2 + H_2$$

There are also other possibilities.

The classic concept for burning of the solid residue considers reaction in two zones (22):

Oxidation zone
$$C + O_2 = CO_2$$

Reduction zone
$$CO_2 + C = CO$$

The description is as follows (13):

1) An oxidation zone which produces CO_2 and consumes most of the oxygen.
2) A reducing zone where CO_2 is reduced to CO by carbon.

Another concept replaces the oxidation zone with

$$C + O_2 = CO$$
$$CO + O_2 = CO_2$$

Thring and Essenhigh (23) advance a three-zone theory for the combustion of carbon:

1) $C \xrightarrow{O_2} CO$

2) $CO \xrightarrow{O_2} CO_2$

3) $CO_2 \xrightarrow{C} CO$

This replaces the older two-zone theory. Other concepts consider all reactions to be simultaneous and question the use of "zones".

For the mechanism of combustion of solid hydrocarbons, three alternatives have been considered (13):

a) CO_2 as the first product.
b) CO as the first product.
c) The first product a complex which breaks down to give CO and CO_2.

The latter is regarded as the more modern approach. For small particles at least, calculations indicate that CO is formed first at the char surface (8). The CO then diffuses away and reacts. Such is not necessarily so for larger particles. In the two-film or two-zone combustion theory, CO_2 which is formed is considered to diffuse back to the surface and react with carbon:

$$C + CO_2 = CO$$

It is concluded that the principal product released at combustion temperatures in the firing of p.f. is CO. In fact the ratio of CO to CO_2 released may be as high as 12. This is subsequently altered.

HEAT BALANCE

The heat balance provides the means to relate the heat absorbed (by the surroundings, which include the heat transfer media) to the heat produced by combustion. The presence of water vapor in the combusting system appears in the latent heat effects. One consequence of high moisture content in coal combustion is that a part of the heat is lost due to evaporation of the moisture in the coal and is not recouped from the combustion products.

Reactions with Air or Oxygen (Combustion)

The heat liberated to the heat transfer surfaces in a furnace is established by a heat accounting. To quote Wilson and Wells (21):

> "The quantity of heat available from combustion is the sum of the total heat in the fuel and air, plus the heat of combustion, minus those amounts of heat carried away in the stack gases and in any ash and unburned fuel."

The heat absorbed is accounted by the enthalpic change of the surroundings, principally the heat transfer media.

The makeup of the heat balance does not incorporate kinetic, potential, or dissipative energy effects, which are relatively small. It is rather an overall statement balancing thermal input versus output for and given furnace performance. A statement of the heat balance has been made as follows (21):

1. Total heat input
    a) Heat of combustion of the fuel.
    b) Sensible and latent heats of air, fuel, and other materials, respectively.
    c) Heat from exothermic reactions, other than combustion, which may enter into the process.
2. Total heat output
    a) Sensible and latent heats in carbonization products.
    b) Heat absorbed by endothermic reactions.
    c) Sensible and latent heats in combustion products (ash and stack gases).
    d) Heat of combustion of unburned fuel.
    e) Heat losses to surroundings by convection, radiation, and conduction.

Though there are latent and sensible heat losses due to the presence of moisture, Gauger (9) makes the following comment:

> "... evaporation of extra moisture increases the superheated-vapor item of the stack losses, but this is more than offset by the improvement in carbon dioxide and reduction in ashpit loss."

The statement was also made that moisture up to 6% in the coal may be without effect on the overall heat requirement, since the sensible heat of the gases vaporizes the water. Above this, there is a noticeable effect on the heat load. With respect to gaseous fuels, the water content is usually so low that it can be neglected in the heat balance (15). It can be argued that the chemical reaction of water (to CO and H_2) followed by the burning of the products to yield water once more, does not contribute or detract from the overall energy balance. In these terms, the effect of water would be catalytic only, to promote the rate of reaction of the carbon. The only energy loss would be the latent heat change of the water from the liquid or absorbed phase to the vapor phase.

Efficiency:

Efficiency is commonly defined as the ratio of the heat absorbed to the heat input when considering the performance of combustion systems, notably boilers. Furthermore, the heat input is usually based on the gross heating value of the fuel, as is stressed elsewhere in this work. This introduces an inherent discrepancy since the latent heat of the moisture in the combustion product gases is not ordinarily recoverable in practice.

The relatively large proportions of the heat absorbed relative to the heat input, as generally encountered, often makes it more desirable to establish the heat absorbed by difference: that is, to subtract the heat losses from the heat input. An example of this procedure is demonstrated in the heat balances of Table 3.1 for the performance of boilers.

This type of determination is shown graphically in Figure 3.1. Other determinations of boiler efficiency are presented in Figures 3.2 and 3.3.

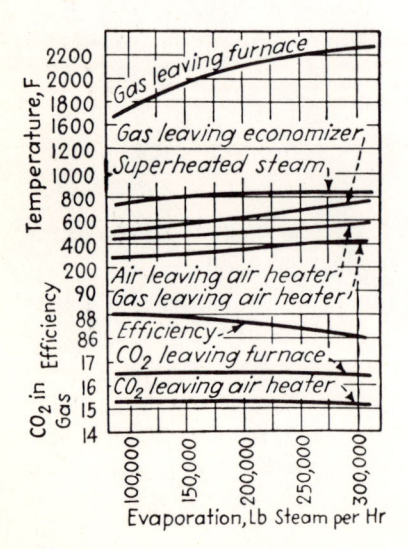

Fig. 3.1. Boiler performance using pulverized coal (from *Mechanical Engineers' Handbook*, fifth edition. Used with permission of McGraw-Hill Book Company).

Reactions with Air or Oxygen (Combustion)

TABLE 3.1. Examples of Heat Balances (from *Mechanical Engineer's Handbook*, fifth edition).

Item	Burning pulverized coal — Plant A, dry bottom furnace	Burning pulverized coal — Plant B, slagging bottom furnace	Burning coal on stoker — Plant C[a]	Burning natural gas — Plant D
Capacity, lb per hour	900,000	450,000	220,000	225,000
Losses, per cent				
Dry gas loss	7.07	5.18	11.3	3.80
Moisture in coal or from combustion of H_2	3.86	5.86	4.2	10.53
Moisture in air	0.11	0.19	....	0.13
Combustible in refuse or unburned carbon	1.24	0.14	2.8	
Unburned gas	....	....	....	0.10
Radiation and unaccounted for	0.60	0.54	0.5	0.51
Heating of water in slag pit	....	0.74		
Heat in slag	....	0.22		
Mill tailings	....	0.08		
Total losses	12.88	12.95	18.8	15.07
Efficiency by difference	87.12	87.05	81.2	84.03

[a]The relatively lower efficiency of this installation was due to the absence of heat-recovery equipment.

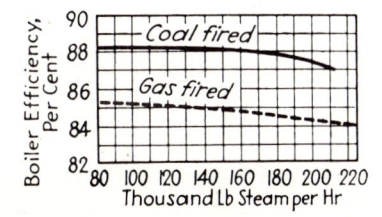

Fig. 3.2. Boiler efficiencies with coal and gas fuels (from *Mechanical Engineers' Handbook*, fifth edition. Used with permission of McGraw-Hill Book Company).

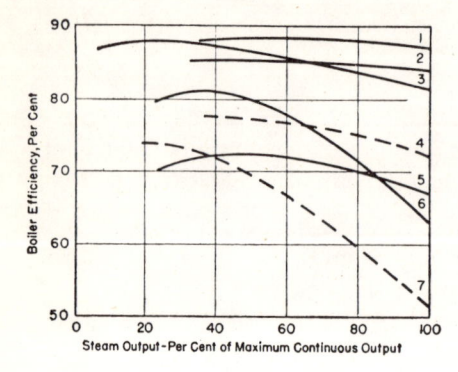

Fig. 3.3. Boiler efficiency curves. 1. Coal-fired drum boiler, economizer, and air heater. 2. Gas-fired drum boiler, economizer and air heater. 3. Oil-fired naval boiler (destroyer). 4. Coal-fired industrial boiler, no air heater. 5. No. 3 buckwheat-fired header boiler, no economizer or air heater. 6. Bituminous-fired four-drum boiler, economizer, no air heater. 7. Coal-fired locomotive boiler. (From *Chemical Engineers' Handbook*, 3rd edition. Used with permission of the McGraw-Hill Book Company).

SOOT FORMATION

Sir Humphry Davy is credited with being the first to recognize that carbon particles produce the luminosity in flames (24). Carbon black production is one aspect of the subject of the formation of dispersed carbon in hydrocarbon diffusion flames (25), and thus even more broadly of combustion. It has been the subject of many works and reviews, but the mechanism is still uncertain.

According to Field et. al. (1) little is known about the limiting requirements for soot formation. Furthermore, soot that is formed may be burned again. Burning times for 500 Å soot particles, at 1600°C, are estimated to be ~ 0.2 sec. in 0.05 atm. partial pressure of oxygen.

It has been noted that conditions for the formation of soot may occur if contact with oxygen is delayed.

In the explosion of methane,

$$2CH_4 + O_2 = 2CO + 4H_2$$

it would be expected that the reaction should proceed stoichiometrically (24). Instead, about 1/5 of the carbon will appear as soot. On increasing the oxygen concentration, the yield of soot is decreased and the yield of CO increased. Also, less water is formed. In the explosion of ethane,

$$C_2H_6 + O_2 = 2CO + 3H_2$$

appreciable soot, some water, and about 10% CH_4 is formed.

Explosion yields were compared as taking place in a large spherical vessel and a long cylindrical tube. It was found that reaction in the sphere

Reactions with Air or Oxygen (Combustion)

gave appreciable quantities of soot, the tube little. Hence the formation of carbon or lack thereof was reasoned to be associated with surface effects.

The channel process for the production of carbon black utilizes an air-deficient diffusion flame. The carbon is collected on cool plates. The main parameter of carbon black or soot formation is believed to be the number of carbon particles formed (25). The size and yield will be a function of the form of the nuclei, and their growth.

Tesner et.al. (25) studied the number of particles formed in a diffusion flame, based on carbon removal. Two methods were checked against each other:

1) Carbon black was deposited on a solid surface, which was introduced into the flame at different heights.
2) Carbon black was removed from the gaseous product with a bag filter. The reaction was stopped at different heights in the reactor by introducing a water-cooled lattice.

There was good agreement between the two methods. Methane was used as the feed gas. A 3 mm. porcelain tube formed the burner. Flame height was 240 mm.

It was found that a tar-like material adhered to the surface of the carbon particles. The dispersion of the particles themselves varied within the flame. The dispersion varied up and down, and across as well, with the largest particles toward the center of the flame.

If nitrogen or hydrogen is added to the methane, the dispersion of the carbon black increases but the yield decreases. If higher hydrocarbons are added, the yield increases but the specific surface decreases.

Paraffins and olefins added to the methane decrease the number of carbon particles. Aromatics increase the number of particles and the yield, up to a point. A further increase lowers the yield and increases the particle size.

In batch oxidations of the light hydrocarbons it is reported that unsaturates do not form as much carbon (soot) as saturates (24).

CONVECTION AND RADIATION

It has been said that there is no combustion problem, only one of heat transfer (1). For the rate of combustion can always be increased by finer grinding of the coal, or by controlling the excess air, etc.

In furnace operation, it has been estimated that 20% of the reaction heat is released directly as radiant energy (7). The remaining heat energy resides in the combustion products, from which about 30% of the energy is then released as radiation.

The presence of water vapor in the combustion gases itself may have some appreciable effect upon the gas emissivity and radiation. Radiation from CO flames is attributed principally to the presence of the CO_2 molecule formed (2).

Hottel (26) makes the following statement with regard to radiant heat transmission involving nonluminous gases:

> "Of the gases encountered in heat-transfer equipment, carbon monoxide, the hydrocarbons, water vapor, carbon dioxide, sulfur dioxide, ammonia, hydrogen chloride, and the alcohols are among those with emission bands of sufficient magnitude to merit consideration."

In addition to creating practical problems such as ash removal and fouling, the ash is a factor in radiation to the walls (7). The presence of soot particles has been noted to raise the emissivity of the combustion gases (1).

Since radiant heat can only penetrate short distances in a fixed bed, the temperature and coefficient of convective heat transfer will determine the intensity of warming up and ignition in fuel beds (18).

A tabulation of theoretical flame temperatures for gases with stoichiometric air indicates temperatures to 3600-3700°F or 2000°C (21). After a point, in burning p.f., there is no way of increasing the rate of heat transfer, except possibly by increasing the flame emissivity (1). Further excess air is inefficient, and grinding costs increase as particle size is decreased. In final analysis, the heat transfer problem becomes one of costs.

The burning time of p.f. is independent of pressure (22). This offers a potential means of increasing capacity; for the feed rate can be increased in direct proportion to pressure.

A steam-jet ignition system has been developed for improving the ignition of low-grade coals (4). The high turbulence of a steam-jet increases convection by drawing hot gases over the bed.

It has been observed in the combustion of p.f. in turbular reactors that convective transfer due to convection is considerably greater than radiation heat transfer — at least in the inlet section (1).

Orning (7) states that the influence of water-cooled walls on combustion has been the subject of considerable speculation but little experimentation. In comment, it would be expected to affect not only the reaction, but convective heat transfer as well.

Reed (27, 28) has noted that recirculation occurring in industrial furnaces may transmit a considerable part of the heat load by means of convective heat transfer. In tube stills, for instance, the magnitude of recir-

Reactions with Air or Oxygen (Combustion)

culation will be affected by such factors as wall-to-tube and tube-to-tube spacing. It is the colder temperature of the tubes which tends to set up the circulation. In comment, it would be expected that the cooler temperature of a furnace wall could also set up recirculation of the combustion gases.

Theoretical calculations are provided by Hedley and Jackson (29) on the influence of recirculation in combustion processes. Adiabatic and non-adiabatic oil-fired systems are idealized to obtain the effects on volumetric combustion rates. Hoffman (30) has used a combustion flow model to introduce the effects of recirculation.

FOULING

The efficiency of heat transfer in furnaces is impaired by deposits and corrosion of the heating surfaces. In this connection fuel-bed conditions which affect the volatilization of Na and K compounds (and their reaction with SO_3) become important (4). Ash deposits are related to Na_2O content, and firing temperatures should be kept below the ash fusion temperature (11).

It has been found that the composition of the ash will vary throughout a furnace (7). For example, the furnace bottom will have a greater concentration of the lower oxides of iron. Since ferrous oxide has greater fluxing power than ferric oxide, this in turn will contribute towards slag formation.

Coal ash is composed mostly of metal oxides (15). Typical ranges of composition are:

SiO_2	40-60%
Al_2O_3	20-60
Fe_2O_3	5-25
CaO	1-15
MgO	0.5-4
$Na_2O + K_2O$	1-4

The composition affects the softening point. Iron oxides are a particular source of trouble. The $CO-H_2$ reducing atmosphere produced by the water gas reaction in the fuel bed, serves to reduce ferric oxides to ferrous oxide and ferrous sulfide and produce clinker.

To minimize fouling, the combustion gases should be reduced in temperature to below where ash will deposit and sinter (1). This temperature is critical at about 1100°C. It is indicated that the principal heat loss will be by radiation. The Babcock & Wilcox Company has devoted considerable effort to the study of fuel-ash effects on boiler design and operation (31), as has Combustion Engineering (15).

Fouling has been related to alkali content and the formation of sulfates (31). It has been observed that coals with total alkali content of less than 0.5 wt. percent as equivalent Na_2O give deposits that can be removed by the action of soot blowers. For coals above 0.6% as equivalent Na_2O, the fouling deposits increase markedly and become a pronounced problem.

EXCESS AIR

Combustion reactions are a balance on the one hand of high reaction or flame temperatures, which favor carbon monoxide at equilbrium, and on the other hand of the use of excess air which drives the conversion to carbon dioxide. Relative to these two counter-acting facets, rates of reaction are generally controlling, manifested by short residence times with rapid heat transfer such that the system temperature is lowered before equilibrium can occur. Hence the consideration that (complete) combustion is a non-equilibrium process.

The above is contrary to gasification by partial combustion, which occurs at lower temperatures and longer residence times without heat transfer (the system remains adiabatic). Here, equilibrium conditions tend to apply.

From a practical standpoint, the amount of excess air can be viewed as the determining feature of combustion. It not only affects combustion rates but flame temperatures and the heat balance as well. These latter aspects are presented in calculational detail in the Appendix.

TABLE 3.2. Excess Air at Furnace Outlet.

	Fuels	Excess Air, %
Solid fuels	Coal	10-40
	Coke	20-40
	Wood	25-50
	Bagasse	25-45
Liquid fuels	Oil	8-15
Gaseous fuels	Natural gas	5-10
	Refinery gas	8-15
	Blast-furnace gas	15-25
	Coke-oven gas	5-10

TABLE 3.3. Excess Air Supplied

Fuel	Type of Furnace or Burners	Excess Air % by Weight
Pulverized coal	Completely water-cooled furnace for slag-tap or dry-ash-removal	15-20
	Partially water-cooled furnace for dry-ash-removal	15-40
Crushed coal	Cyclone Furnace — pressure or suction	10-15
Coal	Spreader stoker	30-60
	Water-cooled vibrating-grate stoker	30-60
	Chain-grate and traveling-grate stokers	15-50
	Undefined stoker	20-50
Fuel oil	Oil burners, register-type	5-10
	Multifuel burners and flat-flame	10-20
Acid sludge	Cone and flat-flame type burners, steam-atomized	10-15
Natural coke-oven, and Refinery gas	Register-type burners	5-10
	Multifuel burners	7-12
Blast-furnace gas	Intertube nozzle-type burners	15-18
Wood	Dutch oven (10-23% through grates) and Hofft-type	20-25
Bagasse	All furnaces	25-35
Black liquor	Recovery furnaces for kraft and soda-pulping processes	5-7

Though thermodynamic and rate calculations may be used to maximize flame temperatures and conversions, in last analysis it is empirical observation and evidence in each situation that will dictate the optimum air-to-fuel ratio — and will depend upon fuel analysis including moisture, air humidity and temperature, and other day-to-day operating variables.

Some representative guidelines from References 15 and 31 illustrate the diversity and are presented in Tables 3.2 and 3.3. Though the values vary from ~ 5% to 50 or 60% excess, 15-25% is perhaps a median range. It depends not only upon the fuel but on the type of combustion system and the control of that system.

3.4 COMBUSTION MECHANISMS

While the exact nature of the combustion process may never be resolved — and is further complicated by the assignment of rate-controlling steps — one statement is as follows (22):

"It is agreed that the burning of coal involves three stages. 1) The release of volatile matter resulting from the heating of the coal. 2) The burning of the volatile matter of the gas phase. And 3) The burning of the solid residue of carbonaceous matter."

There are other points of view. To wit: the oxidation of coal appers to consist of two main reactions: the degradation of hydrogen and carbon (32). In the intial stage, hydrogen degradation outweighs the slow starting carbon degradation. The hydrogen degradation is doubled with oxygen addition and water formation. As the temperature of the coal increases during the oxidation process, the CO/CO_2 ratio decreases (33). The removal of CO and CO_2 is a decomposition process which continues during the entire oxidation process. Decomposition yields mainly CO. The evolution of CO_2 is due to decarboxylation of groups in the oxy-coal.

Within a temperature range of 30-100°C, activated diffusion is the predominant reaction (34). The oxidation speeds of different coals may vary as much as a factor of 10. The speed is a hyperbolic function of time and is independent of grain size below 0.5 mm. The activation energies of CO and CO formation are 10-17 Kcal/mole respectively. The amount of oxygen in CO is between 0.4% and 4% and the amount in CO_2 is between 2% and 26%.

Reactions with Air or Oxygen (Combustion)

Field et.al. (1) note that there is a fundamental difference, between gas phase reactions, such as with coal volatiles, and heterogeneous reactions, involving a gas and a surface. The former type of reaction is homogeneous — involving only one phase; the latter, as the term "heterogeneous" implies, involves two or more phases. In many cases, the rate of diffusion between phases may be rate controlling. Furthermore, reaction rates in the gas phase are usually higher — unless the surface acts as a catalyst.

It is commented by Gaydon and Wolfhard (24) that one difficulty in examining the literature is that many times distinction is made between diffusion and premixed flames. In diffusion flames, the coal and air are not premixed. This type of flame has been characterized as the most important from an industrial standpoint. Gaydon and Wolfhard further state that in a diffusion flame, the chemical reaction rate is usually not controlling. The white-hot part of the flame may be diffusion controlling, but the preheating zones at the side of the flame control the overall rate.

With this introduction we will proceed with the subject in more detail.

STAGES OF REACTION

During combustion in fuel beds, water and volatile material distill off first (15). At the same time, carbon monoxide and hydrogen are produced via the water gas reaction. This gas, plus the volatiles, then burns in the space above the bed.

In the simplest model for combustion, complete devolatilization is assumed to occur first, before ignition, then followed by the rapid burning of the volatiles as compared to the char (1). In the actual combustion process, conceivably, it would be expected that combustion would occur at the surface before volatilization becomes appreciable. This would then be followed by rapid devolatilization such that further oxygen could not reach the surface.

The physical mode of combustion proposed by Gaydon and Wolfhard (24) may initially have a difussion flame of burning hydrocarbons around each particle. This would be followed by surface combustion with the oxygen diffusing to the surface of the particle and the CO and CO_2 away. It is noted that temperature would have little effect on the diffusion rate. Flame propagation would occur by both radiation and conduction.

Lavrov (35) considers the mechanism for the combustion and gasification of a solid fuel to consist of the following stages:

1) Diffusion of reactive gases to the carbon surface.

2) Adsorption.
3) Formation of transitory complexes.
4) Desorption of products.

Field et.al. (1) consider the overall reaction mechanism for the devolatalized coal char particle to be:

>Transport of O_2 to the surface
>Reaction with the surface
>Transport of products away

The rate of transport will be increased by any means of reducing the thickness of the boundary layer.

At 900°C, 20 milliseconds are required for the evolution of 40% of the volatiles. The times for "rapid" decomposition are considered to be on the order of 10-15 seconds. This is still two orders of magnitude longer than for p.f. firing. Field et.al. make the statement that the burning time of a coal is controlled by the diffusion of reactant (diffusion of oxygen to the interface of the volatiles layer).

Contradictory to observations in industrial furnaces, burning times in small-scale experiments are found to increase with increasing volatile content (7).

Field et.al. (1) consider the controlling step of the raction sequence to be the oxidation of CO to CO_2. After the formation of the volatiles, for coal particles in the 100 micron range, at temperatures of $\sim$ 1000°C, it is estimated that $\sim$ 7 milliseconds are required for diffusion of the oxygen. The oxidation of CO to CO_2, which is assumed to be the controlling reaction step, takes $\sim$ 3 milliseconds. Hence it is believed that diffusion of the oxygen is the controlling overall step since it is also generally assumed that the transport of products away will not be controlling.

Mayers (14) states that the rate of combustion in fuel beds is limited by oxygen transport to the reacting surface, and not to the rate of reaction itself. The diffusion rate of the oxygen is also apparently independent of the gas film composition (1).

It is commented by Field et.al. (1) that — though the ordinary diffusion relations are inappropriate for distances on the order of the mean free path of the gas — at combustion temperatures the mean free path is $\sim$ 1 micron whereas the p.f. particle size is $\sim$ 50 micron. Hence the customary equations for diffusion can be used for transport of the reactants.

The statement has been made that diffusion slows down the overall reaction rate at higher temperatures (16). The inference is that diffusion

Reactions with Air or Oxygen (Combustion)

becomes rate controlling. Above 1000-1400°C, diffusion-limited combustion occurs (36).

In the fluidized combustion of coke at 700°C, Khitrin et.al. (16) found CO_2 to be the principle product and chemisorption of oxygen to occur. At higher temperatures, Burdukov and Nakoryakov (37), in experiments involving the combustion of 700μ coal particles in an ultrasonic field, found that diffusion inhibition of combustion occurred when temperatures of 1200-1300°C were reached. The chemical composition of coal was found to have no influence on combustion in an ultrasonic field.

In magnetohydrodynamics, for instance, there is involved combustion at very high temperatures (~ 2700°K), under pressure (1). Air/fuel ratios are essentially stoichiometric, chamber geometry is minimized to reduce radiant heat losses, and pressures of 5-10 atmospheres are encountered.

As the pressure of the system is increased, the mechanism tends toward diffusional control. This is due to the mass transfer or diffusion rate coefficient being inversely proportional to pressure. The overall rate, however, will increase due to the increased oxygen partial pressure at the higher pressures. It is also noted that the CO reaction rate increases, but not the reaction rate of carbon to CO.

Reactions of the other reactants,

$$C + CO_2$$
$$C + H_2O$$

have been found to be diffusion controlling even up to 2800°K (38). Thus it appears to be a clear case all the way around of the heterogeneous reactions being controlled by the rate of diffusion of the gaseous reactants.

HOMOGENEOUS BEHAVIOR AND IGNITION

Field et.al. (1) consider the principal combustion reactions as

$$C + O = CO_2$$
$$H_2 + \tfrac{1}{2}O_2 = H_2O$$
$$\tfrac{1}{2}N_2 + \tfrac{1}{2}O_2 = NO$$
$$S + O_2 = SO_2$$

Usually, equilibrium conditions are not attained due to short residence times. The products also tend to dissociate:

$$CO_2 = CO + \tfrac{1}{2}O_2$$
$$H_2O = \tfrac{1}{2}H_2 + OH$$

$\frac{1}{2}H_2 = H$ (above 2000°C)
$\frac{1}{2}O_2 = O$ (above 2000°C)
$\frac{1}{2}N_2 = H$ (above 3000°C)

At the same time reactions may occur to form CH_4 or other hydrocarbons, NH_3, C_2N_2, HCN, etc. Hence a consideration of reaction behavior requires the further examination of homogeneous reactions (e.g., oxidation and ignition) among the products which may also include hydrocarbon volatiles and oxygenated compounds.

As to hydrocarbons, three theories have been advanced for the combustion of gaseous hydrocarbons (13):

a) Preferential reaction of H.
b) Preferential reaction of C.
c) Hydroxylation:

$$C_xH_y + O \rightarrow C_xH_yO$$

These reactions would involve the volatiles, and would both precede and accompany the reactions of carbon.

The reaction of steam with methane occurs at 500-1100°C, and at 10-20 atm. (39); at lower temperatures, a reforming catalyst is required. Steam-to-methane ratios mentioned for the reforming are 5/1 to 10/1, with the lower ratios giving the higher reaction rate. The steam reforming of methane to CO and H_2 ordinarily would be inhibited by the steam-carbon reaction.

The relationship of gas composition and temperature to distance has been determined for combustion in fuel beds (15). Field et.al. (1) present diagrams giving the yield of combustion products versus excess air and temperature. The basis is low-rank coal containing 10% moisture.

It has been observed that at higher combustion temperatures the yields of CO_2 and H_2O drop off, whereas the yield of CO, [OH] and H_2 increases.

The ignition of coal has been described as occurring in just a few hundredths of a second (20). A p.f. particle goes on to burn in less than half a second. In experiments on a tubular reactor the ignition distance has been observed to be on the order of 1 micron (more or less), with times of the order of 0.3 sec. (1).

The CO formed by reaction at the char suface is said to burn on to CO_2 at distances close to the surface — from 0.05 to 0.4 cm.

Experiments have shown that water is evaporated first, so that ignition propagates through a dry fuel bed (4).

Reactions with Air or Oxygen (Combustion)

The rate of flame propagation is highest for high-volatile, low-ash fuels (20). The highest rate will occur using about 50% of theoretical air, at velocities up to 45 ft./sec. The proportionality rate constant for the burning of volatiles is independent of coal rank (8). It is rather controlled by diffusion.

The low-temperature ignition range for fuels has been characterized as being at temperatures less than 800°C (23). The high-temperature, or flame temperature range, is at temperatures greater than 1200°C.

For coals, the ignition temperature is given to be on the order of 700°C (1292°F), but may be as low as 600°C (1112°F) or as high as 800°C (1472°F) (7), depending on volatiles given off.

Ignition temperatures depend on rank (20) and range from

200-300°C (392-572°F) for lignite

to

450-600°C (842-1112°F) for anthracite

These values range considerably lower than those cited in the previous paragraph.

Ignition temperatures also depend on particle size. Values listed for the ignition temperatures of powdered fuels are as follows (13):

Lignite	150°C (302°F)
Gas coal	200°C (392°F)
Hard coking coal	250°C (482°F)
Anthracite	300°C (572°F)

Kindling temperatures (for sustained reaction) were listed as well:

Cannels	370°C (698°F)
Lignites	450°C (842°F)
Steam coals	470°C (878°F)

At the higher temperatures of combustion, Yavorskii (18) notes that gases formed from the thermal decomposition of coal are much more stable than the free valences and oxygen compounds which react at low temperatures. These gases require a higher temperature for ignition than for evolution.

Combustion is affected by general conditions in the ignition zone (6). Variables are temperature gradient and cloud velocity. An electric field gives greater apparent flame strengths to an opposed-jet diffusion flame, according to Heinsohn and Wolfhorst (40).

Flame propagation itself is considered to occur by both radiation and conduction (24). A maximum will occur in the rate of flame propagation, depending upon the air ratio (13).

Flame propagation in premixed gases may be diffusion controlled (5). If so, the system can be treated as an isothermal "cold" flame, as far as the rate determining step is concerned.

While the normal flammability limits for simple hydrocarbons in air are from 2% to 10-30% by volume of combustible, the presence of H_2 or CO can extent this limit to 70% (1). On the other hand, the presence of inerts can cause the zone to disappear.

Explosion velocity is characterized as follows (13):

1) The velocity of the explosion wave increases to a maximum and then is uniform.
2) An increase in pressure will increase the velocity slightly.
3) Inerts act as diluents and slow down the velocity. A large excess of combustible will also slow down the velocity.
4) Combustion is usually incomplete in explosions.
5) Water vapor, up to 5%, increases the velocity of CO explosions. Higher concentrations decrease the velocity.

Mixtures displaying maximum velocity have excess fuel (2).

SURFACE EFFECTS

The statement has been made that all gaseous reactions are speeded up by contact with hot surfaces, and may be selective at lower temperatures (13). Boudart (5) notes that the problem of reaction initiation in the absence of walls may be severe. For instance, in the H_2-Cl_2 reaction, the container walls ordinarily take care of the generation of Cl atoms to start the reaction. These heterogeneous effects make analysis of the mechanism much more difficult:

> "... the decomposition of diatomic molecules, such as chlorine, is a unimolecular reaction beyond the scope of classical chemical kinetics."

Lewis and von Elbe (41) classify surface materials with respect to chain breaking. Such materials include KCl, $BaCl_2$, Na_2WO_4, $K_2B_2O_4$, $LiCl$ and $RbCl$, and porcelain. Chain breaking depends upon the ability to retain a species in an adsorbed state. As an example, [HO_2] and [H_2O_2] decompose at a surface to break the chain reaction.

Clean quartz breaks the chain in thermal reactions and changes the explosive limits. Pyrex, boric acid and $MnCl_2$ act similarly.

The combustion of coal dust has many similarities to the combustion of metals but is more complicated due to the release of volatiles (24). There is also a similarity to the combustion of oil droplets. For metals, burning with air or oxygen may yield temperatures limited by the boiling points of the metal oxides: e.g., lead, tin, zinc, cadmium, bismuth, antimony, copper. This may also be an important factor in steel production.

Khitrin et.al. (16) make the comment that though there have been numerous theoretical and experimental investigations, there are still some gaps in the understanding of the combustion of coal. For instance, little attention is usually paid to the chemisorption of oxygen, which is related to the formation of surface oxides.

The conditions under which a fuel is ignited and the behavior during ignition will relate to (4):

a) The structure of the volatilized coal.

b) The temperature in the coke-burning stage.

A question to be considered with respect to the mode of combustion is the manner in which the volatile matter is released (1). That is, does the particle burn by first releasing all the volatile matter, or does the so-called fixed carbon burn simultaneously. It is further remarked by Field et.al. (1) that the question has not been resolved, but that it seems unlikely that oxygen would reach the surface in the presence of volatiles. In other words, any oxygen tending to diffuse through the volatiles layer would react instead.

During the combustion of p.f., there are several possibilities for the mode in which the carbon reacts in the particle structure. The carbon may react only from the surface, or uniformly throughout the particle. Alternately, the particle may be regarded as a hollow sphere, with burning occurring on both the outer and inner surfaces.

Unusual as it may seem, there is evidence that coal particles do form hollow spheres during combustion. Such spheres are called cenospheres and are believed formed during volatilization while the coal is plastic, e.g., coking coals in an inert furnace atmosphere. Even non-swelling coals are liable to form cenospheres. In experiments with p.f. flames, a high proportion of particle samples taken were found to be hollow spheres, and the walls themselves often contained bubbles.

Measurements on captive coal particles of different sizes indicate the burning times of both the volatiles and residue to vary as the square of the initial particle diameter (8). This is in accord with the surface area proportionality.

The rate constant for the surface reaction is dependent upon the temperature and upon the fuel type (1). The porous structure of the char will have an effect, as will the fact that the exothermicity of the surface reaction will have raise the particle temperature above the gas temperature. This difference may be several hundred degrees. Field et.al. (1) have provided a model to account for the various effects in the burning of a particle.

Field et.al. also present a mathematical formulation for the burning of a coal particle. The same authors develop a mathematical model for combustion chambers. Plug flow is assumed and recirculation taken into account, along with radiation.

Gray and Kimber (38) have studied the reactions of charcoal particles with CO_2 and H_2O at temperatures up to 2800°K. It was found that at temperatures up to 2300°K and even to 2800°K, both of the reactions

$$C + CO_2$$
$$C + H_2O$$

were diffusion controlling.

Golovina and Khaustovich (36) have investigated the interaction of carbon with CO_2 and O_2 at temperatures up to 3000°K. Most data in the literature on carbon combustion goes up to only about 1500°C. However, it is well known that in the high temperature range (>1000°C) the so-called "secondary" reactions are of importance:

$$C + products\ (CO_2)$$
$$CO + O_2$$

It is also known that with an increase in temperature, the reduction of CO_2 becomes the predominate side reaction.

Above 1000°C, Golvina and Khaustovich found the combustion to be diffusion limited. Above 1400°C, CO was the main product for the reduction of CO_2. However, a breaking of CO_2 reduction occurs in the range of 1400-2000°C. This is attributed to the formation of surface oxides of the type C(O).

COMBUSTION RATES

Field et.al. (1) have done extensive work on the combustion rates of pulverized coal. Generally speaking, residence times may be regarded as ~ 0.5 sec or less, and often ~ 0.25 sec or less. Selected information appears in Figures 3.4, 3.5 and 3.6.

Particle size significantly affects rate as reprsented in Figure 3.7.

Other types of combustion systems may be rate controlled due to the influence of the Boudouard reaction, $C + CO_2 = 2CO$. This is discussed in Chapter 4.

Reactions with Air or Oxygen (Combustion)

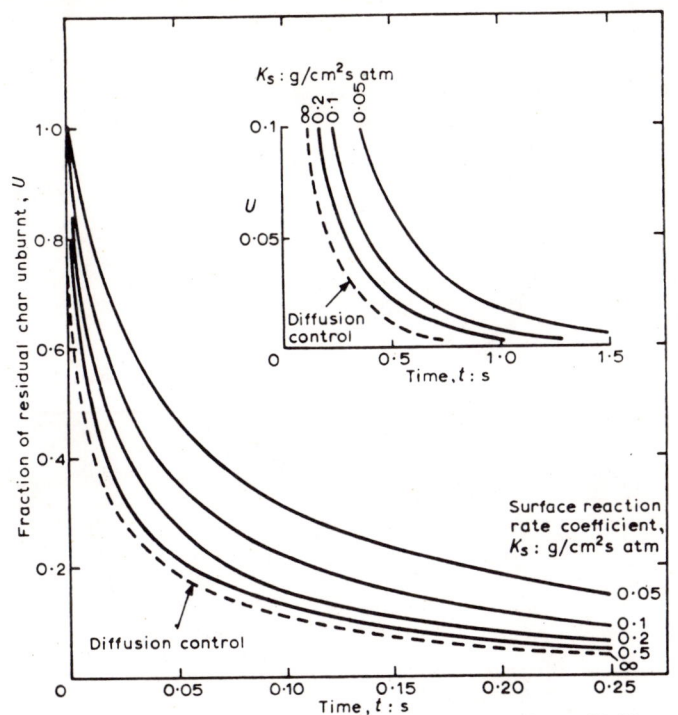

Fig. 3.4. Effect of surface reation rate on burn-out [from Field, *et.al.* (1), courtesy of Inst. of Fuel".

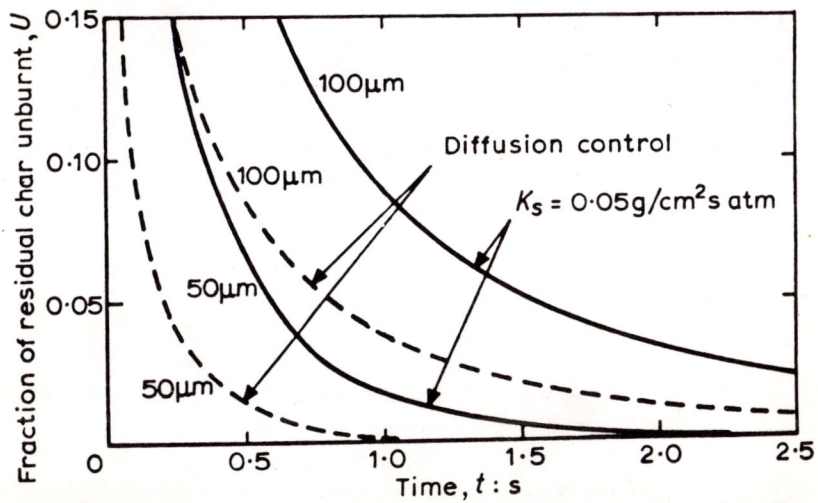

Fig. 3.5. Burn-out times [from Field, *et.al.* (1), courtesy Inst. of Fuel].

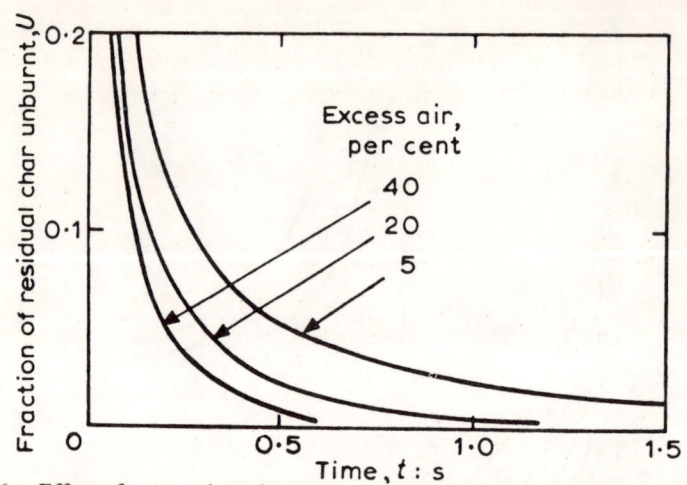

Fig. 3.6. Effect of excess air on burn-out times [from Field, *3t.al.* (1), courtesy of Inst. of Fuel].

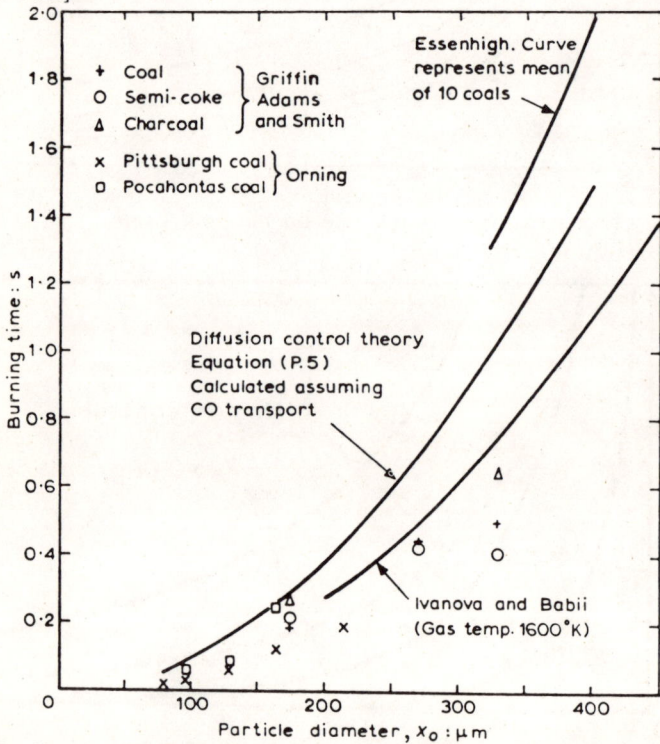

Fig. 3.7. Burning times of particles in air [from Field, *et.al.* (1), courtesy of Inst. of Fuel].

Reactions with Air or Oxygen (Combustion)

COAL/AIR TRANSPORT

The entrained transport of pulverized coal (~ 200 mesh) is accomplished with ratios of about 1.4 lb air per lb of coal, though pulverizer performance may sometimes require twice this ratio. Gas Reynolds numbers are in the fully-developed turbulent region.

If this fuel-air mixture is combusted there is the possibility of flashback unless linear velocities are at ~ 55 ft/sec or greater. Although gaseous flame speeds are only a few feet per second, as denoted in Figure 3.8, the coal volatiles mixed with air can form a combustible gaseous boundary layer which is essentially stationary, allowing flame propagation. This propagation can occur unless the boundary layer thickness is a certain minimum — that is, flames will not progress through tubing, for instance, if the tube diameter is less than a certain minimum, depending upon the combustibles. A bulk velocity of 55 ft/sec or higher assures that the effective boundary layer thickness is less than this minimum.

ADDITIVES AND CATALYSIS

The fact that additives may catalyze or otherwise affect combustion processes has long been indicated. For instance, in times past salt has been thrown into chimneys to assist in getting rid of soot deposits.

Fig. 3.8. Flame Speeds (from *Chemical Engineers' Handbook*, 3rd edition).

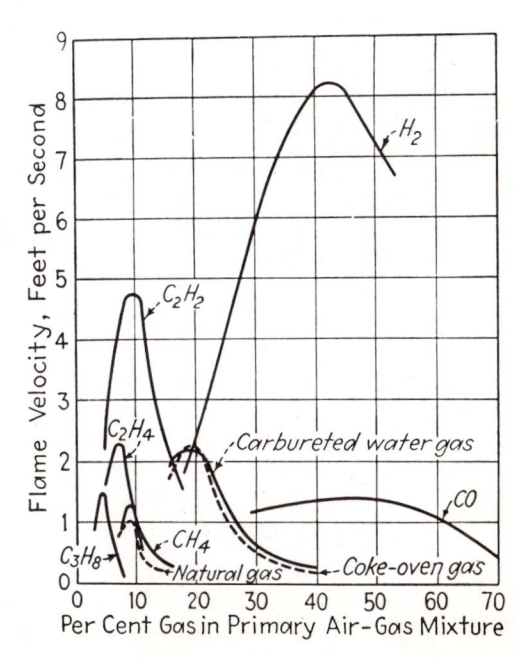

Investigations by Delattre (42) show that solid fuels can be activated by additives. These additives, which serve to catalyze combustion, may be salts or oxides of heavy metals that exist in more than one stage of oxidation, and/or alkali metal or alkaline earth salts. As an example, coal was treated with the following mixture in the study:

BaO_2	2.0%
NaO_2	0.4
CO oxide	0.5
Mn salt	97.1

It is apparently not resolved whether the catalytic effect is with regard to carbon-oxygen reactions, or is more indirect. For this catalytic effect resembles in many respects the catalysis of coal-steam systems. Here alkali salts serve to catalyze the carbon-steam reaction to give $CO-H_2$ synthesis gas. In turn this gas may react to form hydrocarbons and oxygenated compounds in the presence of activated catalyst (two degrees of valency) of the transition elements. The subject is discussed further in Chapters 6, 7 and 8.

Orning (7) has observed that ash may have an effect on combustion. This is in line with the reasoning that combustion is related to the carbon-steam reaction, which in turn is catalyzed by components of the ash such as alkali metal and iron compounds.

The combustibility of briquetted carbon is improved during manufacture by the introduction of certain vapors (43). Recommended additive mixtures are

1) 12-22 grains camphor, 300-400 cc. petroleum ether plus light fuel oil to make one liter.

2) 17 grains camphor plus 350 cc. petroleum ether or gasoline, 27 grains naphthalene, 6 cc. ether, 12 cc. $PhNO_2$ plus light fuel oil to make one liter.

In comment, the above additives are also combustible, and the improvement may be one of volatility. The degree of moisture present was not noted.

Andrzejewski (44) states that, at high temperature, cokes are activated by CO_2, water vapor, air, and other gases. This activation effect has been noted in other references.

The increase in activity appears as an improved gasification and combustion rate with increased reactivity toward oxygen and CO_2. A conclusion is that activation increases the inner surface area of coals, which is manifested as an increased reactivity.

Reactions with Air or Oxygen (Combustion)

The CO-shift,

$$CO + H_2O = CO_2 + H_2$$

is catalyzed by hot carbon (21). It is mentioned by Al'tschuler and Shafir (39) that carbon activates the $CH_4 + CO_2$ reaction. The water content in the flue gas has been mentioned as inhibiting the formation of H_2SO_4 (4). This is due to the fact that water reduces the catalytic oxidation of SO_2 to SO_3 on steel surfaces. Furthermore, there may be an optimum water content to flue gases. In comment, this may be in part due to the condensation of water which then can absorb SO_3 to form the acid.

The water dew point of flue gases is usually around 90-110°F. Water that condenses out would be unimportant in itself. However, acids can condense out up to 350°F. The water content has little effect on the H_2SO_4 dew point, but would serve to dilute the acid. In further comment, condensed water would also absorb SO_3 and SO_2. The entire acid-water equilibrium should be examined.

A small amount of SO_3 in coal gas will cause a premixed flame to become luminous due to carbon formation (24). There is little effect upon a diffusion-controlled flame, which is already luminous. (In a diffusion flame, by definition, the air or oxygen migrates from the surroundings to the body of the flame. Contrarily, for the case of a premixed flame, the fuel and air or oxygen are mixed prior to combustion.) For either premixed or diffusion flames, both SO_2 and H_2S decrease carbon formation, contrary to the action of SO_3.

While the presence of inerts would be ordinarily expected to dilute the reactants and slow the reaction, inerts may sometimes accelerate the reaction (2). The effectiveness of inerts has been given the following order:

$$He:N_2:A:H_2O = 1:3:4:5$$

In fact, at times, the addition of nitrogen is as effective as extra oxygen.

Jost (2) suggests that the accelerating influence of inerts is apparently due to an inhibition of chain agents to the wall. That is, the inerts reduce the probability of chain breaking in the reaction sequence.

According also to Jost, nitrogen plays a remarkable role in CO combustion. The NO formed from the oxidation of the nitrogren serves to act as a catalytic intermediate in the presence of water vapor:

$$2CO + 2H_2O = 2CO_2 + 2H_2$$
$$N_2 + O_2 = 2 NO$$
$$\underline{2NO + 2H_2 = N_2 + 2H_2O}$$
$$2CO + O_2 = 2CO_2$$

Other possible nitrogen reactions include
$$2NO + H_2 = N_2O + H_2O$$
$$2N_2O = 2N_2 + O_2$$

Jost comments that more NO is produced than corresponds to equilibrium. He further states, however, that this is no violation or contradiction of thermodynamic principals — for the NO is formed by another reaction coupling. This statement parallels conclusions reached concerning heterogeneous catalytic reactions, in which the catalyst serves to form new systems with the reactants and products so that the relations for homogeneous reactions no longer apply (see Chapter 8).

Inactive diluents such as CO_2, N_2 and recirculated flue gas reduce carbon formation (24). Adding CO or H_2 to a premixed flame at the point of incipient carbon formation may produce carbon, but may alternately decrease carbon formation.

In studies utilizing hollow cylinders of electrode carbons the following results are reported on the effect of inhibitors (22):

1) Chlorinated compounds inhibit the oxidation of CO to CO_2 in a dry atmosphere.
2) Moisture destroys the inhibiting power of these compounds.
3) Inhibition makes O_2 available over a longer distance to burn carbon.

The above studies on chlorinated compounds as inhibitors to CO oxidation have also shown that the presence of moisture destroys the inhibiting power of these compounds. In comment, under this circumstance, moisture would favor combustion.

The action of chlorinated compounds in interfering with the oxidation of CO has been ascribed by Sherman and Landry (22) to the drying power of these compounds. Since chlorine compounds react with water vapor to form HCl, and this acid has a strong affinity for water, the net result is to dry the mixture. This in turn inhibits the action of water vapor in catalyzing the oxidation of CO. Phosphorus compounds have the same effect as chlorine compounds.

Halogenated hydrocarbons have been found to reduce the flame speed of methane (45). Compounds studied include

Bromotrifluoromethane

Dibromofluoromethane

Methyl chloride

Chlorotrifluoromethane

Dichlorofluoromethane

Reactions with Air or Oxygen (Combustion)

Bromides are more effective than chlorides. The reduction in flame speed was proportional to the amount of inhibitor added.

The addition of 1% formaldehyde or acetaldehyde to the oxidation of ethane (at 310°C and 720 mm) serves to eliminate the induction period (41). It is also noted that ethanol or water shortens the induction period. Acetaldehyde not only shortens the induction period, but causes explosion.

EXPERIMENTAL SYSTEMS

More recent studies on combustion emphasize fluidized combustion, though entrained systems using pulverized fuel lend themselves more readily to theoretical analysis — the point particle or Lagrangean concept of flow as compared to the Eulerian or continuum concept. Work on fluidized combustion for the most part has been described at meetings and appeared in the form of proceedings and reviews, e.g. References 46 and 47. This is the usual state of affairs for a newly-developing subject area.

On-going work in the United States is for the most part sponsored by the Energy Research and Development Administration (now absorbed into the Department of Energy).

Past work on fluidized combustion by the British Coal Utilization Research Association (BCURA) has been presented by Thurlow (48, 49). Unpublished experimental work of the BCURA was concerned with very rapid heating of pulverized coal (1). For intance, 20 micron p.f. has been heated to 1000°C in less than 50 milliseconds, decomposed at temperature for 100 ms., then quenched. Heating was accomplished by injecting the particles in a stream of hot nitrogen.

The results of the foregoing experiments have been recorded in plots of proximate volatile matter vs time for parameters of temperature, and as weight loss vs time for parameters of temperature.

In experiments using a carbon channel (air flow through a packed carbon tube), the relative carbon saturation was determined versus distance (23). Predictably, there was a trend toward equilibrium with distance.

A plug flow system is probably the simplest experimental setup for the study of combustion. The BCURA has used a vertical cylinder of 18" diameter by 20 ft. high, which is lined with refractory bricks (1). The feed was introduced at the top with primary air such that the suspension filled the total crossection and mixing was minimized. Secondary air was injected through jets downstream.

In the BCURA experiments on anthracite coals, representative conditions were 1700°K, 50-150 micron particle size, 2.6% moisture with fuel rich flames. It was found that the rate of combustion for anthracite proved to be lower than that predicted for diffusion controlling. Hence it was assumed that the combustion was surface reaction controlling. In other experiments, the following results were noted:

The rate of combustion decreased with time

The flame temperature showed an optimum with time

An optimum air/fuel ratio existed

The BCURA experiments using bituminous coals were conducted in a horizontal, hot-walled, cylindrical furnace. The dimensions were 3 ft.-6 in. I.D. and 18 ft. long. Feed rates were 200 lb./hr. of coal, with the coal and air premixed and fed through an annular burner. The progress of combustion was followed by both gas analysis and solids sampling.

The bituminous coals ranged from 0.4 to 2.3% moisture. The rate of combustion proved either to be higher than for diffusion control or about the same. Thus it was assumed that diffusion was the controlling step for the combustion of bituminous coals. However, other experiments indicated the combustion to be surface reaction controlling. As with anthracite, it was noted that the burning rate decreased with time and that temperature showed a maximum with time.

The U.S. Bureau of Mines has utilized a 12 ft. diameter spherical vessel for studies on ignition, flammability, and detonation (50). Such a large vessel minimizes wall effects, which catalyze homogeneous reactions. The system is not only suitable for vapors and gases alone, but for solid explosives.

Heterogeneous interaction occurring at surfaces or walls may sometimes be detected by varying the surface to volume ratio (5). Differential calorimetry offers a means of tracing reaction behavior. A convenient means of obtaining combustion rate data as affected by moisture is by means of thermogravimetric analysis: weight loss vs time (4). In the aforemetioned study of Golovina and Khaustovich (36), a high frequency current was used to generate the temperatures used (to 3000°K). A single particle of carbon was employed (15 mm. in diam.), attached to a ceramic holder in a quartz tube. The change in weight was used as the measure of reaction rate as CO_2 or air passed through the tube.

Sherman and Landry (22) describe experiments involving the passing of dry and moist air through beds composed of particles of carbonaceous materials. The system was heated externally to produce

Reactions with Air or Oxygen (Combustion)

temperatures in the range from 647-890°C. At the lower temperatures the CO/CO_2 ratio in the off gas was essentially the same. At the higher temperature, however, the CO/CO_2 was six times greater for the dry air than for the moist.

In perhaps more convenient terms, the CO_2/CO ratio was 6 times greater for moist air than for dry. This result can be interpreted as a greater degree of combustion. (Ordinarily at higher equilibrium temperatures, the CO_2 concentration goes down, the CO concentration up — other things being equal).

The combustion characteristics of the coke residues from peat, coal, anthracite and shale were studied in a fluidized system by Khitrin, et.al. (16). The reactor was a vertical stainless steel tube, 8 mm. I.D. by 800 mm. long, with the feed introduced from a hopper by a screw conveyor. Preheated air was the fluidizing medium. Temperatures to 700°C were used, at which chemisorption was manifested.

In the preceding experiments, CO_2 was listed as the principal product, and the rate does not depend upon particle size (29).

3.5 EFFECTS OF MOISTURE ON COMBUSTION

Lavrov (35) notes that the presence of water has a marked effect on combustion by entering into various combustion reactions. He further notes that experimental evidence was found for the existence of active centers for chain reactions involved in the further combustion of carbon monoxide and hydrogen (which would be reaction intermediates in the combustion of coal).

The conclusion is reached by Haslam and Russell (13) that moisture catalyzes the oxidation of carbon. Water has been noted to shorten the induction period for the oxidation of ethane (41). This produces a faster reaction.

Adsorbed water on walls or packing has been observed to increase the reaction rate (2).

The presence of water is essential for some reactions (13). For example CO and O_2, if dry, will not explode. Anhydrons methane will not ignite until the dissociation temperature is reached (28).

The presence of moisture in coal has been noted to have a beneficial side-effect in firing, as the expansion and evaporation may cause fracture (21).

Gauger (9) notes that "occasionally, firemen add water to coal prior to firing." There is a renewed interest in the use of water conditioning. To quote Whittingham (4):

> "In recent years there has been an increasing awareness of the value of water-conditioning to improve the combustion of low-grade fuels of high fines and ash contents and attention has been re-directed to the use of air humidification particularly as providing a solution to the deposit problems in modern high-pressure boilers."

It can be thus concluded that the more ash in the coal, the more moisture needed during combustion to prevent fouling.

STEAM-CARBON REACTION

Whittingham (4) notes that the endothermic steam-carbon reaction is primarily responsible for cooling effects in furnaces. Gauger (9) states that the presence of moisture is believed to cause heat generation at the surface of the bed and in the combustion space due to the endothermic reaction forming CO and H_2 deep in the fuel bed, and which in turn burn at the surface.

On the other hand, the presence of water vapor adds possibilities to the reaction mechanism (1):

$$H_2O + C = CO + H_2 \quad \text{(at surface)}$$
$$CO + O_2 = CO_2 \quad \text{(in film)}$$

The effect may be as if the water vapor acts as a catalyst for the oxidation of the CO. And indeed, CO is oxidized by the CO-shift:

$$CO + H_2O = CO_2 + H_2$$

The presence of water vapor in p.f. combustion has been noted to reduce CO concentration, whereas the concentration of CO_2 and O_2 is not affected appreciably.

The reaction of product CO_2 and water vapor with the carbon particle will have an effect on the combustion process. In the one case, CO is produced; in the other, CO is regenerated. (These reactions are but part of the set of reactions which may occur, including the approach toward equilibrium involving reverse reactions.) In the presence of CO_2, the reaction rate goes down; whereas in the presence of water vapor, the reaction rate goes up. In fact, the rate may be twice as fast as with air alone.

It is concluded that for CO_2 the reaction effect is small compared to the reaction with O_2.

OXIDATION OF HYDROGEN

It has also been noted that water vapor may accelerate the initial reaction rate between hydrogen and oxygen, though the rate constant for the H_2-O_2 reaction is apparently independent of water concentration (41). In comment, this does not preclude the rate from being dependent on water concentration, depending upon the formulation of the rate equation. In fact, the water vapor product provides self inhibition of rate, particularly near the explosion limits.

Lewis and von Elbe (41) advance H_2O_2 as a product involved in the chain reaction between H_2 and O_2. The sequence is restricted only to wall effects, however.

In reactions of H_2-O_2 systems outside the explosion limit, the surface reaction at low temperatures is also inhibited by the presence of water (2). This is noticeable as the water reaction product forms. It is simultaneously noted, however, that water vapor may also accelerate the reaction.

Conversely, water vapor has been observed to inhibit low-pressure explosions in H_2-O_2 mixtures. It also inhibits wall reactions at low temperatures, $\sim$ 540°C, as noted. In comment, the remark is appropriate that H_2O is a product of the reaction.

CO OXIDATION

Carbon monoxide combustion shows many similarities to hydrogen combustion (2). The most striking facet of CO combustion is the great sensitivity to small amounts of water vapor. Whereas dry CO-O_2 mixtures react very slowly and will not explode by spark, water vapor affects the following properties:

       Ignition temperature

       Flame velocity

       Detonation

       Radiation

There is no effect on the explosion limits. The ignition temperature is lowered, however.

In the other direction, pure, dry CO_2 is not reduced by dry charcoal, even at a red heat (13).

Another variable is vessel diameter, which has been noted to affect the rate of the CO-O_2 reaction (41).

Sherman and Landry (22) conclude that water vapor catalyzes the oxidation of CO. The mechanism for the effect of moisture is but little understood (13). Possibly the H_2O is some sort of catalyst. Thus for the CO-O_2 reaction, starting with the CO-shift,

$$CO + H_2O = CO_2 + H_2$$
$$\underline{H_2 + \tfrac{1}{2}O_2 = H_2O}$$
$$CO + \tfrac{1}{2}O_2 = CO_2$$

Jost (2) also considers the favorable effect of water vapor on CO combustion to be due to the CO-shift as per above, though the reaction may in fact be a chain sequence: e.g.,

$$CO + OH = CO_2 + H$$
$$H + O_2 + M = HO_2 + M$$
$$HO_2 + CO = CO_2 + OH$$

Etc.

Lewis and von Elbe (41) treat the water-catalyzed reaction of CO and O_2 in considerable detail. They note that the reaction is of a steady state type, and that the rate will decrease with increasing initial oxygen concentration (and will become explosive). Plausible chain reactions which are advanced are:

Chain Initiation:

$$H_2O + CO = CO_2 + H_2 \qquad \text{(at surface)}$$

Chain Carriers (dissoc. of H_2O_2):

$$H_2O_2 + M = 2OH$$

Destruction of H_2O_2:

$$H_2O_2 + CO = CO_2 + H_2O \qquad \text{(at surface)}$$

A complete reaction scheme for the oxidation of CO in the presence of water, as indicated by Lewis and von Elbe, would involve the following reactions:

$$H_2O + CO = CO_2 + H_2 \qquad \text{(surface)}$$
$$H_2 + O_2 = H_2O_2 \qquad \text{(surface)}$$
$$H_2O_2 + M = 2\ OH$$
$$H_2O_2 = CO = CO_2 + H_2O \qquad \text{(surface)}$$
$$OH + CO = CO_2 + H$$
$$H + O_2 = OH + O$$
$$H + O_2 + M = HO_2 + M$$
$$O + H_2 = OH + H$$
$$O + O_2 + M = O_3 + M$$

Reactions with Air or Oxygen (Combustion)

$$HO_2 + CO = CO_2 + OH$$
$$HO_2 \rightarrow \text{destruction}$$

Jost (2) states that the presence of water vapor affects both the combustion velocity and the flame velocity of CO-air mixtures. For instance, a maximum in the combustion velocity is found at 9.4% moisture, rising from 16 to 55 cm./sec.

The reaction-accelerating effect of water also occurs in slow oxidation and in detonation. Detonation velocities for $2CO + O_2$ are as follows:

0 % water	1260 cm./sec.
5.6%	1738
38.4%	1266

A maximum occurs at 5.6% water.

The effect of water vapor concentration on the explosion velocity for CO has been noted by Haslam and Russell (13). Up to 5% concentration, water vapor increases the velocity; at higher concentrations, the velocity decreases. In comment, an optimum water vapor concentration may be expected for explosion or combustion reactions. At the higher concentrations the moisture may be acting as a diluent to retard combustion.

The combustion velocity of CO is accelerated by H_2 and organic compounds containing hydrogen, while antiknock compounds have a small inhibiting effect (2).

FURNACE PERFORMANCE

The following statement was made by Gauger (9) in 1945:

"It is generally accepted by firemen and engineers that some moisture is desirable to promote combustion. Until recently, however, there have been few published data indicating the actual effect on combustion of varying amounts of water."

A practice is sometimes made of "tempering" coal with water before firing (13). The effect is to retard the rate at which volatile matter is distilled, and serves to reduce smoke.

Whittingham (4) lists the principal causes of boiler shut-down to be:

1) Overheating and destruction of grate bars.
2) Clinker formation in fuel beds.
3) Slag formation on refractory surfaces.
4) Fouling and corrosion of heating surfaces.

All of the above are associated with the occurrence of excessive temperature at some stage of combustion. Consequently the endothermic reaction of carbon and steam can be of major importance in controlling furnace temperatures.

The introduction of steam to the air supply is a common practice in limiting grate-bar temperatures. The practice can also be uneconomic. The wetting of coal can cause a swelling and coking during ignition which can cause a change in the size distribution. In burning coal on a grate, an increase in the percentage of fines has been noted as lowering grate temperatures.

Among the results related to combustion of adding water to coal are the following (9):

1) Water to fuels with large percentages of fires decreases the bed resistance to air flow.
2) Water to coking coals causes smaller pieces of coke to form, giving a more uniform bed.
3) The initial ignition of the fuel is retarded, and the advance of the ignition point slowed.

Moisture in coals may contribute to fracture during heating due to volumetric expansion and vaporization (21), as previously noted. The effect is to enhance reactivity.

The hydrolysis of coal produces a disintegration effect on coal. In comment, if the presence of water in the coal contributes to hyrolysis in any fashion, then such presumably would contribute to the combustion process.

The adding of flue gas to the inlet air has been reported as effective in producing a more porous and less sintered ash in the fuel bed (4). In comment, this may be caused by a lowering in temperature due to a reduction in reactant concentration.

Moisture in the coal causes more ammonia to be produced in the flue gases (9).

Whittingham (4) has categorized the following effects of moisture on burning in coal beds:

1) More fuel gasification takes place during ignition. Ignition is lengthened at the expense of the burning-out stage.
2) Luminosity of the volatile flame in ignition is decreased.
3) Temperature at the ignition plane decreases.
4) There is a considerably lower bed temperature during burning-out, especially with fine coal (-1/8" + 1/32").

Reactions with Air or Oxygen (Combustion) 125

The insensitivity of ignition temperature to changes in moisture content is explained to be due to a more complete combustion of the primary gasification products. Furthermore, the increased heat released compensates for the latent heat required for evaporating the surface moisture.

Disadvantages listed for the presence of moisture during combustion are (4):

1) Moisture is a noncombustible inert.
2) Moisture absorbs heat due to latent and sensible heat effects.

The presence of water vapor may have a deleterious effect on brick furnace walls, especially in coking practice.

Barshtein (51) notes the following effects of an increase in moisture content in p.f.:

1) Ignition phase prolonged; less time for combustion.
2) Combustible losses in ash increases.
3) Theoretical combustion temperature lowered. Combustion product temperature lowered at exit.
4) Temperature of flue gases thru superheater and economizer will rise somewhat.

In comment, it is difficult to assess the above statements as to whether overall efficiency would improve. Presumably temperatures in the furnace proper would be lower, but the temperatures in the superheater section are raised due to delayed combustion.

The presence of moisture is believed to retard the rate at which volatile matter distills off and cuts down on smoke occurrence (9). In comment, this effect could result in an increase in efficiency.

Studies indicate that surface moisture has an effect upon the combustion of coal (4). For instance, an increase in the surface moisture content of fines from 2 to 8.6% resulted in an increase in boiler efficiency from 54 to 71%.

Whittingham (4) notes that the major source of water is in the combustion air:

> "It is not unreasonable therefore to expect that humidification of combustion air will have a much greater effect on bed temperatures during burning-out that is obtained by initial wetting of the fuel."

And in fact, water vapor averages some 6.7% of the total product gas steam from combustion in furnaces. An increase in humidity (even from winter to summer) is effective in raising this efficiency, as is the introduction of steam.

The fraction moisture in the coal will affect the temperatures at the grate. Some temperatures near the grate varied as follows:

Added moisture	Max. grate-like temp.
1.4%	980°F
2.5	930
3.5	710
6.0	760
10.0	800
12.0	760

Steam may be introduced through jets, though there is some problem with erosion. With 1% steam added, reductions in the link temperature of 40-90°F have been reported. An addition of 20% steam lowered fuel-bed temperatures from 3060 to 2390°. In general, changes in fuel-bed temperature are difficult to measure, though correlation can be made to grate tempertures. Useful information on tempertures and heat release can also be obtained from examination of the ash.

3.6 COAL/WATER SYSTEMS

The slurry pipelining of coal appears on the verge of becoming a fact of life. The subsequent de-watering (and drying to a limited extent, if necessary) brings up the problems of combusting coals with high moisture content. There is even the matter of combusting aqueous suspensions of coal. (The firing of pulverized coal-fuel oil slurries is already an emerging technology — e.g., work underway at the General Motors Corp.) These and related matters are reviewed.

TRANSPORT OF COAL/WATER SUSPENSIONS

Bonnington (52) presents a review of developments in the hydraulic transport of coal. Several investigations indicate economic feasibility with a reduction in transportation costs. Costs for the transportation and firing of slurry fuels have been compared with other systems by Hopkins (53).

Of special note were the efforts of the Pittsburgh Consolidation Coal Company culminating in a 108 mile pipeline transporting 150 T/hr. of coal. A more recent effort is the Black Mesa slurry pipeline in northern Arizona. In the offing is a slurry line from Wyoming.

Reactions with Air or Oxygen (Combustion)

Pneumatic or gas transport of entrained coal is an interesting alternative. Problems of pneumatic coal transport have been treated in papers presented at the first U.S. Symposium on the Pneumatic Transport of Solids (54), sponsored by the Institute of Gas Technology and the U.S. Bureau of Mines in 1966, and more recently at the International Symposium on Freight Pipeline, Washington, D.C., Dec., 1976. Transcontinental Gas Pipe Line Corporation (Transco) and Stone and Webster Engineering Corporation are presently engaged in experimental studies on gas-solids transport with potential application to pulverized coal.

Reference 55 is a review on the state of hydraulic transport in bulk or in containers. Condolios et.al. (56) describe new trends in solids pipelines. Other novel means for transporting solids through pipelines includes the use of paste slugs. Cylindrical paste segments are carried by a liquid hydrocarbon such as crude petroleum. The action is similar to capsules. Presumably a paste or slurry of coal and water could be transported in this manner.

The pipeline transport (and combustion) of mixtures of coal, H_2O and oil has been studied in Germany (57). The proceedings of the aforementioned Symposium on the Pneumatic Transport of Solids is another source on the pipeline movement of coal (58). Of note is the use of natural gas as the suspending medium, though synthetic natural gas (SNG) or low-Btu gas from coal is now advocated (59). The unit train concept is an on-going practice.

Costs have at various times been compared for the various means of coal transport (60): rail, barge, road, pipeline — though overall efficiency is a new consideration.

Laboratory tests have shown that anthracite slurries can be dewatered by atomization followed by mist removal with an air current (61). Recoveries of 67-79% are claimed. More conventional means are the practice.

The injection of coal in the form of slurries also requires a study of optimum velocities and properties (62). Pipileu et.al. (63) at the U.S. Bureau of Mines have established relationships for the hydraulic transport of coal. Bituminous slurries up to 48% solids were used to establish the relations between concentration, velocity, and pressure drop. For maximum capacity, an optimum concentration exists.

COMBUSTION OF COAL/WATER SLURRIES

The possibility of burning coal/water slurries without dewatering or drying is not as far out as would first appear. To wit, it has been proven possible to burn coal directly from pipeline suspension, with no expensive drying process (57). A simple burner design is utilized. Whittingham (4) has reviewed experiments on the effect of adding water to coal and combustion air.

The "Coal Process. Combust. Inf. Bulletin" of the National Coal Board contains reviews or abstracts of Russian articles on the combustion of aqueous suspensions, as does Chemical Abstracts and BCURA Abstracts. The limited amount of research in this area appears to be mostly of Russian or continental origin, though some work referenced is from the United States.

Davydova et.al. (64, 65) have experimentally investigated the combustion of coal-water slurries. In one series of tests the apparatus consisted of a combustion chamber and an air-feed stream, with air heating. The moisture content of the suspension was 50-70%, with combustion air preheated to 300-700ºC. The fuel particle size was 0.8-1.2 mm., and the length of the combustion zone was 0.8-1.2 meters. A slight excess of air was used, with a moisture content of 54-64%.

During combustion the maximum temperature of the chamber was 1230-50ºC. Variation of fuel moisture had no effect on this temperature. Combustion was complete, with no CO, H_2, or CH_4 found in the products.

Delyagin et.al. (66, 67, 68, 69) has studied the combustion of coal-water suspensions in air. In a series of tests, Delyagin and Smetannikov (68) introduced drops (1-1.35 mm.) of a suspension of carbon (50-80% H_2O) into air at 400-1000ºC. The successive phases of warm up, evaporation, ignition and burning were studied. The drop was suspended in a cylindrical chamber, and the temperature measured inside and outside the drop.

At first, during evaporation, the temperature of the drop remained constant while that of the medium decreased. Then as the water content decreased, the temperature inside increased — evidently due to decomposition reactions.

At 350-400ºC the combustible gases evolved and then ignited. This caused the temperature around the drop to go to 1500-1600ºC, and the size of the drop increased $\sim$ 3-fold.

Due to the increased surface, the coke so formed burned faster than a piece of coal of the same size. It was found that the elapsed time for combustion was proportional to drop size and inversely proportional to the temperature of the medium. The processes of ignition and burning were more like those of liquid fuels than of solid fuels.

In another series of tests on the combustion of coal-water suspensions, Delyagin and Smirnova (69) used an apparatus for the investiga-

Reactions with Air or Oxygen (Combustion)

tion of combustion zones and the evaporation of moisture in the suspension layer. The reaction vessel was 80 mm. in diam. by 160 mm. high, and fabricated of ceramic-lined metal. The coal particles were 0.063 mm., formed into a suspension of 45-55% initial moisture at 850°C.

It was found that the rate of ignition of the suspension increased with the temperature of the furnace crown, and was inversely proportional to the moisture content of the coal. Combustion occurred in a thin layer at the surface, while evaporation of the moisture occurred in the deeper layers of the suspension (at tempertures > 100°C).

The rate of removal of the gaseous products did not affect their composition and did not change the characteristics of the fuel layer. It was noted that, at 140-150°C and above, CO, CO_2 and H_2 were obtained.

Delyagin (66) has proposed a mathematical model for the combustion of pulverized coal-water suspensions in an air stream. Reactant (oxygen) and product mass transfer rates are introduced and the effect of moisture included. The theoretical calculations are said to agree with experiment.

Delyagin (67) has also analyzed the combustion of a single drop of an aqueous suspension of coal dust in a hot air stream. Oxidation conditions were assumed, and in correlations derived relating the degree of burn-up, water content and excess air, agreement was obtained with experiment for the combustion of coal suspensions of 50-64% water.

The work of Davydova, et.al. and Delyagin et.al. on the combustion of coal-water mixtures is a direct tie-in with the hydraulic transport (and storage) of coal, including beneficiation by classification or washing.

In the United States, the Babcock & Wilcox Company has fired wet fuel in a cyclone furnace (*Power Engineering,* Jan. 1962). A 70-30 solids water mixture was used, decanted from a 60/40 pipeline mixture. The demonstration was full-scale, conducted at the E. H. Werner Station of the New Jersey Central Power and Light Company at South Arnboy, N.J. The boiler used was fired by two 9-foot cyclone burners, with a capacity of 475,000 lb./hr. at 1500 psi and 1000°F. Test operations were completely satisfactory.

In an article by Frey et.al. (70) of Combustion Engineering, methods are discussed for utilizing pipeline coal. The slurry may be fired directly, decanted, or centrifuged to reduce the water content. In a subsequent article by Jonakin and Caracristi (71), operation of a centrifuge-pulverizer is discussed further.

Kashirskii (72) describes investigations upon the oxidative pyrolysis of peat. A feed of powdered peat containing 15% water was used in a continuous lab setup. The reactor was tubular with external heating, with the peat blown in by air. Reactor conditions were:

Wall temp.	1000°C
	860-880°C at exit
Air	1.5-15% of theo.
Rate of temp. rise of head	900-1700°C/sec.

Exit gas compositions were in the following ranges:

H_2	22.2-30.8 mol %
CO	22.6-30.2
CO_2	11.3-16.3
CH_4	2.6- 4.6
C_2H_4	1.4- 6.8
C_3H_6 - C_4H_8	0.2- 0.8
iC_4H_8	0.2- 0.3
N_2	10 -40

Kashirskii observed that at the lower air rates thermal decomposition predominated and the amount of hydrocarbons increased with an increase in air rate. Conversely, at higher air rates the yield of hydrocarbons *decreased* with an increase in air rate. The diversity was explained by oxidation of the hyrocarbons on to CO_2 and H_2O.

Other properties of the yield from the peat are of interest:

Max. thermal value of the gas
2550 k cal/kg of dry peat

Gasoline
1.1-1.8% of wt. of dry peat

About 52% of the total thermal value of the peat was in the gas yield. The coke residue was satisfactory as fuel. As aqueous condensate was also obtained which contained NH_3, pyridine, and acids and phenols.

Means have been tried for improving blast furnace performance by injecting coal/oil mixtures (73). As an example, coke at the ratio of 71 kg of coke per ton of metal was replaced by a coal/oil mixture of only 45 kg per ton of metal.

INTEGRATED SYSTEMS

A process study has been made between the firing of pulverized coal and of a pulverized coal/water (60% solids) suspension, for steam genera-

Reactions with Air or Oxygen (Combustion)

tion and power plants (74). The following advantages are claimed for using the slurry:

A 10% reduction in installation costs

Simplification of pulverizer installation

Automatic regulation of storage and feed

Reduction of fouling of heating surfaces

Reduction of powder deposits

Reduction of wear on coal-handling equipment

Reduction of personnel

An economic estimate has been made to compare the relative costs of wet firing as compared to dry firing (75). For one thing, the wet firing cuts down the cost of ash removal.

Schwartz and Merton (76, 77) have investigated the firing of boilers with coal/water mixtures. Experimental data and operating experiences were reported for the grinding, transport, and firing of coal/water suspensions (76). The test unit was rated at 100 T/hr. of steam.

The commercial conversion of a boiler at a power station has also been reported (77). Originally using pulverized fuel, the boiler was converted to slurry firing.

Lukes et.al. (78) describe the oxidation of coal in an aqueous medium. Coal was ground with water in the absence of oxygen to a particle size less than 1 mm, and then further diluted with water to form coal/H_2O slurries of 1:3 or 1:4. The coal slurry was introduced to the reactor countercurrent to air. Reactor temperatures were 130-350°C.

Meier zu Kocker and Huning (57, 79) have demonstrated that coal/H_2O oil mixtures of high coal/H_2O content are directly combustible. The oil provides the energy for evaporating the water.

The discharge end of pipelines handling coal-water suspensions ordinarily would require a separation. One means proposed involves centrifugation, followed by movement of the partially dewatered coal by a non-clogging screw conveyer (80).

A means for the recovery of cleaned coal from aqueous slurries involves the use of 5-25% of an organic liquid mixed into the slurry (81). Purposely, the boiling point of the liquid is kept below 100°C, and the liquid must also show a greater surface affinity for coal than for water. Petroleum hydrocarbons fill this requirement. By holding the slurry temperature intermediate between the boiling points of the hydrocarbon and water, the coal dust is brought to the surface and removed.

Coal/oil suspension have been separated thermally in a fluidized bed (82). A mixture of 30% coal and 70% crude was heated to temperatures of 700-900°F to give a product of char, oil, and gas. It was concluded that such a means was feasible.

The order of coke replacement in blast furnaces has been listed as follows (83):

1) Coal-tar fuel, fuel oil, dried, low-ash coal.
2) Coal of higher ash and moisture.
3) Natural gas and coke-oven gas.

Research on the injection of coal/oil slurries into blast furnaces shows a considerable savings in coke (62, 83). A unit installed in 1963 has a capacity of 85T/D using a slurry density of 50% coal fines. The use of the slurry has cut down on the cost of pig-iron production.

CONCLUSIONS

Moisture appears to play a more integral part in the combustion of hydrogen-deficient carbonaceous fuels (such as coal) than is commonly realized. The empirical evidence on air humidification and the addition of steam or water to coal-fired furnaces attests to this fact. Moreover, water injection is a commonly recognized method for inproving the power performance of the internal combustion and jet engines — and has been put into practice, particularly for aircraft.

This role of water is also born out in a consideration of the chemistry of the reactions associated with combustion. The carbon-steam reaction to produce carbon monoxide and hydrogen, which are then oxidized on to the end products, is an important step in the combustion sequence, as is the CO-shift to produce hydrogen and CO_2 as an end product.

Since the entire system involves reaction and equilibria involving C, H_2O, CO, H_2 and CO_2, and reaction speeds are fast, it will be difficult if not impossible to say which reactions are directly controlling in the mixture. This is complicated by the heterogeneous nature of the system, whereby the rate of mass transfer of a reactant(s), or of the products, becomes controlling.

Resolution of the matter would probably be in the form of a mathematical analysis to establish a criterion of which step is controlling.

Work in this direction already indicates the film diffusion of oxygen to be the overall controlling rate step, and that the oxidation of CO to CO_2 is the most important surface (or film) reaction. Whether the source of the CO is from the carbon-steam reaction or the $C-O_2$ reaction has not been conclusively decided, as far as this report is concerned.

Reactions with Air or Oxygen (Combustion)

Furthermore, the increasing importance of combustion in fluidized as opposed to fixed-bed systems is recognized. For these reasons, it is indicated that there is ample reason to further experimental studies of the combustion of coal/water mixtures in fluidized systems at a pilot-plant level of investigation.

Some work has already been noted in this connection, in particular References 42, 57, 64, 66, 67, 68, 69, 74, 76, 77, 78 and 79. These works establish or at least infer that coal/water slurries or suspensions containing appreciable amounts of water can be burned.

The connotation of burning mixtures containing water is not only that low-grade coals of high-moisture content can be successfully utilized, but, more importantly, that coal can be transported in slurry form, and then burned, without costly recovery or drying operations.

For all these reasons, it is desirable to establish the economies of the combustion of coal/water systems at a level whereby useful design information can be obtained. More specifically, such information should be directed toward the energy balance: what efficiencies of heat transfer can be maintained and at what temperature levels. This will depend to a great extent upon reactor or furnace design and upon determining optimum operating conditions. Other information, while useful, will be secondary to these overriding factors.

3.7 COMBUSTION SYSTEMS

Coal, the fueld of primary interest in this study, may be combusted as pulverized coal (p.f.) in air entrained systems, as sized particles ($\sim$ 1/4 in.), or as larger pieces in fixed or slowly-moving beds. The latter involves using some kind of stoker to feed the coal and some form of grate to support the coal particles and admit combustion air.

Pulverized fuels, which are on the order of 200 mesh particle size, are entrained in what is known as primary air. If additional air is required for combustion, it is added to the combustion zone as secondary air. This supplementary air may also be required for the other types of systems.

While coal combustion systems are categorized in terms of the type of bed, there is also the matter of which side of the heat transfer surface is traversed by the combustion reactants and products, as well as other items of configuration and juxtaposition. Thus, in the case of boilers, there are water-tube boilers and fire-tube boilers. In the former, water and steam are on the tube side, with the combustion gases moving across and around the tubes. In the latter, the combustion gases move through the tubes, with the water and steam occurring on the shell side.

The fire-tube boiler is characteristic of steam engines or locomotives, for instance, while the water-tube boiler is used in larger applications, notably central power stations.

The two principle types of coal-burning systems have been designated as layer and chambered (16). The former refers to fixed beds. The latter is more specifically for pulverized fuel. The various feeding methods have been classified as follows (15):

> Overfeed
>
> Frontfeed
>
> Underfeed

In tube furnaces, combustion of the pulverized fuel is along the axis (1). Downshot furnaces are used with low-volatile bituminous coals and anthracite, where ignition instability poses a problem.

Specifications for fired heaters for the thermal processing of fluids and solids are presented by Ellwood and Danatos (84). Sherry (85) discusses equipment and methods for burning low-grade fuels such as lignite.

It is more or less agreed that the burning of coal involves three stages, a topic already encountered:

1) Release of volatile matter.
2) Burning of the volatiles in the gas phase.
3) Burning of the solid residue.

The mechanism that determines the release of heat energy is primarily physical (14). It depends in part on the geometry of the system and the flow properties of gas-solid mixtures. For convenience, we will here refer to the systems as being either fixed or fluidized with respect to the combustible.

FIXED OR SLOWLY-MOVING BEDS

For fuel-bed burning on a grate, a distillation effect occurs (21). The result is that liquid components which are formed will volatilize before combustion temperatures are reached. Cracking may also occur.

The ignition of green coal is almost entirely by radiation from hot refractory arches and from the flame-burning of volatiles (4). In fixed beds, the radiant heat above the bed can only penetrate a short distance into the bed (18). Consequently, convective heat transfer determines the intensity of warming up and ignition. Convective heat transfer also plays an important part in the overall flame-to-surface transmission (27, 28, 29).

Reactions with Air or Oxygen (Combustion)

The reaction of gases is greatly accelerated by contact with hot surfaces (21). While the reaction away from the walls may proceed slowly, reaction at the surface proceeds much more rapidly.

In fuel beds, a high bed resistance precludes the distribution of air and flow of combustion products (4). Studies indicate that the presence of water reduces this resistance. Evidently wetting causes the adhesion of fines to large particles giving a more permeable structure. Whittingham (4) comments:

> "The marked improvement in bed structure and higher burning rates consequent upon moisture conditioning more than compensates for the diluent effects of the increased content of water vapor in the flue gases."

Fixed or slowly-moving beds may be designed and operated so as to produce a minimal amount of fly-ash carryover. The sulfur problem still remains, however. If suitably simple, efficient and economical pollution control systems can be devised, then direct-fired coal heating systems will again find a place for household heating and for the heating source of small industrial plants (see Chapter 10).

Examples are shown in Figures 3.9 and 3.10.

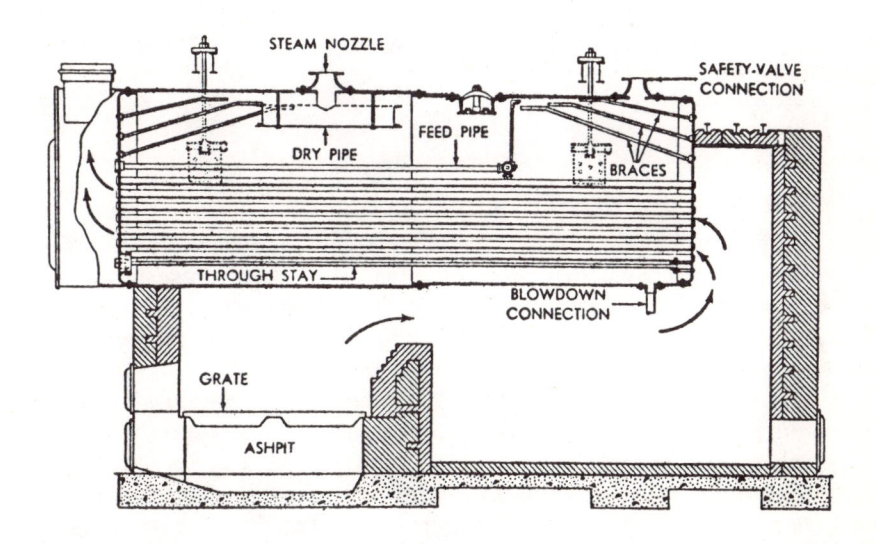

Fig. 3.9. Stationary fire-tube boiler (Babcock & Wilcox).

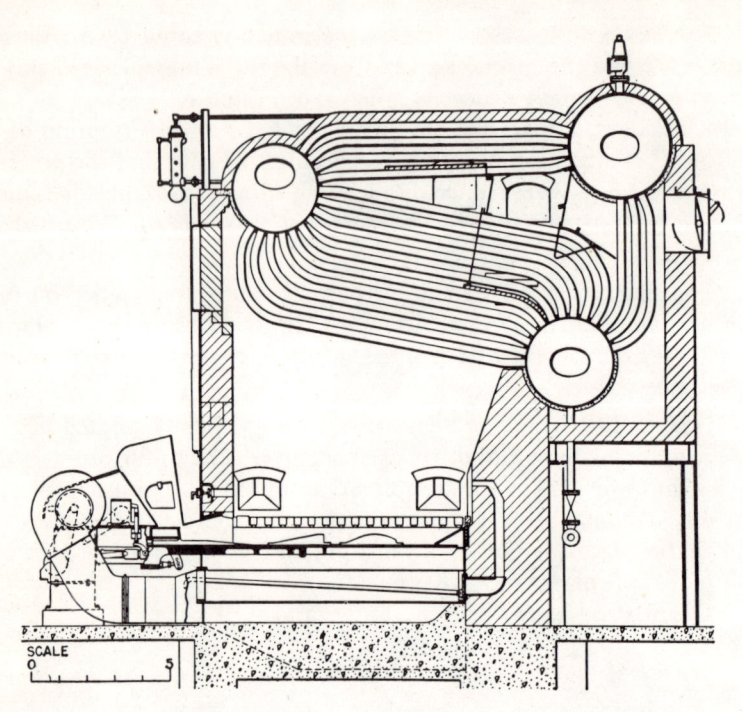

Fig. 3.10. Water-tube boiler with motor-driven stoker (Combustion Engineering).

FLUIDIZED COMBUSTION

There is a surge of interest and work in fluidized combustion as a means for providing high heat transfer rates, controlling sulfur, and reducing nitrogen oxide emissions due to the low temperatures in the combustion zone — which, incidentally, also favor the carbon dioxide equilibrium. Much of the early work originated in England, notably at the BCURA (British Coal Utilization Research Association). It is being continued in the United States under government sponsorship as previously mentioned.

In simplest terms, fluidized combustion occurs in expanded beds as diagrammed schematically in Figure 3.11. Reaction occurs at lower temperatures ($\sim 1700^\circ$F), but high convective heat transfer rates exist due to the bed motion. In fact, higher heat loads can be effected than in comparably-sized radiation furnaces. Or, put in a different way, smaller

Reactions with Air or Oxygen (Combustion)

chambers produce the same equivalent heat load. Moreover, fluidized systems can operate under substantial pressures, permitting effective cleanup of the off-gases — though cleanup can also be effected in the combustion zone. More is said on the subject in Chapter 10.

The compressor or blower requirements are transferred to the combustion air, which is at a lower temperature than the stack gases, and which can lead to some savings in compression requirements. More interestingly, under sufficient pressure, the combustion gases can be used to drive a gas turbine.

Other scaled-up versions are shown in Figures 3.12 - 3.14.

There remain the mechanical problems associated with fluidized combustion. Difficulties are encountered with the feeding of coal and particularly the withdrawal and separation of the sidestream. This is for the purposes of separating the ash from the char or unreacted coal, for recycle back to the combustion chamber.

There are also problems with pollution control. While the sulfur may be captured downstream with suitable add-on gas controls, the sulfur may also be captured in the bed proper — which adds to the separations and recycle problems. Capture during combustion, however, is recognized as the ideal and is a source of high hopes for fluidized combustion.

Ideally, the ash would agglomerate, permitting more ready separation and reducing bed carryover. Attrition does occur, however, and fly-ash particulates in the off-gases require controls.

Thus there remain problems in sulfur and particulate control along with problems with feeding, withdrawal, separation and recycle which await resolution. Nevertheless, fluidized combustion presents an intriguing prospect for direct firing.

ENTRAINED SYSTEMS

The combustion of pulverized fuel is vastly different than combustion in a fuel bed (22). For one thing p.f. is ignited mainly by radiation from the adjacent flame. It is strongly dependent upon

    Air/fuel ratio

    Percent volatiles

    Percent ash

    Particle size

No such limitations exist in, say, overfeed beds. Furthermore, the burning of p.f. requires that the volatiles burn first, the carbon residue last. This is not so with fuel beds.

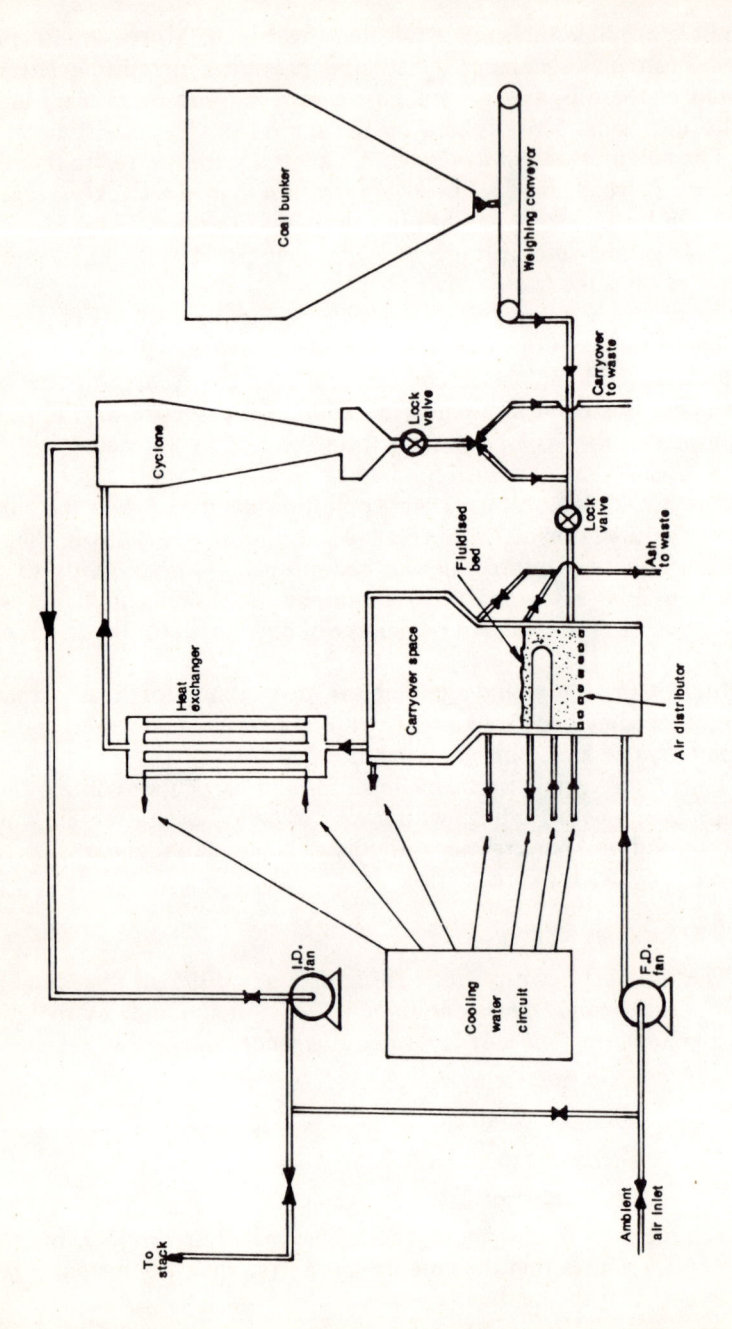

Fig. 3.11. Fluidized combustion layout.

Reactions with Air or Oxygen (Combustion)

BOILER SIZE	100 k lb./H rating, 125 psig, saturated steam
BOILER TYPE	Fluidized bed, watertube, atmospheric pressure, shop fabricated and assembled
FUEL	High sulfur (3.6 wt. %) Illinois coal, ¼" nominal size
FUEL PREP'N	No further prep'n required

ENVIRONMENTAL PROVISIONS:

SO_2	Fluidized bed system meets sulfur dioxide emission criteria
NO_x	Fluidized bed system meets nitrogen oxides emission criteria
Particulates	Electrostatic precipitator (ESP)

OVERALL RATES

Coal 12,000 lb/H = 144 T/D
Limestone 4,500 lb/H = 54 T/D
Total Dry Solids to Disposal
(Ash & Spent Stone) 2.28 T/H = 55 T/D

Fig. 3.12. Fluidized bed boiler system with sulfure removal (ERDA).

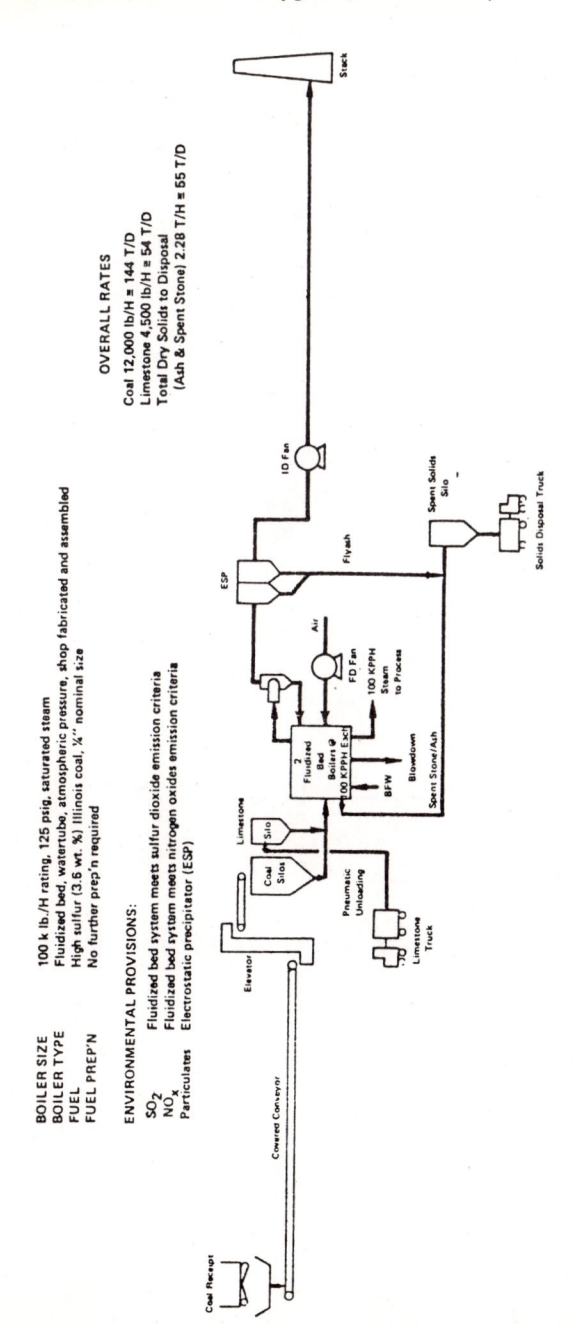

Note: Limestone may not be required when low sulfur "compliance" coal is burned in a fluidized bed. In this case, provision is made for receipt, storage, and handling of small quantities of inert makeup material (e.g. crushed cinder, alumina) for the bed.

140   Coal Conversion

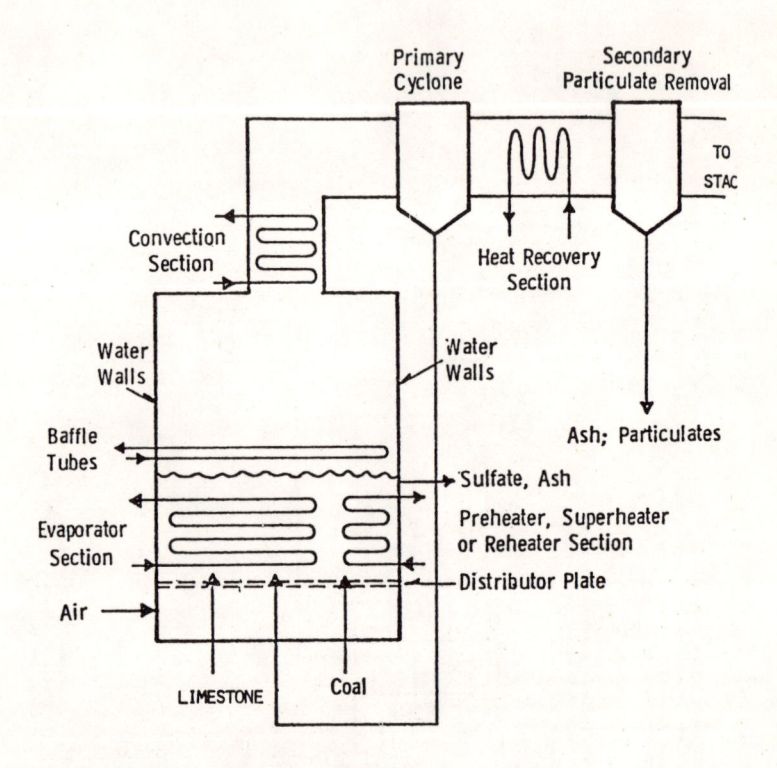

```
Combustion Pressure - close to atmospheric (usually balanced draft)
    Bed Temperature - 1400 to 1600°F
    Gas Velocity    - 2 to 12 ft/sec
    Bed Material    - limestone or dolomite, or inert material for low sulfur fuels
                      not requiring sulfur capture
    Fuel            - coal, fuel oil, bark and wood wastes, coke, char, etc.
```

Fig. 3.13. Atmospheric FBC boiler (ERDA).

Reactions with Air or Oxygen (Combustion)

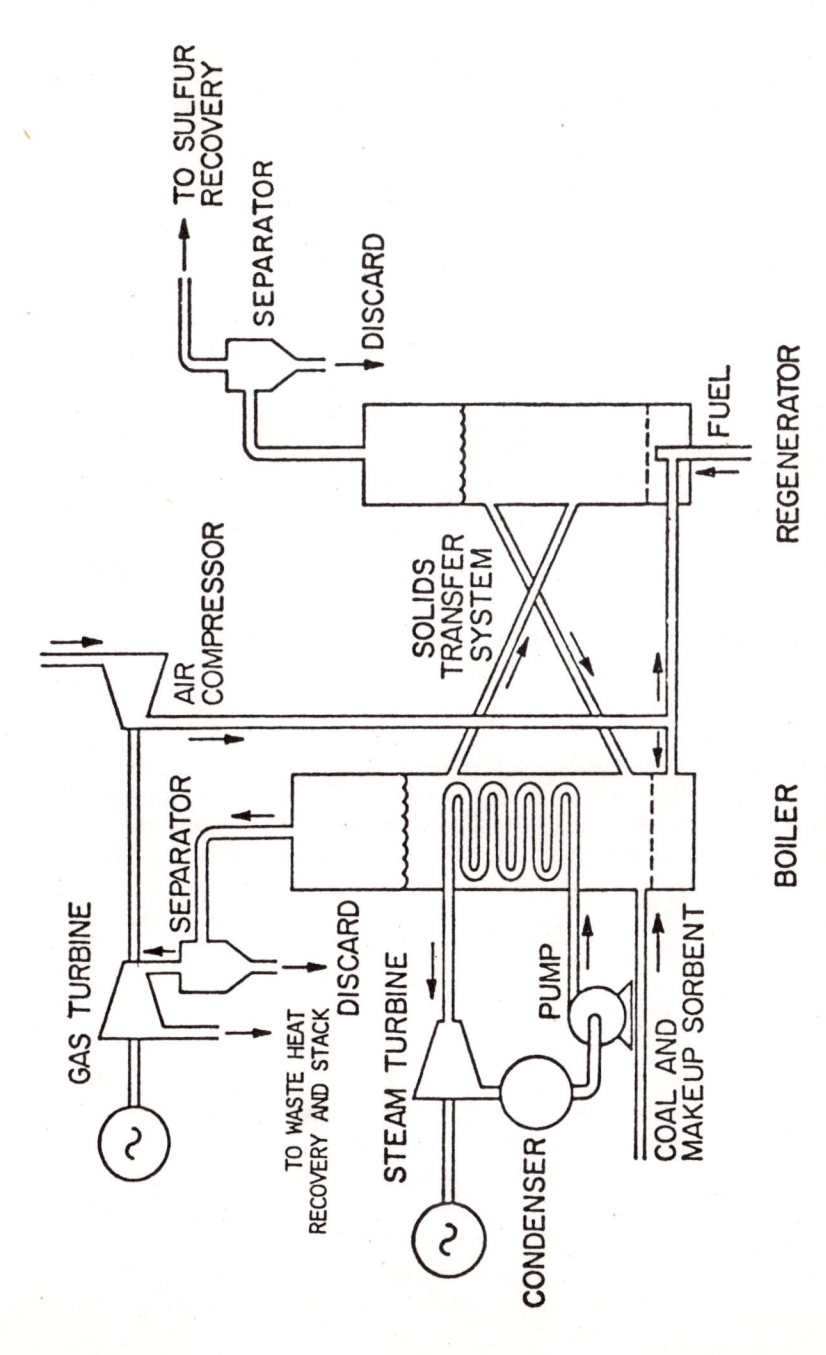

Fig. 3.14. Pressurized FBC boiler with sorbent regeneration (ERDA).

High-moisture content adversely affects pulverizer performance (11). In addition to reducing the Btu content, it requires a higher air temperature to the pulverizer for evaporating moisture to meet ignition requirements.

In entrained systems fine grinding and increased retention times intensify combustion (18). The temperature of the carrier and degree of dispersion are also important. In present day practice, for the burning of pulverized fuel, the fuel is introduced at high velocities (1). These velocities may be greater than 100 ft/sec, and involve expansion from a jet to the combustion chamber.

The pulverized coal is usually suspended in a stream of primary air (about one-fourth of the total), the remainder of the air for combustion being introduced as secondary air. The secondary air may enter at an inlet region surrounding or adjacent to the primary air duct, or at some distance away.

The introduction of pulverized fuel poses problems in flashback and in deposition of coal deposits in the lines. For these reasons, the fluidized velocity is kept at values greater than about 55-60 ft/sec using primary air/coal ratios greater than 1.4 or 1.5/1.

The temperature of the primary air should be regulated within limits. A lower temperature reduces the possibility of flashback. A temperature of at least 60°C is required to prevent the condensation of moisture in the lines; but at temperatures of 80-130°C bituminous coal particles, for instance, will soften and stick.

For protection against flashback into the feed pipe, the primary air should be limited to 25% or less of the stoichiometric amount for combustion, and the velocity should be at least 55-60 ft/sec as has been noted.

The entrained reactor provides a model for the experimental study of the combustion of pulverized coal-air mixtures, as encountered in industrial furnaces. Other types of entrained systems include cyclone furnaces, which have been used for lignite (11). Systems have been developed and utilized for the injection of coal-oil slurries into blast furnaces (86)., and for the burning of coal-water slurries as previously presented (57, 64, 65, 66, 67, 68, 69, 74, 76, 77, 79). The latter trades off latent heat losses against enhanced reactivity and convenience.

Prior to 1945, few data were available for conditions within p.f. flames. This has been supplanted since by work by the BCURA (British Coal Utilization Research Association) and others. Two pilot plant projects at BCURA were concerned with the combustion of coal in gaseous suspension (48, 49). One was a large-scale operation involving pulverized flames; the other the burning of coal in fluidized beds. Unfortunately, the

Reactions with Air or Oxygen (Combustion)

BCURA now exists in name only with the research undertaking discontinued.

Commercial types of p.f. combustion systems are shown in Figures 3.15 and 3.16, along with an example of a type of burner used (Fig. 3.17).

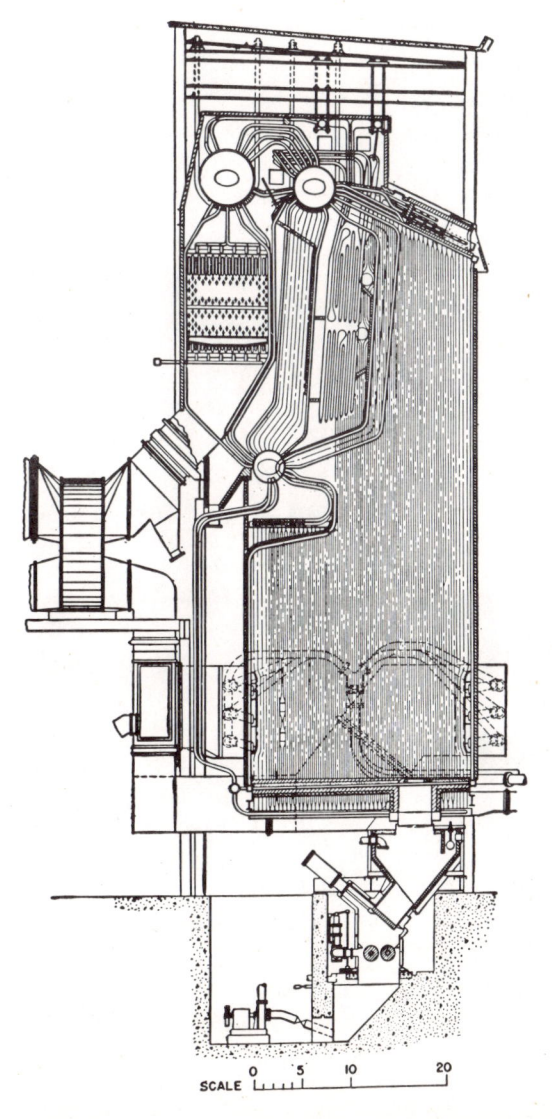

Fig. 3.15. Steam generating unit with slag tap (Combustion Engineering).

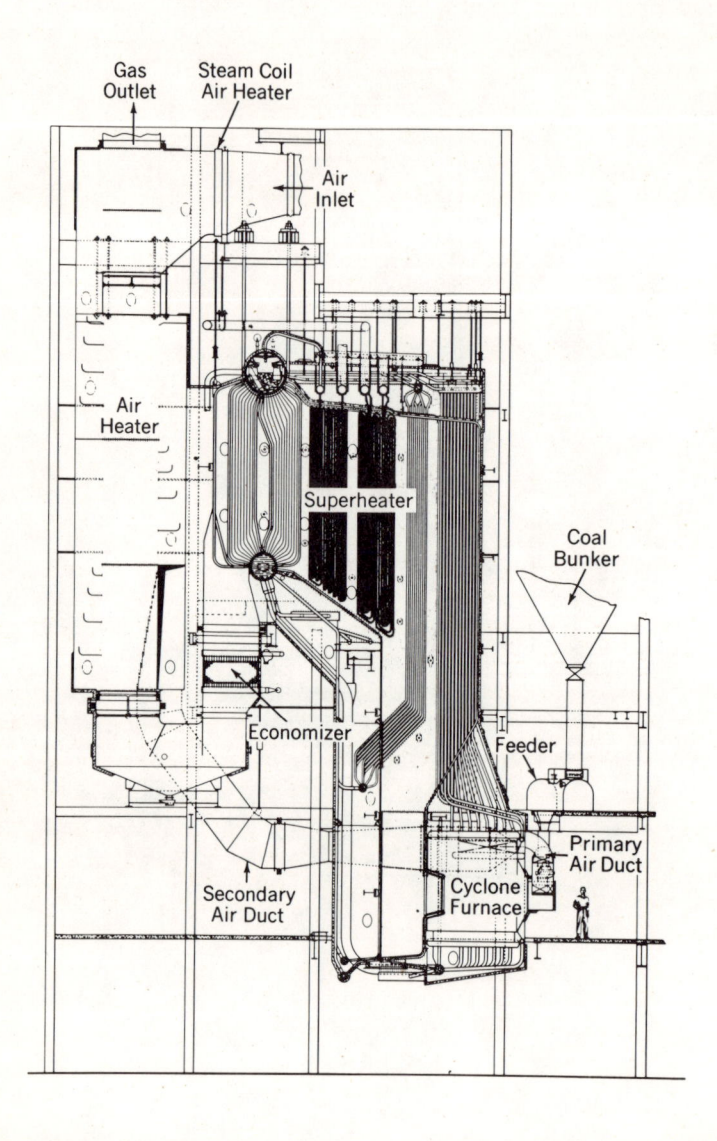

Fig. 3.16. Steam generating unit fired by cyclone furnace (Babcock & Wilcox).

Reactions with Air or Oxygen (Combustion)

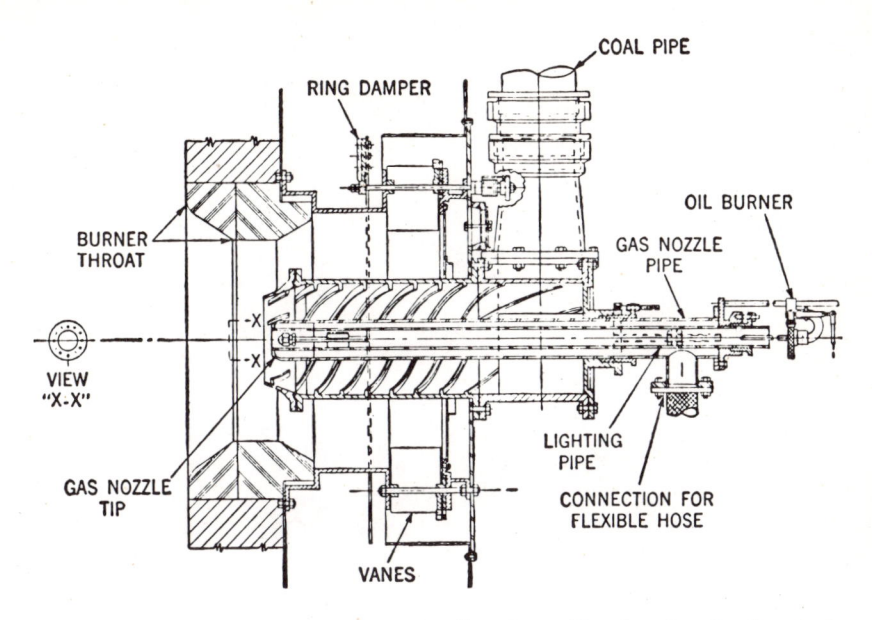

Fig. 3.17. Burner for pulverized coal, oil and gas (Combustion Engineering).

REFERENCES

1. Field, M. A., D. W. Gill, B. B. Morgan and P. G. W. Hawksley, *Combustion of Pulverized Coal,* British Coal Utilization Research Association (BCURA), Leatherhead, Surrey (1967).
2. Jost, W., *Explosion and Combustion Processes in Gases,* McGraw-Hill, New York (1946).
3. Start, P. L., "Combustion of Coal," *J. Inst. Fuel, 35,* 496-502 (1962).
4. Whittingham, G., "Some Effects of Moisture on the Combustion of Coal," British Coal Utilization Research Association, Bulletin *27* (6), 229-236 (June, 1953).
5. Boudart, M., "Chemical Kinetics and Combustion," in *Eighth Symposium on Combustion,* **Williams and Wilkins,** Baltimore (1962), pp. 43-50.
6. Essenhigh, R. H., "Constitution, Rank Index, and Flammability of Coals," *Fuel, 34,* 497-501 (1955).

7. Orning, A. A., "The Combustion of Pulverized Coal," in *Chemistry of Coal Utilization*, H. H. Lowry Ed., Wiley, New York (1945), pp. 1522-1567.

8. Essenhigh, R. H., "Influence of Coal Rank on the Burning Times of Sigle Captive Particles," *J. Eng. Power, 85* (3), 183-190 (1963).

9. Gauger, A. W., "Condition of Water in Coals," in *Chemistry of Coal Utilization*, Vol. I, H. H. Lowry Ed., Wiley, New York (1945, pp. 600-626.

10. Schmidt, L. D., "Changes in Coal During Storage," in *Chemistry of Coal Utilization*, Vol. I, H. H. Lowry Ed., New York (1945), pp. 627-676.

11. Markley, G. F., "Progress in Lignite Firing," *Power Engineering, 71* (11), 55-57 (Nov., 1967).

12. Sondreal, E. A., "Low-Tempeature Oxidation of Lignite as Related to Storage," U.S. Bur. Mines Information Circular No. 8234, 1964, pp. 59-73.

13. Haslam, R. T. and R. P. Russell, *Fuels and Their Combustion*, McGraw-Hill, New York (1926).

14. Mayers, M. A., "Combustion in Fuel Beds," in *Chemistry of Coal Utilization*, H. H. Lowry Ed., Wiley, New York (1945), pp. 1482-1521.

15. *Combustion Engineering*, O. deLorenzi Ed., Combustion Engineering, Inc., New York (1957) . . .

16. Khitrin, L. N., M. B. Rakvitch, and L. L. Kotova, "The Characteristics of the Combustion of the Coke Residue of Solid Fuels," in *Eighth Symposium on Combustion*, Williams and Wilkins, Baltimore (1962), pp. 793-800.

17. Kiebler, M. W., "The Action of Solvents on Coal," in *Chemistry of Coal Utilization*, Vol. I, H. H. Lowry Ed., Wiley, New York (1945), pp. 677-760.

18. Yavorskii, I. A., "Peculiarities in Warming Up of Coals and Ways for Intensifying Their Ignition During Stratified Combustion," Tr. Transp.-Energ. Inst. Akad. Nauk SSSR, Sibirsk, Otd. 1959, No. 8, 5-17, CA 58:3398d.

19. Walker, T. K., "The Role of Water in Spontaneous Combustion of Solids," *Fire Res. Abstr. Rev., 9* (1), 5-22 (1967). BCURA Abstract No. 1298 (1967).

20. Griswold, J., *Fuels, Combustion and Furnaces,* McGraw-Hill, New York (1946).
21. Wilson, P. J., and J. H. Wells, *Coal, Coke, and Coal Chemicals,* McGraw-Hill, New York (1950).
22. Sherman, R. A. and B. A. Landry, "Combustion Processes," in *Chemistry of Coal Utilization,* Supplementary Volume, H. H. Lowry Ed., Wiley, New York (1963), pp. 773-819.
23. Thring, M. W. and R. H. Essenhigh, "Thermodynamics and Kinetics of Solid Combustion," in *Chemistry of Coal Utilization,* Vol. III, H. H. Lowry Ed., Wiley, New York (1963), pp. 754-772.
24. Gaydon, A. G. and H. G. Wolfhard, *Flames: Their Structure, Radiation and Temperature,* Chapman and Hall, London (1953).
25. Tesner, P. A., H. J. Robinovitch and I. S. Rafalkes, "The Formation of Dispersed Carbon in Hydrocarbon Diffusion Flames," in *Eighth Symposium on Combustion,* Williams and Wilkins, Baltimore (1962), pp. 801-806.
26. Hottel, H. C., "Radiant Heat Transmission," in *Heat Transmission,* by W. H. McAdams, McGraw-Hill, New York (1954).
27. Reed, R. D., "A New Approach to Design for Radiant Heat Transfer in Process Work," John Zink Company Process Heating Seminar, Tulsa, Oklahoma, May 1951. Also refer to "Radiant Heat Transfer versus Wall-to-Tube Spacing," Technical Paper 53-15, Western Petroleum Refiners Association, 1953.
28. Reed, R. D., *Furnace Operation,* Gulf Publishing Co., Houston (1973).
29. Hedley, A. B. and E. W. Jackson, "The Influence of Recirculation in Combustion Processes," *J. Fuel Heat Tech., 15,* 7-11 (Sept.-Oct., 1966). BCURA Abstract No. 826 (1967).
30. Hoffman, E. J., "Convection and Radiation Heat Transfer in Tubestills and Furnaces," Combustion Institute, University of West Virginia, Morgantown, October, 1969.
31. *Steam,* The Babcock & Wilcox Company, New York (1955, 1960, 1963, 1972).
32. Schuett, E., "Oxidation Kinetics of Coal. I," *Erdoel v Kohle, 17* (12), 985-9 (1964). CA 62:6303d.
33. Van Loon, W. and H. H. Smeets, "Combustion of Carbon," *Fuel, 19,* 119-21 (1950. CA 44:5563[a].

34. Muenzer, H. and W. Peters, "Kinetics of Coal Oxidation Between 30-100°C," *Brennstoff Chem., 46* (12), 399-407 (1965). CA 64:7926ª.

35. Lavrov, N. V., "The Mechanism of Reactions Taking Place During Combustion and Gasification of a Solid Fuel," *Dkl. Akad. Nauk Uz. SSR, 1961* (3), 10-14. CA 61:411g.

36. Golovina, E. S. and G. P. Khaustovich, "The Interaction of Carbon with Carbon Dioxide and Oxygen at Temperatures up to 3000°K," in *Eighth Symposium on Combustion,* Williams and Wilkins, Baltimore (1962), pp. 784-792.

37. Burdukov, A. P. and V. E. Nakoryakov, "Combustion of Coal Particles, Injected as Dust, in an Ultrasonic Field," *Kinetika Goreniya Iskop. Topliv, Akad. Nauk SSSR, Sibirsk, Otd., Khim.-Met. Inst., 1963,* 87-96. CA 61:13088d.

38. Gray, M. D. and G. M. Kimber, "Reaction of Charcoal Particles with Carbon Dioxide and Water at Temperatures up to 2800°K," *Nature, 214,* 797-798 (May 20, 1967).

39. Al'tshuler, V. A. and G. S. Shafir, "Reaction of Methane with Steam and Carbon Dioxide in the Presence of Carbon," *Gazifikatsiya i Piroliz Topliv, Akad. Nauk SSSR, Gos. Kom. po Toplivn, Prom. pri Gosplane SSSR, Inst. Goryuch. Iskop., Sb. Statei, 1964,* 58-63.

40. Heinsohn, R. J. and D. E. Wulfhorst, "The Effects of an Electric Field on an Opposed-Jet Diffusion Flame," *Combust. Flame, 11* (4), 288-296 (Aug., 1967).

41. Lewis, B. and G. von Elbe, *Combustion Flames and Explosions of Gases,* Academic Press, New York (1951).

42. Delattre, J., "Catalyzing Combustion," Fr. Patent 854, 277, April 9, 1940.

43. Gerard, P. A., "Additive to Improve Combustion of Briquetted Carbon," Belg. Patent 671, 411, Feb. 14, 1966. Assigned to Societe Anon. Antipollutant Ariosan. CA 65:P10383d.

44. Andrzejewski, R., "Activation of Cokes with Gas and Its Significance in Combustion and Gasification Processes," *Koks, Smola, Gas, 6* (4), 121-129 (1961). CA 60:5242a.

45. Halpern, C., "Effect of Some Halogenated Hydrocarbons on the Flame Speed of Methane," *J. Res. National Bur. Standards (Phys. Chem.), 70A,* 133-141 (March-April, 1966).

46. "Proceedings of the Fourth International Conference on Fluidized-Bed Combustion," Mitre Corporation, McLean, Virginia (1975).
47. Skinner, D. G., "The Fluidized Combustion of Coal," British Coal Utilization Research Association, Leatherhead, Surrey (1970).
48. Thurlow, G. G., "The Combustion of Coal in Suspension. Part 1," *Ind. Process Heat, 7,* 6-20 (July, 1967).
49. Thurlow, G. G., "The Combustion of Coal in Suspension. Part 2." *Ind. Process Heat, 7,* 4-7 (Aug., 1967).
50. Bartkowiak, A. and J. M. Kuchta, "A Large Spherical Vessel for Combustion Research," *Ind. Eng. Chem. Process Design Dev., 5,* 436-439 (Oct., 1966).
51. Barshtein, I. K., "Influence of Moisture Content in Fuel on Boiler Performance," *Eng. and Boiler House Rev., 64,* 177-180. CA 43:5921e.
52. Bonnington, S. T., "Developments in the Hydraulic Transport of Coal," *Coal Prepar., 2,* 219-223 (Nov./Dec., 1966).
53. Hopkins, W. E., *Combustion, 33,* 30-35, Nov., 1961.
54. Kane, L. J. and J. D. Spencer, "Pneumatic Coal Transport: Difficult but Promising," *Power Engineering, 70,* 58-59 (Aug., 1966).
55. "Transport of Materials in Bulk or in Containers by Pipe-Line." U.N. Dept. Econ. Soc. Affairs, New York, 1966.
56. Condolios, E., E. E. Chapus and J. A. Constans, "New Trends in Solids Pipelines," *Chem. Eng., 74* (10), 131-138 (May 8, 1967).
57. Meier zu Kocker, H. and R. Huning, "The Combustion Behavior of Coal/Water/Oil Mixtures," *Brennst.-Chem.,* 48, 1-6 (Jan., 1967). BCURA Abstract No. 666 (1967).
58. Proc. Symp. Pneumatic Transport. Solids, Morgantown, 1965. U.S. Bur. Mines Information Circular No. 8314, 1966.
59. Nichols, D. G., B. S. Patel, C. Lin, K. C. Chaung and C. Y. Wen, "An Analysis of the Pneumatic Pipeline Transmission of Coal with Fuel Gas," Proceedings of the International Symposium on Freight Pipeline, Washington, D. C., December, 1976.
60. Smith, F. L., "Pipeline Movement of Bulk Materials," *Pipeline Engr., 37,* 48-52 (Sept., 1965).
61. Brady, G. A., H. H. Griffiths and J. W. Eckard, "Dewatering Anthracite Slurry," U. S. Bur. of Mines Report of Investigation No. 7012, 1967.

62. Morinaga, K. and Y. Jomoto, "Coal Slurry Injection into the Blast Furnace," *Iron Steel, 40,* 92 (March, 1967).
63. Pipilen, A. P., M. Weintraub and A. A. Orning, "Hydraulic Transport of Coal," U. S. Bur. Mines Report of Investigation No. 6743, 1966.
64. Darydovia, I. V., C. N. Delyagin, B. V. Kantorovich and V. S. Levanevskii, "Experimental Investigation of Combustion of Coal-Water Suspensions," *Novye Metody Szhiganiya Topliv i Vopr. Teorii Goreniya, Akad. Nauk SSSR, Inst. Goryuch. Iskop., 1965,* 140-145. CA 64:13967h.
65. Davydova, I. V., G. N. Delyagin, B. V. Kantorovich and V. S. Levanevskii, "The Combustion of an Aqueous Suspension of Coal," *Teplo-i Massoperenos, 4,* 213-220 (1966). CA 65:15109f.
66. Delyagin, C. N., "Regulrities of the Combustion of Pulverized Coal-Water Suspensions in an Air Steam." *Novye Metody Szhiganiya Topliv i Vopr. Teorii Goreniya, Akad. Nauk SSSR, Inst. Goryuch. Iskop., 1965,* 72-83. CA 64:13967g.
67. Delyagin, G. N., "Mass Transfer During the Combustion of an Aqueous Suspension of Coal in an Air Steam," *Teplo-i Massoperonis, 4,* 221-223 (1966). CA 65:15109c.
68. Delyagin, G. N. and B. N. Smetannikov, "The Ignition of an Aqueous Suspension of Carbon in a Stationary High-Temperature Medium" *Teplo-i Massoperenos, 4,* 234-40 (1966). CA 65:15109d.
69. Delyagin, G. N. and Z. V. Smirnova, "Interaction of Coal with Water in Combustion in a Coal-Water Suspension Layer," *Norvye Metody Szhiganiya Topliv i Vopr Teorii Goreniya, Akad. Nauk SSSR, Inst. Goryuch. Iskop., 1965,* 126-130. CA 64:12967f.
70. Frey, D. J., J. Jonakin and V. Z. Caracristi, *Combustion, 34,* 20-24 (June, 1962).
71. Jonakin, J. and V. Z. Caracristi, *Combustion, 34,* 35 (Oct., 1962).
72. Kashirskii, V. G., "Oxidative Pyrolysis of Peat," *Izv. Vysshikh Uchebn. Zavedenii, Energ., 7* (1), 80-85 (1964).
73. Benesch, R., "Improving the Blast Furnace Process by Injections of Coal/Oil Mixtures," *Hutnik, Katowice, 33* (3), 73-77 (1966). BCURA Abstract No. 933 (1967).
74. Schwarz, O., "Combustion of Powdered Coal and of Coal/Water Suspensions in Water-Tube Boilers," *Journees Intern. Combust. Conversion Energie,* Paris, 1964, 383-392. CA 65:15109b.

Reactions with Air or Oxygen (Combustion)

75. Schroeder, K., "Possibilities to Reduce Power Generation Costs," *Combustion, 38* (4), 43-50 (Oct., 1966).
76. Schwarz, O. and H. Merten, "Production, Transport of Coal/Water Suspensions," *Brennst.-Warme-Kraft, 18,* 474-478 (Oct., 1966). BCURA Abstract No. 146 (1967).
77. Schwarz, O. and H. Merten, "Direct Combustion of Coal/Water Slurries in Power Station Boilers," *Gluckauf, 103* (5), 225-231 (March 2, 1967). BCURA Abstract No. 930 (1967).
78. Lukes, J., J. Kucera, J. and M. Zizka, "Combustion of Coal in an Aqueous Medium," Czech. Patent No. 104, 245, July 15, 1962. CA 64P:9453e.
79. Meier Zu Kocker, H. and R. Huning, "Combustion: Problems of Coal/Water/Oil Suspensions," *Chemie-Ingr-Tech., 39* (1), 7-12 (Jan. 13, 1967). BCURA Abstract No. 521 (1967).
80. "Pipeline Coal Discharge Conveyer," Br. Pat. Spec. 1 054 196, Appl. date (U.S.A.), May 9, 1963. Assigned to Combustion Engineering Ltd. BCURA Abstract No. 739 (1967).
81. "Recovery of Cleaned Coal from Aqueous Slurries," Br. Pat. Spec. 1 041 547, Appl. date June 11, 1964. Assigned to Esso Res. and Eng. Co. BCURA Abstract No. 284 (1967).
82. Round, G. F., J. Kruyer and G. W. Hodgson, "Thermal Separation of Coal/Oil Suspensions in a Fluidized Bed," *Canad. J. Chem. Eng., 44,* 43-46 (1966.
83. Morinago, K., "Research on Coal-Oil Slurry Injection into the Blast Furnace," *Tetsu-to-Hagane', 52,* 107-113 (Feb., 1966). BCURA Abstract No. 932 (1967).
84. Ellwood, P. and S. Dnaatos, "Process Furnaces," *Chem. Eng., 73,* 151-175 (April 11, 1966).
85. Sherry, J. L. "Steam Raising with Low Grade Fuels," *Steam Heat Engr, 35,* 6-13 (Nov., 1965). *35,* 42-45 (Dec., 1965); *35,* 26-32 (Feb., 1966); *35,* 18-21 (March, 1966).
86. Limpach, R. et.al., "Injection of Coal-Oil Slurry into the Blast Furnace," CNRM, 3-6 (July, 1965). BCURA Abstract No. 3674 (1966).

Chapter 4

PARTIAL COMBUSTION

The conversion of solid carbonaceous material to gaseous mixtures containing CO has been of large industrial significance. The result of the gasification, called producer gas, represents a gaseous product which has further heating value. This low-Btu gas, a producer gas, is generated by means of the partial combustion of carbonaceous fuels with air, or with air and steam.

In principle, the reacting system is the same for partial combustion as for so-called complete combustion. Only the proportions and reacting conditions vary. The same components are involved, but include carbon monoxide and hydrogen in more significant proportions.

Partial combustion is necessarily a slower process than complete combustion since reaction may occur at lower temperatures — or, at higher temperatures, equilibrium conditions are reached. Complete combustion involves higher concentrations of the reactant air or oxygen, which is ~ 15-25% in excess of the stoichiometric amount for complete conversion, depending upon the fuel. The heat of reaction to produce CO_2 is greater than to produce CO, resulting in higher reaction temperatures and rates. The excess air drives the reaction to CO_2, a reaction which is also faster than the reaction to CO. Thus there is the matter of relative rates, though equilibrium favors CO at higher temperatures.

The proportions of CO depend upon the mode of contact between the fuel charge and the air or oxygen (1). Lower space velocities encourage higher CO formation, for example. For fixed or moving beds the allowable temperature of a fixed fuel bed is most usually limited by the ash fusion point of the charge, which normally is from ~ 1800 to 2800°F. An exception is operation at slagging conditions. The temperature may be controlled by the addition of steam, which also generates CO and H_2. In the analysis of the reaction zones which exist in countercurrent flow (slowly-moving beds) with the coal charge introduced at the top, oxidation to CO_2 occurs at the bottom followed by reduction to CO, then a distillation zone for the volatile matter.

Fluidized and entrained systems call for still other criteria, particularly the degree of conversion that can be attained.

The reaction mechanisms are similar if not the same as for more complete oxidation and, depending upon what source consulted, may involve surface oxides and their further reaction and decomposition (1), and may be diffusion controlling (2).

It is the purpose here to discuss some of the aspects in more detail, employing further references with the accent on carbon monoxide as the reaction product.

4.1 REACTIONS PRODUCING CARBON MONOXIDE

The basic producer mixture is essentially carbon monoxide and nitrogen, but may also contain other components, notably hydrogen; even hydrocarbons are added to produce a "town gas". Producer gas is traditionally obtained by blowing air or a mixture of air and steam through a fuel bed of coal or coke. The air reacts to produce CO and some CO_2, while the steam reacts to yield CO and H_2. A representative analysis is ~ 5% CO_2, 27% CO, 15% H_2, and 52% N_2.

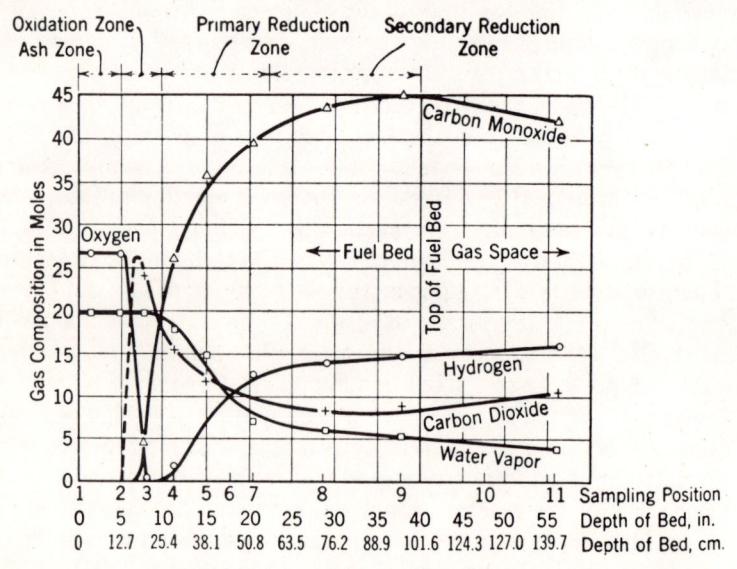

Fig. 4.1. Gas compositions in the fuel bed of a producer (from *Chemistry of Coal Utilization*, Vol. II). [From B. J. C. van der Hoeven (1). Adapted from R. T. Haslam and R. P. Russell, *Fuels and Their Combustion*, McGraw-Hill, New York (1926).]

Partial Combustion

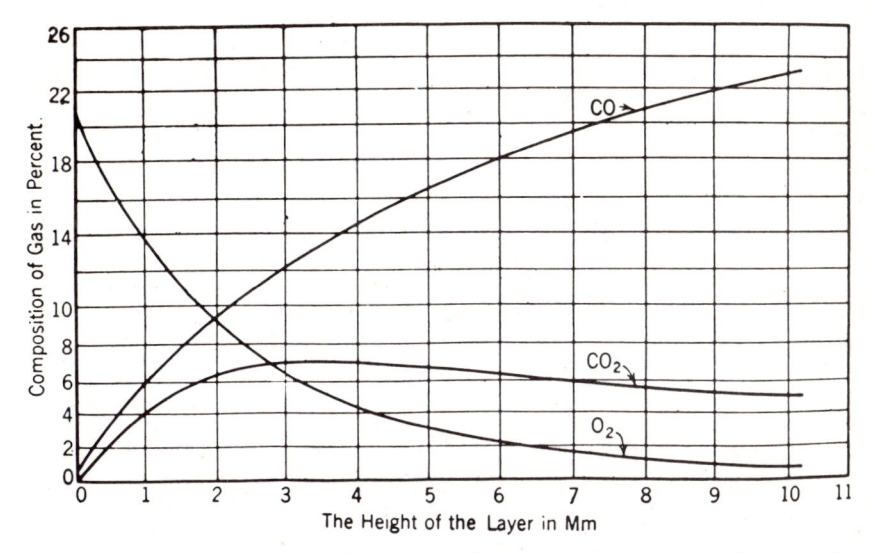

Fig. 4.2. Gas compositions in a layer of charcoal [From N. A. Karzhavina, *Fuel, 19*, 220-5 (1940), by permission of the publisher, IPC Business Press, Ltd. Consult *Chemistry of Coal Utilization,* Vol. II, Wiley, New York (1945), p. 1594.]

The composition of the product will depend upon the composition of the carbonaceous material used: the carbon, hydrogen, oxygen and moisture content. It will also depend upon the proportions of air and steam used. Furthermore, the product composition will depend upon the reactor configuration: whether fixed or slowly-moving bed, fluidized, or entrained. There is also the matter of residence time and the temperature profile.

Historically, the most predominate system in use has been the fixed or slowly-moving bed. Illustrative compositional profiles are shown in Figures 4.1 and 4.2. (Other profiles are shown in Figures 4.7 and 4.8).

The production of fuel gas from solid carbonaceous materials is a complicated process, in the same manner as is combustion. It has one saving grace, however, in that the reaction products tend to equilibrium. Equations fundamental to the conversion are reiterated as follows in cgs units, with free energy changes also listed:

		ΔH (291°K)	ΔF (298.1°K)
1.	$C + O_2(g) = CO_2(g)$	− 94,390 Cal.	− 94,260 Cal.
2.	$C + \frac{1}{2}O_2(g) = CO(g)$	− 26,490	− 32,510
3.	$C + CO_2(g) = 2CO(g)$	+ 41,410	+ 29,240
4.	$CO(g) + \frac{1}{2}O_2(g) = CO(g)$	− 67,900	− 61,750
5.	$H_2(g) + \frac{1}{2}O_2(g) = H_2O(g)$	− 57,830	− 54,507
6.	$C + H_2O(g) = CO(g) + H_2(g)$	+ 31,340	+ 22,000
7.	$C + H_2O(g) = CO_2(9g + 2H_2(g)$	+ 21,270	+ 14,760
8.	$CO(g) + H_2O(g) = CO_2(g) + H_2(g)$	− 10,070	− 7,240

The carbon is β-graphite, a solid phase. The values of ΔH and ΔF here have units of gram cal./gram mole of reactant. A negative value for ΔH indicates an exothermic reaction while a negative value for ΔF indicates equilibrium in the forward direction.

Heats of reaction are a function of temperature (and pressure) and may be adjusted for the effect of temperature by means of the heat capacities of the reactants and products as shown in the Appendix. Tabulated values for several of the reactions are given in Table 4.1. There will be some variation in the data from one source to another.

The reaction equilibria for the more significant reactions are plotted in Figures 4.3 and 4.4 as the log of the reaction equilibrium constant versus temperature (1). Carbon is given an activity of unity. Data are also provided in Table 4.2 (2).

It should be noted that coals exhibit different reactivities than carbon (β-graphite). This departure is exhibited in Figure 4.5 for the methane-hydrogen equilibrium. The discrepancy is readily explained by the fact that coals have a molecular structure that includes other elements and are not simply representable as "carbon". Thus hydrogen and oxygen are also involved in the structural makeup of coal as has been emphasized previously — not to mention inorganic constituents of the ash which may have catalytic or other effects.

An examination of the equilibrium data will show that CO is favored at higher temperatures, while CO_2 is favored at lower temperatures. Overall, hydrogen is favored by higher temperatures. The CO-shift reaction *per se* (Reaction 8), favors hydrogen at lower temperature, however. Thus generalizations can sometimes be misleading, and each circumstance may require analysis.

Partial Combustion

TABLE 4.1. Heats of Reaction (from *Chemistry of Coal Utilization*, Supplementary Volume). All compounds as gases; carbon as β-graphite.

Temp., °K	C + ½O₂ → CO	C + O₂ → CO₂	H₂ + ½O₂ → H₂O	C + CO₂ → 2CO	C + H₂O → CO + H₂	CO + H₂O → CO₂ + H₂	C + 2H₂ → CH₄
0	−27.2019	−93.9686	−57.107	39.5648	29.905	−9.660	−15.987
298.16	−26.4157	−94.0518	−57.7979	41.2204	31.382	−9.838	−17.889
600	−26.330	−94.123	−58.493	41.463	32.163	−9.300	−19.893
700	−26.407	−94.167	−58.696	41.353	32.289	−9.064	−20.401
800	−26.511	−94.215	−58.882	41.193	32.371	−8.822	−20.823
900	−26.635	−94.268	−59.055	40.988	32.420	−8.578	−21.166
1000	−26.768	−94.318	−59.214	40.782	32.446	−8.336	−21.43
1100	−26.909	−94.364	−59.360	40.546	32.451	−8.095	−21.65
1200	−27.056	−94.410	−59.497	40.298	32.441	−7.857	−21.79
1300	−27.212	−94.456	−59.622	40.032	32.410	−7.622	−21.92
1400	−27.376	−94.505	−59.735	39.753	32.359	−7.394	−22.00
1500	−27.545	−94.555	−59.841	39.465	32.296	−7.169	−22.06
1750	−27.99	−94.68	−60.067	38.70	32.08	−6.62	—
2000	−28.46	−94.83	−60.258	37.91	31.80	−6.11	—
2250	−28.94	−94.98	−60.426	37.10	31.49	−5.61	—
2500	−29.45	−95.14	−60.588	36.24	31.13	−5.10	—

*Negative sign designates exothermicity.

TABLE 4.2. Reaction Equilibrium Constants (from *Chemistry of Coal Utilization*, Supplementary Volume). Atmospheric partial pressure units*; [C] = 1.

Temp., °K	C + ½O₂ → CO	C + O₂ → CO₂	H₂ + ½O₂ → H₂O	C + CO₂ → 2CO	C + H₂O → CO + H₂	CO + H₂O → CO₂ + H₂	C + 2H₂ → CH₄
298.16	1.1163(24)	1.237(69)	1.1149(40)	1.0100(−21)	1.0013(−16)	9.9126(4)	7.916(8)
600	2.1676(14)	2.5167(34)	4.2922(18)	1.8669(−6)	5.0500(−5)	2.7050(1)	1.000(2)
700	9.2232(12)	3.1826(29)	3.8309(15)	2.6729(−4)	2.406(−3)	9.0072	8.972
800	8.5483(11)	6.7080(25)	1.9433(13)	1.0893(−2)	4.3988(−2)	4.0380	1.413
900	1.3352(11)	9.2570(22)	3.1429(11)	1.925(−1)	4.2483(−1)	2.2059	3.250(−1)
1000	3.0034(10)	4.7511(20)	1.1474(10)	1.8985	2.6176	1.3787	9.829(−2)
1100	8.8006(9)	6.3451(18)	7.6012(8)	1.2206(1)	1.1578(1)	0.94850	3.677(−2)
1200	3.1462(9)	1.7376(17)	7.8762(7)	5.6966(1)	3.9945(1)	0.70120	1.608(−2)
1300	1.3111(9)	8.2517(15)	1.1497(7)	2.0831(2)	1.1404(2)	0.54742	7.932(−3)
1400	6.1659(8)	6.0484(14)	2.2057(6)	6.2856(2)	2.954(2)	0.44473	4.327(−3)
1500	3.1946(8)	6.2903(13)	5.2538(5)	1.6224(3)	6.0805(2)	0.37478	2.557(−3)
1750	8.441(7)	6.744(11)	2.9604(4)	1.0565(4)	2.851(3)	0.2699	—
2000	3.064(7)	2.241(10)	3.3934(3)	4.189(4)	9,029(3)	0.2155	—
2250	1.375(7)	1.575(9)	6.2664(2)	1.200(5)	2.194(4)	0.1828	—
2500	7.153(6)	1.892(8)	1.6127(2)	2.704(5)	4.435(4)	0.1640	—

*Numbers in parentheses designate multiplication by that power of 10.

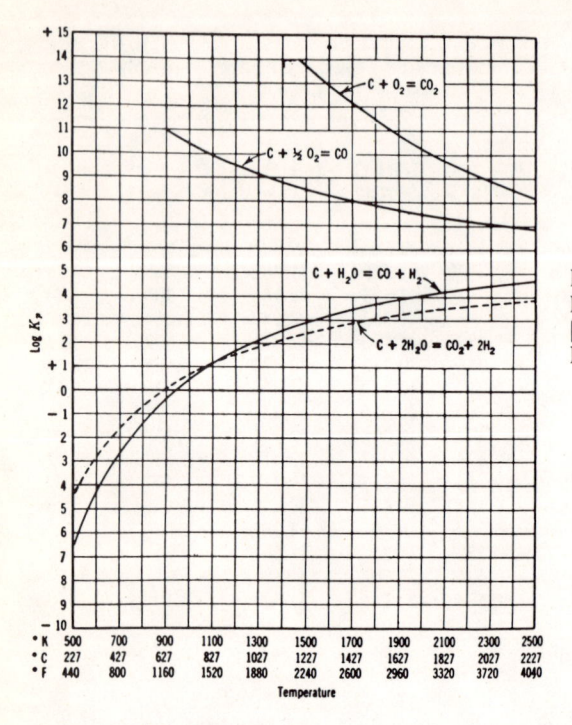

Fig. 4.3. Variation of log K_p with temperature. [From B. J. C. van der Hoeven (1).]

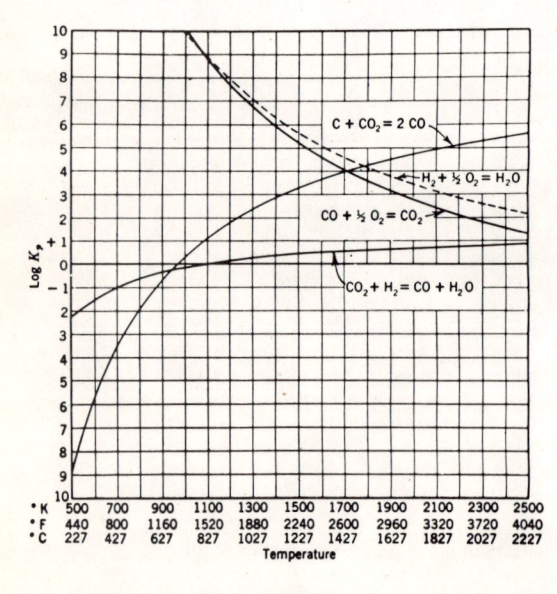

Fig. 4.4. Variation of log K_p with temperature. [From B. J. C. van der Hoeven (1).]

Partial Combustion

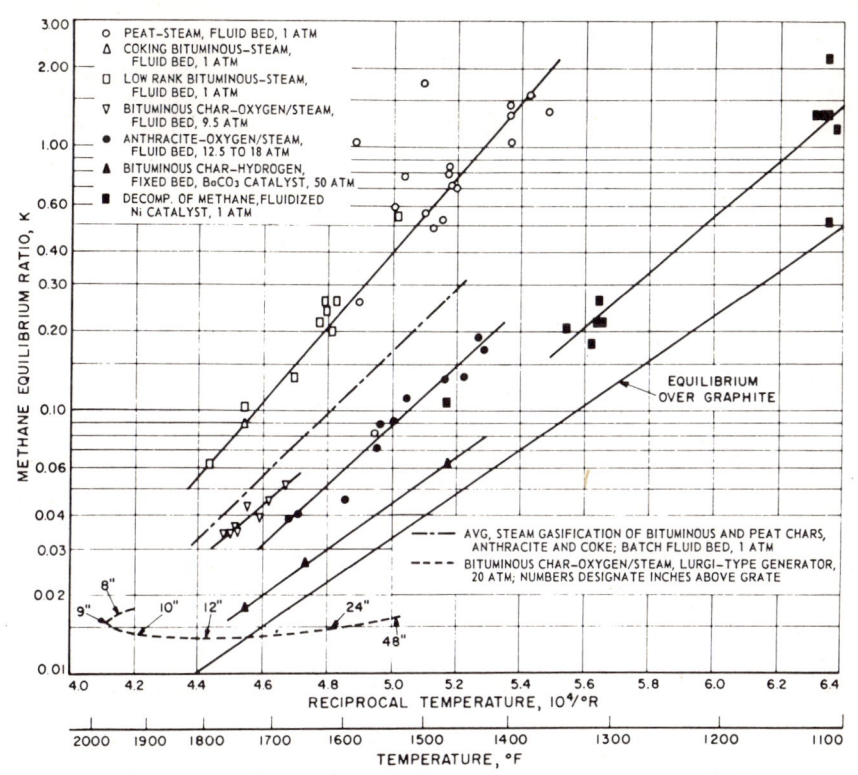

Fig. 4.5. Methane-hydrogen equilibrium (from *Chemistry of Coal Utilization*, Supplementary Volume). [From A. M. Squires, Inst. Chem. Engrs. Meeting, London, Feb. 2, 1960. Consult *Chemistry of Coal Utilization*, Supplementary Volume, Wiley, New York (1963), p. 895.]

The equations

1) $C + O_2 = CO_2$
2) $C + \frac{1}{2}O_2 = CO$

are quite significant since they represent the combustion of carbon. Hottel et.al. (3) in their work in the kinetics of combustion, determined values of the reaction constant as a function of temperature for various gas velocities. At low temperatures the curve was linear and rose sharply to about 1000°K where it broke smoothly into a leveling-off region. At temperatures below 1100°K, the temperature is the rate controlling factor. At higher temperatures, the gas velocity controlling diffusion to and from the carbon surface predominated. In other experiments using thicker fuel beds and more representative coke samples, the findings of Hottel have been confirmed.

The exact mechanism of combustion with regard to the primary product is in doubt, as set forth in the previous chapter. Thus if reaction (1) is assumed, CO_2 is the primary product. It is subsequently reduced to CO by passage through the coal. The steps in the formation of CO may be written as

$$C + O_2 = CO_2$$
$$CO_2 + C = 2CO$$

If equation (2) is assumed, the formation of CO as the primary product is short lived and in the presence of oxygen burns immediately to CO_2. The CO_2 formed is then reduced to CO. Equations for the sequence may be written as

$$2C + O_2 = 2CO$$
$$2CO + O_2 = 2CO_2$$
$$CO_2 + C = 2CO$$

Lavrov (4, 5, 6), for example, suggests that atomic oxygen formed by the dissociation of molecular oxygen, CO_2 or water reacts with carbon to form intermediate products which are thermally decomposed in the reduction zone to liberate oxides of carbon. The dissociation occurs as a result of: A) chemisorption and heterogeneous catalysis; B) chain reactions; and C) thermal dissociation. The ultimate composition of the combustion products depends upon the structure of the intermediate complexes.

Authur (7) reasoned that if CO_2 resulted from the oxidation of the primary product (CO), inhibitors of the oxidation process should increase the percentage of CO in the product. His experimental work showed that with the addition of 0.2% $POCl_3$ to dry combustion air at 850°C the CO/CO_2 ratio was about 6:1 with residual oxygen at about 3%. Similar results were produced by the addition of 0.28% CCl_4, even with the air moisture content up to about 0.2%. However, with an increase of moisture content to about 0.4% the ratio CO/CO_2 decreased to about 1:6. Similar results were experienced by Lavrov (4).

Yamauchi et.al. (8) have shown that the effect of chlorine at partial pressures of greater than 3.5 torr was inhibiting on the oxidation rate of graphite by CO_2. In the same oxidation reaction, partial pressures of bromine up to 0.9 torr caused a very rapid increase in the oxidation rate. The rate then decreased with additional bromine. The rate is always higher with CO_2 containing bromine than with pure CO_2. In comparison, the oxidation rate reached a maximum at an iodine partial pressure of 0.3 to 10 torr.

Partial Combustion

Wicke (9) explained the increase in the CO/CO_2 ratio in the presence of dry air as being due to the ability of chlorine to change the structure of the burning carbon.

Hiles and Mott (10) note a marked increase in the production of CO with an increase in the rate of combustion air supply. Kriesinger (11) predicted that this increase in air supply increases the combustion rate and thereby provides the necessary heat for the endothermic reaction

$$CO_2 + C = 2CO$$

In a new method for investigating the combustion of powdered coal, Khitrin et.al. (12) determined that the main process is first, sorption of oxygen by carbon, then combustion of the carbon (C-O reaction). Results show that the consumption of oxygen in the reaction with carbon and the sorption process are kinetically equal.

Long and Sykes (13) carried out extensive studies of the mechanism of the steam-carbon reaction. The studies were made under static and dynamic conditions, and covered a pressure range of 10 to 760 mm and a temperature range of 680 to 800°C. They concluded that steam first reacts with carbon to give oxygen and hydrogen atoms separately adsorbed on neighboring sites. Only about 2% of the total surface takes part in the reaction. The rate of the first step is accounted for by assuming that the reaction occurs in those collisions in which the combined energy of the incident steam molecules and the two active carbon atoms exceeds 75 kcal. Adsorbed hydrogen evaporates readily, but in the steady state much remains on the surface. Adsorbed oxygen reacts more slowly to form gaseous CO. This CO has no retarding effect and is not appreciably adsorbed by the sites accessible to steam. The activation energy for the conversion of an adsorbed oozgen atom into gaseous CO is 55 kcal. It is suggested that active carbon atoms that can form a single bond should react with steam, whereas the relatively few atoms capable of forming a double bond should react with CO. The effect of catalysts was investigated in a later paper (14).

A set of five simultaneous equations involving the partial pressure of H_2, CO_2, CO, CH_4 and undecomposed steam were written by Majewski and Spurrier (15) to describe the effect of elevated pressures on the carbon-steam reaction. The steam composition is given for temperatures of 750, 800, 850, and 900°K. Their findings indicate that:

1. The amount of methane is reduced by increasing the temperature and/or reducing the pressure, but the pressure effect up to 100 atm. is insignificant.
2. The behavior of CO_2 is similar to that of CH_4.

3. Steam decomposition is considerably favored by an increase in temperature but not by an increase in pressure.
4. Low temperatures favor high hydrogen formation while at high temperatures there is more CO.
5. The CO and hydrogen content increases as the temperature increases.
6. The CO and hydrogen content decreases as the pressure increases.

Investigations of the catalysis of the oxidation of carbon by metallic impurities indicate that the general form of the reaction is not changed though the rates of certain stages are increased (16). Specific findings are:

1. The activation energy for the transfer of oxygen to carbon is decreased.
2.. The decrease in the activation energy for conversion of adsorbed oxygen atoms into gaseous CO is counteracted by an opposing change in the non-exponential factor.
3. The number of active sites is not appreciably altered by the removal of impurities.

The relevancy of catalysis, both in oxidation and in the steam-carbon reaction, is becoming of increasing notice. Additional information of a more practical nature is presented in Chapter 6. Suffice to say, the presence of alkali catalyzes the reaction of carbonaceous materials and oxygen, and/or of carbon and carbon dioxide, as well as carbon and steam in one way or another (1). This may be in part caused by surface effects (i.e., fluxing) or by a more active participation of the alkali component (i.e., in oxidation and reduction reactions).

REACTION TIMES

Considerable study has been made on reaction rates or burning times for the Boudouard reaction (1):

$$CO_2 + C = 2CO$$

It is generally assumed that CO_2 concentration is controlling rather than carbon particle size or concentration — though no doubt the latter are contributing factors.

Residence times of 2 to 10 seconds have been observed, and a characteristic plot is shown in Figure 4.6. Though complete conversion is indicated, an equilibrium actually occurs which depends on temperature, as does rate.

Partial Combustion

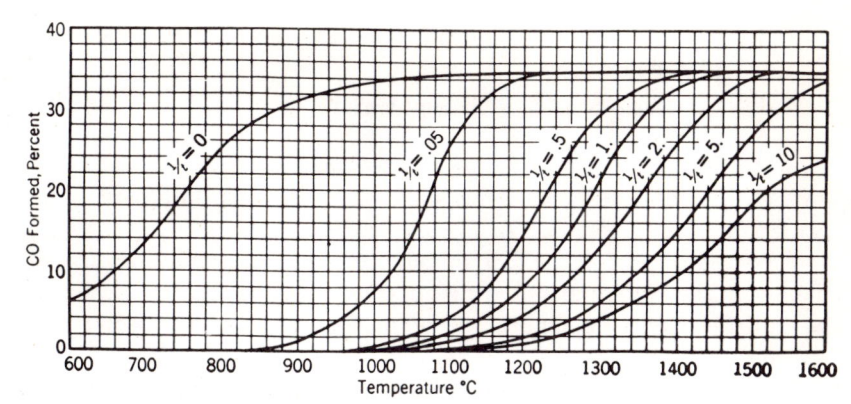

Fig. 4.6. Effect of contact time t in seconds and temperature on the reaction $CO_2 + C$ in the presence of nitrogen. [From B. J. C. van der Hoeven (1). Adapted from J. K. Clements, L. H. Adams and C. N. Haskins, *U.S. Bur. Mines Bull.* 7 (1911).]

If the Boudouard reaction in effect controls the formation of CO (that is, is the slower reaction), then the total residence time for CO formation is dictated by the sequence

$$C + O_2 \rightarrow CO_2$$
$$CO_2 + C \rightarrow 2\,CO$$

As indicated in the previous chapter, complete combustion has residence or reaction times of 0.5 sec or less, based on pulverized coal. Hence, in p.f. systems at least, the second reaction is evidently controlling.

THE BOUDOUARD REACTION

The reaction betweeen carbon and carbon dioxide to produce carbon monoxide is very possibly the controlling factor in gasification by partial combustion as is indicated in the preceding section. As such, the reaction merits further examination from the standpoint of mechanisms, catalysis and other considerations.

In an investigation of the factors which influence the reduction of CO_2, Fedoseev et.al. (17) passed a stream of 100% CO_2 over electrode carbon at rates of 1.5 - 18.8 liter/minute. Their findings indicate that the rate of conversion was affected as follows:

1. Increased with an increase in carbon dimensions from 0.25-8 mm.
2. Was independent of space velocity under near isothermal conditions.

3. Increased with space velocity in the temperature range 1400-1600 degrees.
4. Increased linearly with temperature.

Hottel et.al (3) found that a decrease in carbon particle size from 9 mm to 0.16 mm caused a 12-fold increase in the reaction rate of carbon with CO_2.

In a study of the effect of particle size on conversion of CO_2, Kokurin and Rozental (18) determined that reduction was dependent upon the size of the fuel particles and independent of the structure of the carbon.

Kokurin and Rozental (19) further showed that the maximum yield of CO was obtained with a bed thickness of 25-40 mm. Under dynamic conditions, (contact times of 1.5-2.0 sec.) it was also found that:

1. The CO/CO_2 ratio of the gas product was independent of time and initial CO_2 content.
2. The CO/CO_2 ratio increased with an increase in temperature (600-1100°C).
3. The gas product always contained unreacted carbon dioxide.

In a study of the reduction of CO_2 in a fluidized bed of petroleum coke, it was found that, in the temperature range of 500-800°C at contact times of 3-4 seconds, greater bed heights and lower gas velocities caused a notable increase in reduction of CO_2 in the layer of coke (20). When air was used as the fluidizing agent, the greatest CO content in the exit gas (14%) was experienced at the highest temperature.

Harker et.al. (21), using selected carbon samples, have shown that in the C-CO_2 reactions studied at atmospheric pressures and temperatures of 650-900°C there are two simultaneous equilibria established. These are

$$CO_2 + C \rightleftarrows C(O) + CO$$

and

$$CO_2 + C \rightleftarrows 2CO$$

It was established that the formation of CO_2 in the reaction

$$C + C(O) \rightarrow CO_2 + C$$

leads to an apparent retardation of the CO_2-C reaction. Surface oxide was formed in amounts that approached a maximum and was not desorbed under the conditions of the experiment.

When reacted with equilibrium mixtures of CO and CO_2, the rate of surface oxide formation was considerably less with CO than with CO_2.

Partial Combustion

When CO reacted with carbon previously reacted with oxygen there was an increase in the amount of combined oxygen in the gas phase.

Vastola and Walker (22) studied the reaction of CO_2 with carbon under static conditions. The reaction was carried out at pressures of 2.7-16 cm of Hg. and temperatures of 400-700°C. Carbon used was graphite wear-dust containing about 5% iron. They report the reaction to be first order with respect to CO_2 pressure. Slightly less than 2 moles of CO were formed for every mole of CO_2 consumed.

Blackwood (23) studied the reaction of graphite with CO_2 and hydrogen in a dynamic system in the temperature range of 650-870°C and at pressures of 1-30 atm. The reaction in this temperature range appears to be much simpler than at lower temperatures and the rate of formation of CO may be represented by the equation

$$R = k_1 P_{CO_2}/(1 + k_2 P_{CO})$$

Blackwood and Ingeme (24, 25) studied the reactions of purified carbon with CO_2 at pressures up to 40 atm. and in the temperature range of 790-870°C. They found that at high partial pressures of CO and CO_2 the rate of formation of CO was greater than might be expected from theory. The increase may be explained by proposed additional steps involving the interaction of a CO_2 molecule with an absorbed CO molecule to produce absorbed O_2. The reaction rate decreased with the addition of CO but was not affected by the addition of hydrogen.

Kokurin et.al. (26, 27) found that under static and dynamic conditions the reaction of oxygen, CO_2 and CO with charcoal was governed by a chemisorption mechanism. Maximum absorption of oxygen occurred at 250°C, of CO_2 at 300°C, and of CO at 700°C. At temperatures up to 400°C the reaction product in the gas phase was chiefly CO_2. However, as the temperature increased to 1100°C the concentration of CO was increased to a maximum yield of 95%. Similar results were experienced when charcoal was subjected to CO_2 at temperatures greater than 300°C. At greater than 900°C, the product was almost entirely CO.

In further regard to the mechanism, Long and Sykes (28) postulated that gasification would be enhanced if electrons were transferred from graphite to the catalyst. The transition elements (iron, etc.) accept electrons because of the non-stoichometric character of their oxides. The transition metals also exhibit unfilled d-orbitals. They would therefore accept electrons and enhance the deposition of CO.

Seith and Kubaschewski (29) showed that dissolved carbon migrated electrolytically in austentite and therefore must be in an ionic state.

Another possibility for the mechanism is suggested by the work of Pettit et.al. (30) concluding that the CO_2 dissociates over iron to give CO and adsorbed oxygen atoms. The oxygen atoms, being mobile, diffuse to a free carbon interface and react to produce gaseous CO.

Mikhailov et.al. (31) showed that a mixture of CO_2 and hydrogen could be converted to CO, glyoxal and organic acids when subjected to bombardment with fast electrons (115 kev.). It was suggested that the reaction took place through the decomposition of CO_2 and participation of molecular hydrogen.

Reactivity can be enhanced by other means. One possibility is the activation (excitation) of the reacting oxygen (air) thermally (by preheat) or by the use of an electric field — for either partial or complete combustion. In any case, the method must be energy efficient for commercial application. Electro-magneto-chemical interaction is a still-emerging science.

Rakszawski et.al. (32) investigated the use of iron as a catalyst for the carbon-carbon dioxide reaction. Iron in the form of carbonyl iron powder, Fe_2O_3, or $Fe_2(C_2O_4)_2$ as solution was added to pure graphite powder and moulded into rods. Reaction rates with CO_2 were measured for temperatures from 900 to 1200°C. Their findings indicate that:

1. Iron produced a marked increase in their reactivity for samples without previous heat treatment.
2. Heat treatment at 1400 and 1635°C sharply decreased the catalytic efficiency of iron.
3. Reactivity was further decreased by heat treatment at temperatures of 1910 and 2990°C.
4. Reactivity may be restored by secondary treatment in hydrogen or oxygen.

Rakszawski et.al. contend that the extent to which iron catalyzes the C-CO_2 reaction depends upon the previous heat treatment given the sample. It is indicated that during the reaction, CO_2 reacted with the carbon in the iron to remove it. They also mention that CO_2 is a poor carburizer but that a mixture of CO and CO_2 is a good carburizer.

In a patent to Esso (33), CO is produced by passing CO_2 over a catalyst mixture of CoO and MoO_3 supported on Al_2O_3, at 1100-1400°F in the presence of carbon or gaseous hydrocarbons such as methane. The process may be carried out in two stages by first passing methane over the catalyst to form hydrogen and deposit carbon on the catalyst, then feeding CO_2 over the catalyst to produce CO. A sample product gas composition is listed as CO_2-10%, CO-85%, and H_2-29%.

Partial Combustion

In the case of supported iron catalysts, Swamy (34) has shown that CO_2 adsorbed on the catalyst can be desorbed by evacuation. Between 0-200°C, however, some CO_2 is adsorbed irreversibly. Above 300°C uptake of CO_2 was continuous and no equilibrium was established after a period of 36 hours. Upon evacuation some CO was desorbed along with CO_2. Reduction of the adsorbent by hydrogen liberated water and methane, presumably from the iron oxides and carbides formed during chemisorption.

Carbon monoxide also undergoes a conversion to carbon and CO_2 in the presence of fused iron catalysts. Kagan et.al. (35) report that the reaction is accompanied by the carbonization of the catalyst. Carbonization in the initial stages of the reaction was attributed to the autocatalytic nature of the reaction

$$2Fe + 2CO = Fe_2C + CO_2$$

The decomposition of CO with the formation of carbon begins after the maximum rate of carbide formation is attained.

The activity of iron catalysts with respect to carbide formation is increased by the addition of Al_2O_3, K_2O, and V_2O_5 (36). The formation of CO_2 is also accelerated in the presence of Al_2O_3 and V_2O_5. Carbide, formation is markedly decreased by the introduction of Cr_2O_3, FeCr and especially SiO_2 and B_2O_3. MgO, MnO and CaO accelerate the rate of formation of CO_2. At higher temperatures Na_2HPO_4, talc and WO_3 decrease the rate of the reaction. Aluminum silicates will affect the CO_2 formation depending upon the amount used and their composition.

Pettit et.al. (30) indicate that for oxide thicknesses between 4×10^{-4} and 1.8×10^{-3} cm the oxidation rate was a linear function of the mole fraction CO_2 in the gas phase and the partial pressure of CO_2 and CO. It appeared that the rate determining step is the dissociation of CO_2 into CO and adsorbed oxygen atoms in accordance with the equation

$$CO_2 + \alpha - Fe = \alpha - FeO + CO$$

In studies of the action of catalysts of the spinel type ($ZnAl_2O_2$, $ZnCr_2O_4$, $ZnFe_2O_4$), Yoneda et.al. (37) proposed two reactions for the exchange of oxygen between CO_2 and CO. At temperatures less than 250°C, CO_2 exchanges its oxygen atom with the oxide through carbonate formation,

$$CO_2 + O^= \rightarrow CO_3^=$$

without its dissociation with CO and an oxygen atom, and without the participation of CO in the reaction. At higher temperatures oxygen is transferred to CO from CO_2 through the oxide catalyst by

$$CO + K(O) = CO_2 + K$$

In a study of the oxygen exchange between CO and CO_2, Bank and Verdurmen (38) used oxygen-18 as a tracer in a transparent quartz vessel. The exchange of oxygen between the O_2 and the quartz was observed in the packed vessel. The CO/CO_2 exchange appears to be at least partially heterogenous. Small amounts of water enhance the exchange rate. The exchange rate is dependent upon the CO_2 pressure to the first power, and at 820-850°C is independent of the CO pressure. The rate of exchange appears to follow the equation

$$\text{Rate} = \frac{1}{t} \frac{P_{CO} \cdot P_{CO_2}}{P_{CO} + P_{CO_2}} \ln(1-F)$$

where F is the exchange fraction. Dependence of the exchange rate upon the reactant concentration is given by

$$\text{Rate} = k \, P_{CO}^m \cdot P_{CO_2}^m$$

In studies using carbon-13 as a tracer, Brandner and Urey (39), found the reaction to be largely heterogenous and the rate of exchange to be independent of the total pressure and composition of the gas mixture.

Norris and Ruben (40) studied the exchange reaction using C-14 as a tracer and determined that the rate followed the law

$$\text{Rate} = k \, (P_{CO})^{0.73} (P_{CO_2})^{0.85}$$
$$k = 1.2 \times 10^{-4} \text{ at } T = 850°C$$

The reaction appeared to be bimolecular between the reactants adsorbed on the surface. The data, however, was poorly reproducible.

Hayakawa, (41) using C-13 as a tracer, obtained the rate law

$$\text{Rate} = k \, P_{CO_2} \frac{P_{CO}}{1 + b \cdot P_{CO}}$$

The reaction appeared to be a bimolecular reaction between the two reactants adsorbed on the quartz container. He found CO to be moderately adsorbed and CO_2 weakly adsorbed.

No allowance is made in the above formulas to include carbon activity.

Partial Combustion 169

4.2 PRODUCER GAS

Carbon monoxide has long been generated on a commercial scale by the partial reaction of air (or oxygen) with carbonaceous materials such as coal or coke:

$$C + \tfrac{1}{2}O_2 = CO$$

It is also generated in the formation of water gas by the presence or addition of moisture (steam):

$$C + H_2O = CO + H_2$$

Since the former reaction is exothermic (gives off heat) and the second-mentioned is endothermic (requires heat), partial oxidation with air (or oxygen) has a distinct advantage if carbon monoxide is the principal product desired. Hence this route will be of predominate interest here.

The low-Btu gas obtained by partial combustion of coal or coke with air contains chiefly carbon monoxide and greater amounts of nitrogen. It is generally referred to as producer gas. Ordinarily, pure oxygen would not be considered as a reactant due to expense or other reasons. An exception occurs, however, in the steel industry. Here, the basic oxygen furnace (BOF) generates carbon monoxide as a by-product of significant note. Other gasifiers, such as the Lurgi, are specifically aimed at producing a gas higher in hydrogen by using significant proportions of steam — and which require additional heat input, either by indirect heat transfer or by oxygen-blowing.

VARIABLES

Profiles of fixed- or slowly-moving bed gasifiers are shown in Figures 4.7 and 4.8. Though the temperature is lower at the top of the bed, this is in part the result of the endothermic reduction of CO_2 to form CO by the Boudouard reaction. The bottom of the bed is at a higher temperature since the air is in temporary excess as it passes up through the bed.

The inspection may be placed upon a more formal basis. Thus consider the first four equations of previous enumerations:

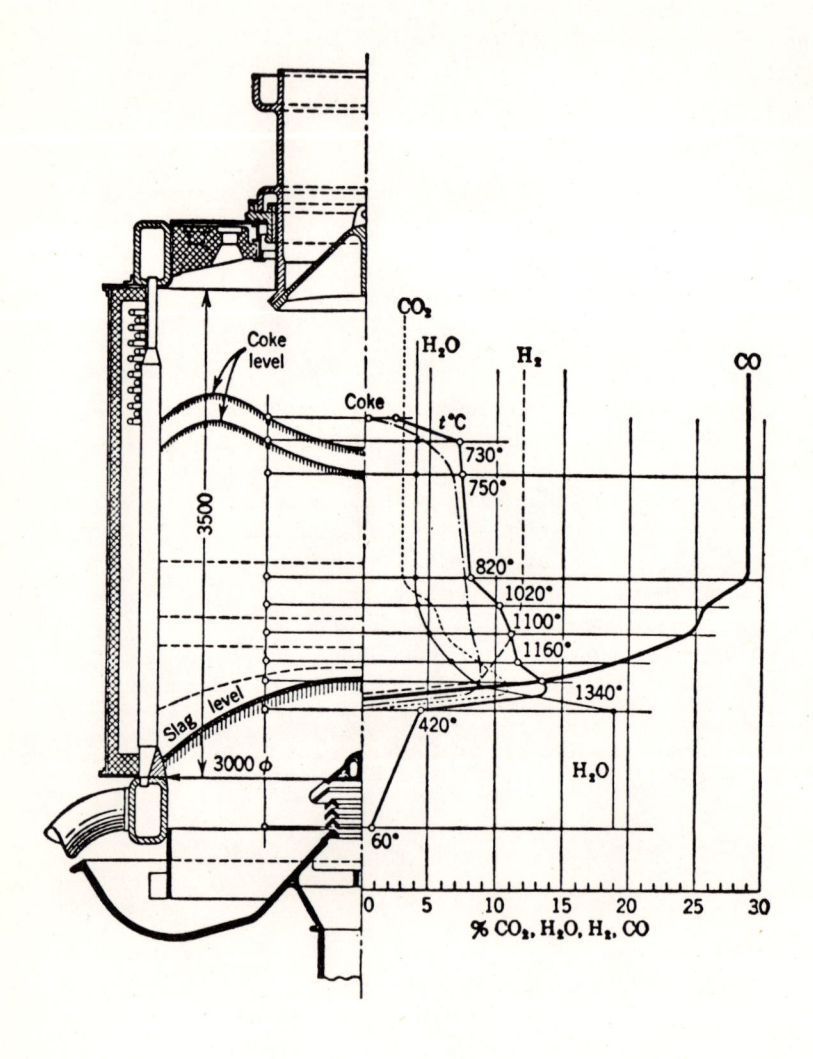

Fig. 4.7. Gas composition vs. temperature in a gas producer [From W. Gumz, *Gas Producers and Blast Furnaces*, Wiley, New York (1950). Adapted from W. Horak, *Z, Oester Ver. Gas - u. Wasserfachm, 74,* No. 11-12, 170-180, 194-203 (1933).]

Partial Combustion

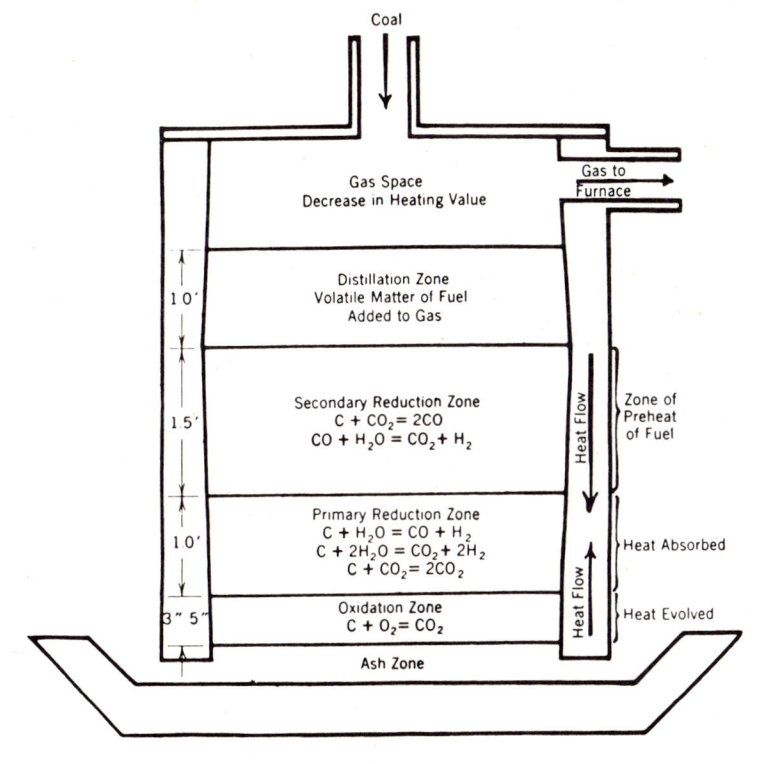

Fig. 4.8. Reaction zones in the fuel bed of a gas producer. [From B. J. C. van der Hoeven (1). Adapted from R. T. Haslam and R. P. Russell, *Fuels and Their Combustion*, McGraw-Hill, New York (1926).]

1) $C + O_2 = CO_2$ $-94,390$ cal/g-mole
2) $C + \frac{1}{2}O_2 = CO$ $-26,490$ cal/g-mole
3) $CO_2 + C = 2CO$ $+41,410$ cal/g-mole
4) $CO + \frac{1}{2}O_2 = CO_2$ $-67,900$ cal/g-mole

The composition of the product stream depends upon the conditions of contact between carbon and oxygen. In fuel beds thicker than about 4 inches, reaction (2) is dominant. In addition to (2) the hot fuel beds resulting from (1) react with the CO_2 according to (3) for the production of CO. Reaction (3) is accompanied by the absorption of heat, thereby lowering the temperature of the gas. Side reactions involving volatiles in the coal and water also take place during gasification. It is common practice to add steam to the oxygen containing air blast. Equations for the accompanying steam-carbon reactions are, as noted previously:

6) $C + H_2O = CO + H_2$
$$+31,340 \text{ cal/g-mole}$$

7) $C + 2H_2O = CO_2 + 2H_2$
$$+21,270 \text{ cal/g-mole}$$

Both reactions are endothermic and cause the lowering of the gas temperature.

(Note that there are only four independent equations in the set of six as listed above. Two more dependent equations may also be formed.)

Early work by Kreisinger et.al. (11, 1) on producers led to the following conclusions:

1. Oxygen is consumed in the first four inches above the grate. CO_2 concentration is 10%-16% in this region of maximum temperature.
2. A temperature of 1300°C and a fuel bed thickness of two to three feet is required for the proper reduction of CO_2.

Conclusions reached by Mott and Wheeler (42, 1) with respect to the use of steam in producer operations were:

1. Steam lowers the temperature of the fire zone preventing heat loss and fusion of the ash.
2. The CO/CO_2 ratio is lowered because of the addition of hydrogen.
3. Excessive steam lowers efficiency.
4. Small amounts of methane are formed.

Haslam (1), using a fuel consisting of coke and anthracite in a bed 1.5 - 4.5 feet thick and a throughput of up to 70 lb-hr per ft² of grate area, found that:

1. The percentage of undecomposed steam is the controlling factor of the CO/CO_2 ratio.
2. A steam to fuel ratio above 0.4 results in a decrease of CO content.
3. An increase in bed depth caused an increase in decomposition.
4. High gasification rates (70 lb/hr/ft²) promote increased temperature and greater steam decomposition.

Partial Combustion   173

Terres and Schierenbeck (43, 1) suggest that a fuel-steam ratio of 0.5 - 0.6 is most effective for gas production.

Most effects of operating variables can be assessed by considering the gasifier configuration and by examining the reaction equilibrium — though the latter can become very complicated due to the number of equations involved (four independent equations).

SULFUR

For the reducing atmosphere of partial oxidation, the sulfur in the gaseous phase appears principally as H_2S whereas in the oxidizing atmospheres of complete oxidation the sulfur appears principally as SO_2. More appropriately it is a matter of the H_2S - S - SO_2 equilibrium, though other compounds also make an appearance. As a rule of thumb, however, H_2S and SO_2 do not coexist; rather it is a matter of H_2S - S or else S - SO_2.

For sulfur (and particulate) removal and additional commentary, consult Chapter 10.

4.3 *IN SITU* GASIFICATION

Most of the reserves of fossil fuels in the United States are located deep underground. For instance, in Wyoming the strippable reserves of coal are nominally about 23 billion tons; down to 3,000 feet the reserves are 120 billion tons; and down to 6,000 feet, an estimated 550 billion tons. Oil shale is another example of an energy reserve which occurs at deeper levels. And on the average, only about 30% of the petroleum reserves are recovered by conventional production methods.

These deeper reserves, which are beyond the scope of conventional mining practices, require development of *in situ* recovery techniques. One such is the gasification (or liquefaction) of these solid materials, in place, followed by retrieval as fluid-flow systems. Before *in situ* gasification can be effected, however, permeability and porosity must be established by fracturing the formation, preferably by chemical explosives or hydraulic means. Channelization is to be avoided.

Gasification in place can be effected by two principal routes: partial combustion with air (or oxygen), or reaction with steam supported by combustion with air or oxygen. The first route is of perhaps more interst here — namely, the techniques by which carbon monoxide concentrations can be maximized so as to produce a low-Btu gas of sufficient Btu-rating for a fuel, or for recovery of the CO followed by further processing.

A recovery operation is diagrammed in Figure 4.9 as proposed for Project Thunderbird, a venture for *in situ* gasification to be located in the Powder River Basin region of northeastern Wyoming.

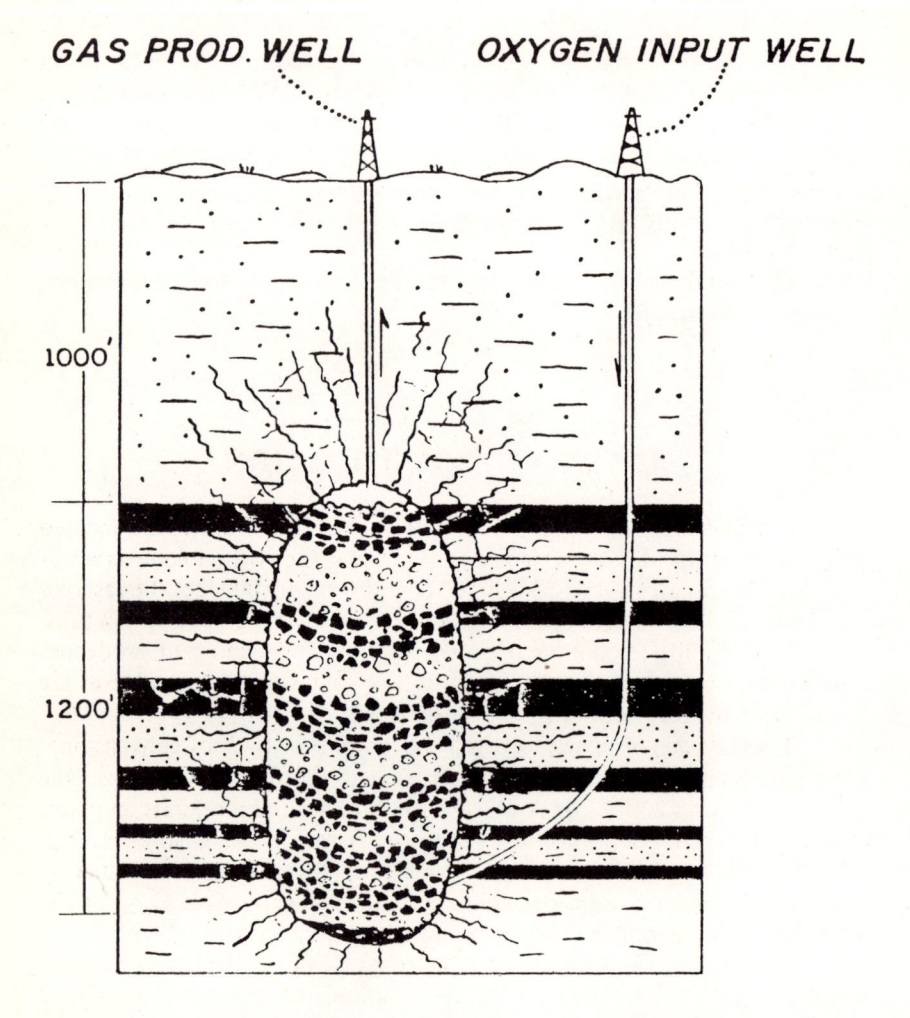

Fig. 4.9. Diagram of gasification operation for Project Thunderbird. [From J. S. Wold and P. Jenkins, Casper, Wyoming. Consult Hoffman (48).]

Projects presently underway include tests near Hanna, Wyoming conducted by the Laramie Energy Research Center (Department of Energy) on the underground gasification of coal utilizing horizontal hydraulic fracture between input and output wells. Other tests proposed or underway are located near Gillette, Wyoming and are directed by Lawrence Livermore Laboratories (Department of Energy). Explosives are to provide fracture. Past efforts in this country were by the U.S. Bureau of Mines in Alabama.

Other announced efforts for coal gasification involve the *in situ* gasification of steeply-sloping seams utilizing techniques developed in Russia, to be supported by the Department of Energy. A private effort in Texas is underwritten by Texas Utilities and purportedly also utilizes techniques developed in Russia. The firm of Booz, Allen and Hamilton has performed studies for ERDA on steeply-sloping seams as well as other aspects of the total problem of coal and oil shale development, including interfacing with the environment. Gulf R & D is working with DOE to test *in situ* gasification of steeply-sloping seams in northern Colorado and southern Wyoming.

Occidental Petroleum is engaged in a large-scale demonstration in western Colorado for the *in situ* retorting of oil shale using underground mining techniques. The Laramie Energy Research Center (DOE) also continues to be active in the underground gasification or retorting of oil shale.

Methods advanced for the extraction of oil shale (kerogen) are potentially applicable to coal. These include hot natural gas (methane), hot hydrogen, and superheated steam injection.

As a fuel, low-Btu gas is suitable for on-site power generation either via conversion to steam or by the more direct use for gas-fired turbines, with increased efficiency.

The use of nuclear explosives has been advanced for fracturing and rubblizing or fragmenting oil shale and coal formations. Experiments have been tried for using nuclear devices to stimulate production from low-permeability gas formations. The safety has been challenged; additionally, the economics may be such that conventional explosives are as cheap, or more so, and just as effective. Moreover, the overall efficiency of using chemical explosives may be greater.

Of other significance is the fact that the environmental impact of strip-mining practices can be side-stepped by going on to *in situ* methods. This fact, on top of the fact that most of the fossil fuel reserves are at deeper levels, is a compelling argument for instituting *in situ* methods.

Much of the work on coal combustion has been concerned with pulverized fuel (p.f.). The nature of the fracture and fragmentation for underground gasification, however, would be related more closely to fixed or moving-beds, such as on a grate. The depth of bed that will be involved would correspond to that for conversion to producer gas — a mixture principally of carbon monoxide and nitrogen. At the same time pyrolytic and distillation effects are produced — in part from the moisture of the coal. Some of these factors are examined.

FRACTURE

As a prerequisite to the underground gasification of coal, oil shale or other solid fuels, permeability and porosity must be established by a fracturing technique. Earlier work in the field was limited by channelization and leakage.

There has been an interest already established for oil shale and also coal, utilizing nuclear explosives. The attendant hazards, economics and efficiency, however, are now resulting in another look at chemical explosives.

Some of these aspects have been publicized by the concern over the use of nuclear devices to stimulate gas production, e.g. Project Rio Blanco in Colorado and Project Wagonwheel in Wyoming. The issue has been debated on the campus of the University of Colorado and elsewhere. Studies at the University of Colorado have also indicated that nuclear stimulation is not economical.

At least two organizations have worked on non-nuclear processes for underground fracturing. These include the Dowell Division of the Dow Chemical Company with the Stratoblast service, and a joint venture of Hercules, Inc. and Petroleum Technology Corp. Dowell has used solid and slurry explosives. The latter firms employ a two-component explosive in their Astro-Flow process, with the components mixed underground.

PROBLEMS OF IN SITU GASIFICATION

Past work done on the *in situ* gasification of coal in the United States has been discussed by Elder (44), and in Russia by Golger, et.al. (45). A common problem has been the obtaining of gas of sufficiently high Btu-rating. Evidently heat leak to the surroundings causes a disproportionate conversion to carbon dioxide to sustain conversion temperatures. This may result in a Btu-rating of less than 100, whereas a rating of 150-160 Btu/SCF is more desirable if the gas is to serve as a fuel (46). Work underway by the Laramie Energy Research Center (LERC) near Hanna, Wyoming, proves that such a Btu rating is attainable, if not necessarily sustainable.

Partial Combustion

Another problem is leakage of gas to the surface through fissures caused by thermal expansion of the overburden. Presumable this could be solved by gasification of deeper formations. The proposed development in underground gasification called Project Thunderbird embraces formations at depths of 1000-2000 feet (47, 48).

The attainment of a product gas of sufficiently high concentration of carbon monoxide may be more of an exercise in adjusting operating conditions than anything else. There may be a limiting moisture condition for conversion. The presence of alkali and mineral constituents in the coal may have a contributing effect. The point and rate of injection of air will affect the conversion. A pretreatment of the coal-bearing formation (possibly with air) is a possibility.

Though air is generally considered the oxidant for gasification, there may be occasions when pure oxygen can be utilized, at least in part: for example in O_2-steam gasification to produce CO and H_2, which can be used as a fuel as is, or reacted further to form hydrocarbons, e.g. SNG (48).

Alternately, superheated steam has been proposed as the medium for gasification (49, 50). In addition to furnishing heat for pyrolysis, it has a distillation effect by lowering the gas-phase partial pressure of the distillates produced. Furthermore, part of the steam can itself react with the coal.

CONVERSION

Though not necessarily within the scope of this section, the use to which the product gas is to be put will have bearing: that is, will the gas be used directly as a fuel — even supplemented with richer fuels — or will the carbon monoxide be separated as a chemical intermediate for further processing?

The gasification of coal to produce a low-Btu gas is regarded as one significant means of gaining a clean, pollution-free fuel. The processes of gasification would be under the necessary and sufficient pressure to permit removal of all particulate matter and sulfur compounds. A working pressure somewhat above ambient is required for effective cleanup, for the processes of separation and recovery require some pressure-drop. This is a problem in the atmospheric combustion operations in central power stations: no appreciable pressure drop can be taken without compressing the stack gases.

A low-Btu gas of sufficient rating can be used as a fuel in a boiler for the steam-power cycle or similar industrial application — or used as the fuel for gas turbines. These gas-turbine operations possess the potential of high efficiency: on the order of 60% (46). Linked gas and steam turbines are considered to have maximum effectiveness by recovering and utilizing the waste heat from the gas turbine exhaust.

Alternately, the carbon monoxide can be selectively absorbed — by solutions of copper ammonium salts, for instance. Raising the temperature level of the CO-rich solution recovers the carbon monoxide and regenerates the solution. More will be said of this in Chapter 9.

A part (or all) of the carbon monoxide can be converted to hydrogen by the CO-shift:

$$CO + H_2O = CO_2 + H_2$$

A number of catalysts favor the water gas shift, including those of iron.

A synthesis gas of varying proportions can be made of hyrogen and carbon monoxide to proportioning the CO-shift. In turn this synthesis gas can be converted by hydrocarbons and/or oxygenated compounds by Fischer-Tropsch reactions. For example, a nickel catalyst will yield methane:

$$CO + 3 H_2 = CH_4 + H_2O$$

Reaction conditions are usually ~ 300 psi and 750°F, though conditions can vary. Carbon dioxide may also be produced as a product. The reaction is highly exothermic and heat must be removed.

Iron catalysts tend to produce hydrocarbon liquids predominately, along with oxygenated compounds. Cobalt catalysts are intermediate. Still other Group VIII transition metals show catalytic activity. Zinc catalysts produce alcohols, chiefly methanol.

The methanation reaction is a topic of current, vital interest in producing high-Btu gas from coal, and research and development is ongoing. The complex at SASOL in South Africa produces both liquid fuels and chemicals as noted in Chapter 1. The production of methanol from synthesis gas is a viable process in commercial operation. These catalytic processes are further discussed in Chapter 7.

REFERENCES

1. van der Hoeven, J. C., "Producers and Producer Gas," in *Chemistry of Coal Utilization,* Vol. II, H. H. Lowry Ed., Wiley, New York (1945).

2. von Fredersdorff and M. A. Elliott, "Coal Gasification," in *Chemistry of Coal Utilization,* Supplementary Volume, H. H. Lowry Ed., Wiley, New York (1963), pp. 892-1022.

3. Hottel, H. C., G. C. Williams and P. C. Wu, "The Reaction of Coke with CO_2," *Am. Chem. Soc. Div. of Fuel Chem. Reprints, 10* (3), 58-71 (1966).

4. Lavrov, N. V., "Combustion of Carbon in Dry and Humid Media," *Dokl, Akad, Nauk, Uz SSR, 22* (4), 32-4 (1965). CA 63:7888a.

5. Lavrov, N. V., "Peculiarities of the Mechanism of Carbon Combustion," *Dokl. Akad. Nauk. SSSR., 156* (3), 662-5 (1964). CA 61:4111g.

6. Lavrov, N. V., "Physical-Chemical Rules in the Process of Combustion of Fuels," *Dokl. Akad. Nauk. USSR., 22* (6), 26-9 (1965). CA 63:9707f.

7. Authur, J. R., D. H. Bangham and H. G. Crone, "Topochemistry of Fuel Beds," *Ind. Eng. Chem. 43*, 525-33 (1951).

8. Yamauchi, S., T. Makaibo and I. Ikeda, "Effect of Chlorine on the Oxidation of Graphite by CO_2," *Bull. Chem. Soc. Japan, 41* (3), 755-6 (1968). CA 68:9913q. See also Yamauchi, Makaibo and Hirano, "Effect of Bromine on the Oxidation of Graphite by CO_2," *Carbon, 5* (3), 243-4 (1967).

9. Wicke, E., *Fifth Symposium on Combustion,* Pittsburgh, 1954. Reinhold, New York (1955), pp. 250-1.

10. Hiles, J and R. A. Mott, "The Mode of Combustion of Coke," *Fuel 23*, 154-71 (1944). CA 39:403[7].

11. Kreisinger, H., C. F. Augustine and F. K. Ovitz, "Combustion in the Fuel Bed of Hand-Fired Furnaces," U.S. Bureau of Mines Tech. Paper No., 137 (1916), pp. 31-34. CA 11:1740.

12. Khitrin, L. N., M. B. Ravich and L. L. Kotova, "Kinetic Combustion Characteristics of Powdered Fuel in a Stream," *Inzh. Fis. Zh. Akad. Navk. Belorussk Ssr., 5*, (1), 7-12 (1966). CA 57:10118e.

13. Long, F. J. and K. W. Sykes, "The Mechanism of the Steam-Carbon Reaction," *Proc. Roy. Soc.* (London), *A 193*, 377-99 (1948). CA 43:3694e.

14. Long, F. J. and K. W. Sykes, "The Effect of Specific Catalysts on the Steam-Carbon System," *Proc. Roy. Soc.* (London), *215A*, 100-110 (1952). CA 47:4180f.

15. Majewski, E. A. and P. L. Spurrier, "Thermodynamics of the Steam-Carbon Reaction," *J. Inst. Fuel, 35*, 258, 296-8 (1962).

16. Long, F. J. and K. W. Sykes, "The Catalysts of the Oxidation of Carbon," *J. Chim. Phys., 47*, 361-78 (1950). CA 44:7132c.

17. Fedoseev, S. G., N. N. Rusinvskaya and A. Rodv, "Influence of Some Factors on the Reduction of Carbon Dioxide and Steam by Carbon," *TR. Mosk. Khim-Tekhnal. Inst.*, No. 48, 180-4 (1965). CA 65:14492d.

18. Kokurin, A. D. and D. A. Rozental, "Gasification of Solid Fuels," *Tr. Leningn. Tekhnal, Inst. im Lensoveta*, No. 63, 139-43 (1964). CA 62:15954f.

19. Kokurin, A. D. and D. A. Rozental, "Influence of Fuel Bed Thickness on Reduction of CO_2," *Tr. Vses. Nauchn-Issled. Inst. Pererabotki i Ispol'z Topliva*, No. 10, 121-17 (1961). CA 57:16977a.

20. Ustimeva, N. M. and E. V. Smidovich, "Reduction of CO_2 in a Fluidized Bed of Petroleum Coke," *Neftepererabotka Neftekhim Navcha. Tekhn*, sb. 11, 15-17 (1965). CA 64:9484c.

21. Harker, H., H. Marsh and W. F. K. Wynn-Jones, *The Carbon and Graphite Papers Conf.*, London, 1957, 291-301. CA 53:10924c.

22. Vastola, F. J. and P. L. Walker, Jr., "Mass Spectrometric Study of the Carbon-CO_2 Reaction," *Am. Chem. Soc. Div. Gas Fuel Chem. Reprints, 1,* 124-32 (Sept. 1959).

23. Blackwood, J. D., "Reaction of Tiraphite with CO_2 and Hydrogen," *Avst. J. Appl. Sci. 13,* (3), 199-206 (1962). CA 57:117h.

24. Blackwood, J. D., "Reaction of Carbon with Hydrogen at High Pressures," *Avst. J. Chem., 12,* 14-28 (1959). CA 53:12158d.

25. Blackwood, J. D. and A. J. Ingeme, "The Reaction of Carbon with CO_2 at High Pressure," *Avst. J. Chem. 13,* (2), 194-209 (1960). CA 54:16146d.

26. Kokurin, A. D., D. A. Rozental and Yu. P. Eudokimov, "Interaction of Oxygen, CO and CO_2 with Charcoal Under Static Conditions," *Tr. Leningr. Technol. Inst. im Lensoveta*, No. 59, 101-6 (1961). CA 62:14391d.

27. Kokurin, A. D., D. A. Rozental, V. P. Suslina and N. I. Tishina, "The Interaction of CO_2 with Carbon in a Fuel Under Dynamic Conditions," *Tr. Liningr. Tekhnol. Inst. im Lensoveta*, No. 59, 107-112 (1961). CA 62:14387d.

28. Long, F. J. and K. W. Sykes, "The Catalysis of the Oxidation of Carbon," *J. Chim. Phys., 47,* 361-78 (1950). CA 44:7132c.

29. Sieth, W. and O. Kubaschehewski, "Electrical Transfer of Carbon in Steel," *Z. Electrochem., 41,* 551-8 (1935). CA 30:415⁹.

30. Pettit, F., R. Yinger and J. B. Wagner, Jr., "Mechanism of Oxidation of Iron in CO-CO_2 Mixtures," *Acta Metallurgica, 8,* 617-23 (1960). CA 55:4302c.
31. Mikhailov, B. M., V. S. Bogdanov and V. G. Kiselev, "Reduction of CO_2 with Hydrogen Under the Action of FAst Electrons," *Izv. Akad. Nauk. SSSR., Ser. Khim 7,* 27-3 (1965). CA 63:1253b.
32. Rakszawski, J. F., F. Rusinko and P. L. Walker, "Catalysis of the Carbon-Carbon Dioxide Reaction by Iron," *Proc. Fifth Conference on Carbon,* Vol. 2, 243-50 (1963). Pergamon Press, New York.
33. Esso Research and Engr. Co., Brit. Patent No. 873,213. Appl. Aug. 8, 1958. U. S., Sept. 11, 1957.
34. Swamy, C. S., "Dissociation of CO_2 on Supported iron Catalyst," *Current Science (India) 34,* No. 17, 505-6 (1965). CA 63:15609f.
35. Kagan, Yu B., A. N. Bashkorov, E. V. Kanzolkina and A. Ya Rozovskii, "Kinetics for Conversion of CO in the Presence of Iron Catalysts," *Zhur. Fis. Khim., 33,* 2706-11 (1959). CA 55:12008g.
36. Bashirov, A. N., E. V. Kamzolkina and Yu. B. Kagan, "Catalyst for the Decomposition of CO," *Navch. Doklady Vysshei Shkoly, Khimi Khim Technol.* No. 1, 162-5 (1959). CA 53:14384d.
37. Yoneda, Y., S. Makishima and K. Hirasa, "Exchange of O atoms among CO_2, CO, and Oxide Catalysts of the Spinel Type," *Am. Chem. Soc. J., 80,* 4503-7 (Sept. 5, 1958).
38. Bank, C. A. and E. A. th. Verdurmen, "Oxygen Exchange Between CO and CO_2," *J. Inorg. Nuc. Chem., 25,* 667-75 (1963).
39. Brandner, J. D. and H. C. Urey, "Kinetics of the Isotopic Exchange Reaction Between CO and CO_2," *J. Chem. Phys., 13,* 351-62 (1945).
40. Norris, T. H. and S. Ruben, "Kinetics of the Isotopic Exchange Reaction Between CO and CO_2," *J. Chem. Phys., 18,* 1595-1600 (1950).
41. Hayakawa, T., "Kinetics of the Isotopic Reaction Between CO and CO_2," *Bull. Chem. Soc. Japan, 26,* 165-72 (1953). CA 48:5617c.
42. Mott, R. A. and R. V. Wheeler, "The Quality of Coke," *Symposium on Solid Fuels,* Am. Chem. Soc. Meeting, Boston (1939), pp. 95-107. Chapman & Hall, London (1939), p. 464. CA 34:1837[5].
43. Terres, E. and J. Schierenbeck, "Coke Gas Producers for Heating Gas Retorts," *Gas v Wasserfach, 67,* 257-63, 278-82, 296-9, 314-4, 325-7, (1924). CA 18:3469.

44. Elder, J. L., "The Underground Gasification of Coal," in *Chemistry of Coal Utilization,* Supplementary Volume, H. H. Lowry Ed., Wiley, New York (1963).

45. Golger, S. P., D. M. Derman, N. V. Lavrov, I. L. Farberov and N. V. Federov, "Production of Technological Gas in Underground Gasification of Lischansk Coals," *Underground Processing of Fuels,* Translated from the Russian and Published for the National Science Foundation, Washington, D.C. Israel Program for Scientific Translations, Jerusalem, 1963.

46. Squires, A. M., "Clean Power from Dirty Fuels," *Scientific American, 227* (4), 26-35 (Oct., 1972).

47. Wold, J. S. and T. C. Woodward, "Coal Makes Bid to Join Plowshare Nuclear Program," *World Oil,* May, 1968.

48. Hoffman, E. J., "Project Thunderbird: Gassification of Coal Via Nuclear Fracture," American Gas Association, Second Synthetic Pipeline Gas Symposium, Pittsburg, November, 1968. Published in the Proceedings.

49. Lyon, L. B., "Primary Extraction, Conversion and Upgrading Hydrogen-Deficient Fossil Hydrocarbons," from Proceedings of the First Intermountain Symposium on Fossil Hydrocarbons, Salt Lake City, October, 1964. Sponsored by Brigham Young University.

50. Lyon, L. B., "Liquid Fuels from Coal, Other Fossil Hydrocarbons," *Coal Age, 69,* No. 12, 70 ff. (Dec., 1964).

Chapter 5

REACTIONS WITH HYDROGEN

The reactions of hydrogen with coal can be classed according to degree of severity. For the range of conditions of interest, a spectrum of materials is produced varying from the appearance of free carbon, through asphaltic materials, tars and oils, naphtha fractions, and on to the more volatile hydrocarbons and methane.

The proportions and characteristics of the products relate to the reaction conditions and hydrogen activity, and to the residence time. With regard to hydrogen activity, a facile hydrogenation occurs if carbon monoxide and steam are used as reactants, serving to generate a very active form of hydrogen (possibly nascent hydrogen) via the CO-shift.

At relatively mild hydrogenation conditions, the production of asphalt-like bitumens prevails. At progressively more severe conditions, liquids in the gasoline-boiling range are produced. And at the most severe conditions methane is produced in predominance, and the operation is sometimes referred to as hydrogasification. Though the complete range of products may appear for any given set of hydrogenation conditions, the distribution varies with the severity.

Most hydrogenation studies for the reaction of coal and hydrogen have involved a liquid carrier or vehicle to form a paste or slurry with the coal. Ideally, the carrier is a part of the product stream, and when recycled, undergoes further reaction. In this type of reaction, at the conditions required, decomposition and rearrangement as well as solvation action accompany hydrogenation. Such reactions in the presence of a carrier are referred to as the liquid-phase hydrogenation of coal. The subsequent further hydrogenation of vaporized product is distinguished as vapor-phase hydrogenation.

The reaction between coal and hydrogen to yield gases without the presence of a carrier or solvent liquid is also called hydrogasification. Formerly this conversion had not received a great deal of attention, but is

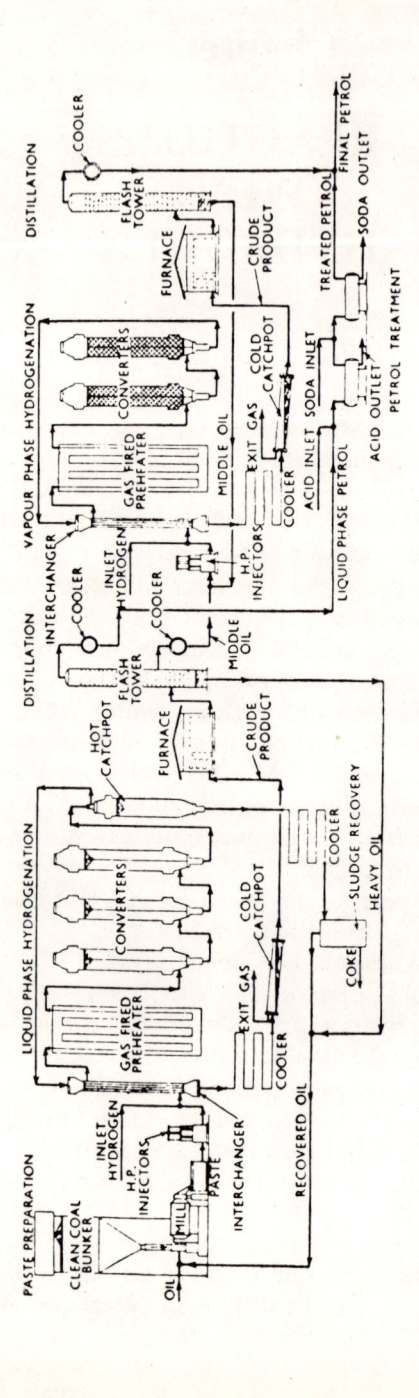

Fig. 5.1. Flow diagram for liquid- and vapor-phase hydrogenation. [From K. Gordon, *Trans. Inst. Mining Eng.* (London), 82, 348-59 (1931). Consult *Chemistry of Coal Utilization*, Vol. II, Wiley, New York (1945), p. 1754.]

Reactions with Hydrogen

coming into more prominence with revived interest in gasification processes. Depending upon less severe conditions, liquid products may also be predominantly produced.

While not emphasized *per se,* hydrogenation is also called for in upgrading the liquids from pyrolysis. Rather here the emphasis is on hydrogenation during pyrolysis — by virtue of the fact the reacting system is brought to hydrogenation temperatures.

Methodology and results are reviewed in this chapter for various hydrogenation routes and conditions. A representative flow diagrm for a commercial embodiment is shown in Figure 5.1 as a point of reference. Other processes are presented in the Supplement to Chapter 1.

5.1 HYDROGENATION WITHOUT CARRIER

The direct reaction between hydrogen and coal without carrier takes place at higher temperatures and lower pressures than used in liquid phase hydrogenation (1, 2). The nature of the reaction under these conditions is also different. Evidently there is a direct reaction with carbon and hydrogen to form methane and other homologs. Both anthracite and chars react readily. The very first stage of reaction is considered to be with the oxygen of the coal. Pure hydrogen need not be used.

Lower-rank coals tend to react more readily than higher-rank coals or coke (3). While methane may result from the direct reaction of carbon and hydrogen (see above), it may also result from the decomposition and hydrogenation (hydrocracking) of higher molecular weight compounds formed in the earlier stages of hydrogenation. The overall rate of methane formation can be regarded as first order with respect to hydrogen partial pressure.

Reaction conditions that have been used in direct hydrogasification reactions are (2, 4, 5):

1500°F	and	600 psi
1400-1700°F		1000-2000 psi
1100-1750°F		300-3000 psi

The hydrogasification reaction is suited to fluidization, and is generally so studied. No catalyst need be employed, as contrasted to liquid phase reactions. In dry hydrogenations to yield liquids, however, catalysts have an appreciable effect (6). At lower temperatures, liquids are produced (5).

In hydrogasification reactions, the conversion to methane is favored by high pressure and relatively lower temperatures (3). Above 900°F the reaction becomes very rapid. While coal is more reactive initially than char, char can be converted at lower temperatures than can coal: 1100°F versus 2300°F, for example.

It has been found that sodium carbonate increases the rate of reaction of hydrogen and coal or coke in the range 1500 to 1650°F. Pretreatment with N_2, air, CO_2, or steam also enhanced the reactivity.

For catalysts added as a powder, stannous chloride is regarded outstanding (7). For catalysts added by solution impregnation, ammonium molybdate is superior, followed by stannous chloride, nickelous chloride, and ferrous sulfate.

The role of catalyst distribution clouds the issue of the effect of a vehicle oil. It is possible that a principal effect of the vehicle is to disperse and distribute the catalyst. If the same object can be accomplished without the vehicle, then the vehicle serves no purpose other than a means of introducing the feed as a slurry — but which is an important purpose for flow processing.

Gases containing as much as 85 percent methane can be produced by the direct dry hydrogenation of pulverized untreated coal (3). Operating conditions are from 450- 500 psi at 900-1700°F with a coal residence time of one-half hour using 40-60 mesh.

Representative studies on the dry hydrogenation of bituminous coal have been made in a batch autoclave to determine the effects of catalyst, hydrogen pressure, temperature, and residence time (6). The catalysts used were either mixed in the dry state or impregnated from solution. Operating ranges were 650-960°F and 500-4000 psi cold, with residence times of 30 minute to three hours at reactor temperature.

The conversion products were evacuated and distilled off to recover gases and distillable liquids. The non-volatile matter remaining was extracted with benzene and the insoluble matter considered as unreacted. The unreacted material was used to calculate the total conversion. The benzene solubles were recovered and extracted with n-hexane. The n-hexane insolubles were measured and classed as asphaltenes. The n-hexane solubles were added to the distillable oils and classified as oil. All conversion yields were computed on a moisture and ash free basis.

Tin and its compounds showed uniformly high catalytic activity, both promoted and unpromoted, and mixed as a powder or impregnated. Ammonium molybdate was inactive as a powder, but very effective when impregnated from solution. Metal naphthenates were the most effective, especially molybdenum naphthenate. The less expensive catalysts of iron

Reactions with Hydrogen

and its compounds were less effective. Whereas an effective catalyst would give 80-90 percent conversion, iron catalysts would give maybe 50 percent (but with notable exceptions), which is on the order of uncatalyzed reactions. As a rule, the impregnated catalysts were more effective.

Some typical results are as follows for the aforementioned dry hydrogenations:

Dry Catalyst Tumbled with Coal

Conditions: 1000 psi cold
2500 psi hot
1 hour at 840°F

	Conv.	Asph.	Gas
No cat.	54%	21%	14%
Tin oxalate	85	45	16
Fe(SO$_4$)	56	27	11

Impregnated Catalysts

Same conditions

	Conv.	Asph.	Gas
No cat.	33		11
1% Mo (as ammonium molybdate)	90	30	14
1% FE (as FeSO$_4$)	85	39	15

The effect of temperature in catalyzed reactions is to cause an increase in total conversion, and an increase in oil and gas produced over the range studied. However, there is an optimum of about 50 percent asphaltene produced at 705°F, dwindling to 10 percent at 930°F. Below 660°F, there is only slight conversion. It should be noted that asphaltenes have been described as an intermediate to the formation of oil (7).

The effect of temperature on dry uncatalyzed reactions is to produce an increase in conversion and proportional increases in gas and oil. With a Pittsburgh coal, the asphaltenes show a decrease (6). With a Wyoming coal, however, the asphaltenes level off to only a few percent.

The effect of increasing pressure in the catalyzed reaction is to increase the total conversion and the asphaltene product. The amount of gases produced is comparatively unaffected.

The effect of increased residence time is to increase the overall conversion and the conversion to asphaltenes in the catalyzed reaction. At the test conditions, 700°F and 750°F, there was little conversion to gas. However, at these lower temperatures it took three hours to accomplish a conversion that would only take one hour at 850°F, say.

Laboratory experiments in a flow reactor with a fixed bed at 1470-1740°F and 300-750 psi yielded 85 percent conversion to gas and 15 percent to liquids, for coals up to and including semi-anthracite (2). Cokes were more resistant but reacted readily in the presence of sodium carbonate. Scale-up caused difficulty due to heat dissipation problems, and only 30-40 percent conversions were obtained. In addition the plastic properties of the coals caused difficulties.

In laboratory experiments passing hydrogen up through a fixed bed which could be heated and cooled quickly, from 35 to 95 percent gas and 5 to 30 percent liquid were produced depending upon the coal charge and residence time (8). Temperatures of 1470°F and pressures as high as 6000 psi were used. The reaction was most rapid the first three minutes, then leveled off. Coal residence times of 0-15 minutes were used. The gas composition was 80-90 percent C_1 with the percentage of C_1 going up with coal rank. Hydrogen residence times were only a few seconds in the reactor (9). With an increased residence time for the hydrogen, the liquid yield would decrease.

In fluidized bed studies with hydrogen, 80-90 percent conversion was obtained at 1350°F and 2500-3500 psi (2). The gas was 70-80 mole percent C_1+C_2. Conditions used by other investigators were 1470°F and 6000 psi with ammonium catalyst giving ~ 70 percent conversion, and also 930-1740°F and 300-600 psi.

A lab process using a fluidized bed of dry coal at 1000 psi and 930°F with no catalyst was reported to give conversions of 20-27 percent oil and also 20-27 percent gases.

Experiments with a low temperature bituminous char in a fluidized bed at 1400-1500°F and 500-2000 psi did not reach 30-50 percent char conversion (9). By increasing the residence time of both the coal and hydrogen, the conversions were increased. The reaction was started by preheating the −16 to +20 mesh charge to reactor conditions and then dumping the feed into preheated hydrogen in the reactor.

The coal hydrogasification process (Hygas) developed by the Institute of Gas Technology calls for direct dry hydrogenation in fluidized beds. Reaction is at 1200-2000° F and ~ 1000 psi at two temperature levels operating countercurrently. A flow diagram is shown in the Supplement to Chapter 1.

Reactions with Hydrogen 189

A difficulty encountered in fluidization reactions is the agglomerating tendency of the particles (2). It has been found that pretreatment with nitrogen, air, carbon dioxide or steam reduces this tendency (3). The agglomerating tendencies may also be reduced by first heating the coal rapidly through its plastic range (10).

In studies with 65-100 mesh char at 1500-1700°F, it was found that the addition of steam accelerated the hydrogenation reaction to produce methane (3). The effect decreases with pressure and carbon burnoff (evidently if O_2 is present). The reaction rate may be 2 to 100 times as great over the pure hydrogen rate, in the range from 1 to 30 atm. The pressure above which steam has little effect is around 30 atm or 450 psi (9).

A purpose of adding steam to the hydrogen is the further generation of hydrogen by the carbon-steam reaction (3). This latter reaction is endothermic and can be used to counterbalance the exothermicity of the hydrogenation reaction. An estimate made shows the H_2 requirements for gasification can be reduced 70 percent by the use of steam (5).

It has been noted that the addition of sodium carbonate or other alkalies increases the gasification rate of coals or cokes with hydrogen (3). This effect is also produced in the steam-carbon reaction (Chapter 6). Pretreatment with nitrogen, air, carbon dioxide, or steam has also been demonstrated to increase the hydrogen reactivity of coal.

5.2 LIQUID-PHASE HYDROGENATION

The liquid-phase hydrogenation of coal first investigated by Bergius, circa 1913, has been studied extensively since. The process has been applied commercially, and continues to be of considerable interest. Research has been directed toward cheaper sources of H_2, better catalyst, less severe operating conditions, and simpler operating procedures — and there is emphasis on elimination of the carrier, i.e. toward dry hydrogenation (11). The successful separation and use of the liquid products has also been of concern (1). Examples of present developments are shown in the Supplement to Chapter 1.

The liquid-phase hydrogenation of coal may take place at around 750°F and 3000-4500 psi with a retention time of 2.75 hours (2). Tin or other compounds may act as a catalyst, and the vehicle or carrier may be some coal tar fraction or other solvent. The operating conditions will depend largely upon the product sought. For instance, mild hydrogenation to produce high molecular weight asphaltic material will require different conditions (12, 13, 14). The carrier oils used to form the paste with the coal are chiefly various recycle oils or tar oils, though hydrocarbon oils and other solvents have been used. Yields of the order of 70 to 95 percent conversion are obtainable (2).

Other ranges of operating conditions listed that have been used to convert coal essentially to liquid products via liquid-phase hydrogenation are:

Temp.	Pressure	Reference
750°F	1500 psi	(15)
900°F	3000 psi	
750-950°F	3500-10,000 psi	(15)
880°F	4500-10,000 psi	(16)
840-1025°F	7500 psi	
910°F	5000 psi	
780°F	1500 psi	(1)
825°F	3000-4500 psi	(2)
[Sn(OH)$_2$ catalyst]		
700-1000°F	3000-10,000 psi	(17)
900°F	3000-10,000 psi	(17)

Pressures of 3000 psi or higher are desirable for liquid-phase hydrogenation at temperatures of 840°F (15). At these conditions, coal can be 80-90 percent dissolved in tetralin-phenol mixtures without the addition of hydrogen.

Increasing temperature (in this case from 790 to 860°F) increases the yield of gasoline and gases, and somewhat the yield of middle oil above the gasoline boiling range, but severely reduces the yield of heavy oil.

Residence times are on the order of two hours. Increasing the retention or reaction time (say from one-half to four hours) increases the yield of gasoline, gas, and middle oil (400-570°F), but severely reduces the yield of heavy oil.

The effect of increasing pressure is to increase the reaction rate. An increase from 1200 to 1800 psi will almost double the conversion. Catalysis serves the same function. It has been suggested that hydroaromatic compounds are first formed with the hydrogen, then react to furnish hydrogen to the unsaturated compounds. This is apparently

Reactions with Hydrogen

the role of hydrogen donors such as tetralin. Increasing the hydrogen pressure increases the rate of regeneration of the hydroaromatic compounds.

Some further observations on the effect of different conditions are as follows. The higher the rank the coal, the more difficult to convert to liquid products (17). The higher the pressure, the better the oil yield (1). Above 10,000 psi pressure is required for high-rank coal. Higher pressures favor direct hydrogenation to liquids.

Above 930°F, there is greater gas formation with attendant hydrogen consumption. Higher temperature may account for more ring structures formed. A temperature of 860°F is mentioned as the upper limit for the production of gasoline and kerosine (17).

The hydrogenation of coal-derived aromatic compounds shows a minimum reaction temperature followed by simultaneous reaction and decomposition (2).

A lower oxygen content in the coal will give a higher liquid yield (17). It has already been mentioned that a high hydrogen content increases yield. No liquid is produced from anthracite.

Luxmasse, an alkaline iron oxide byproduct of the aluminum industry, was originally added by Bergius to remove sulfur from the reaction (7). It has since been shown to be a specific catalyst for hydrogenation and has served to demonstrate the importance of catalytic activity to the reaction.

Tin compounds are possibly the most active group of catalysts used in liquid-phase hydrogenations (18). Molybdenum compounds are very active as well. Iron catalysts are also effective (1, 17). In fact ferric oxide is regarded as quite successful (15, 17). Ferrous sulfate has also been used (14, 17) as well as FeS and mixtures of iron compounds. Iron compounds possess the virtue of being inexpensive. The catalyst may be introduced as a powder or impregnated from solution.

A pertinent note at this point is that Fe_2O_3 and TiO_2 used separately are less active than when mixed together, the synergistic effect.

Iron catalysts work successfully on low rank coals at 3000-4500 psi pressure, but not on higher rank coals (17). However, by increasing the pressure to 10,000 psi, the iron catalysts serve effectively for the hydrogenation of higher-rank coals.

In general metallic sulfides are considered more active than the oxides (15). Thus the presence of hydrogen sulfide can have a beneficial effect. Catalysts of the metalloids phosphorus, arsenic, selenium, and tellurium and their compounds have been reported to have catalytic activity.

Compounds such as $ZnCl_2$, $AlCl_3$ and zeolites are sometimes referred to as solid acid catalysts. An extensive report reviewing these and other materials, particularly oxides and sulfides, has been prepared by Catalytica Associates, Inc. ("New Catalytic Materials for the Liquefaction of Coal," prepared for the Electric Power Research Institute, M. Boudart, J. A. Cusumano and R. B. Levy, Principal Investigators, Catalytica Associates, Palo Alto, California, October, 1975. The entire subject is also reviewed in a later report, "Scientific Resources Relevant to the Catalytic Problems in the Conversion of Coal," prepared for the Energy Research and Development Administration, by J. A. Cusumano, R. A. Dalla Betta and R. B. Levy, Catalytica Associates, Palo Alto, California, October, 1976.)

An application of the use of zinc chloride in catalytic hydrogenation is a process pilot-plant study supported by the Department of Energy, the Conoco Coal Development Company and the Shell Development Company. Coal and hydrogen are reacted in the presence of molten zinc chloride at 650-825°F and 1500-4500 psi. The ratio of zinc chloride to coal is about one-to-one. Fuel gas, gasoline and oil are the products. The HCl is neutralized with caustic. Hydrogen sulfide produced is converted to zinc sulfide (and HCl), water (steam) produced forms zinc oxide (and HCl) and ammonia produced is converted to a complex. The spent melt is regenerated with air in a fluidized sand bed at 1700-1900°F, vaporizing the zinc chloride which is recovered. The sulfur is converted to SO_2 and the ammonia to nitrogen. Zinc converted to the oxide is regenerated with HCl. High conversions are reported in the initial work.

The ash content of the coal may itself have catalytic or anti-catalytic properties, depending upon the composition and nature of the ash (17).

The sequence of hydrogenation and thermal liquification of coals dispersed in a solvent has been described as follows (17):

(1) Hydrogenation

(2) Isomerization

(3) Cracking

followed by

(4) Hydrocracking

(5) Depolymerization

The reaction has also been described as involving both hydrogenation and decomposition in the following steps (2):

(1) Removal of oxygen.

(2) Hydrogenation and decomposition:

Reactions with Hydrogen

    (i) Hydrogen addition
   (ii) Cracking and further H addition
  (iii) Cracking of the coal molecules
  (iv) Recombination of fragments

The continuation leads eventually to gases. Depolymerization and repolymerization effects are also cited, which accompany pyrolysis and are affected by hydrogen chain termination (13).

Among the coal-derived products obtainable in significant quantities from hydrogenation are the aromatics benzene, toluene, xylene, and naphthalene, and the tar acids phenol, the cresols, and the xylenols (8). Disposal of other products for which there is no market could be a problem unless converted, say, by further hydrogenation. Condensed crystalline aromatic structures such as pyrene and corene have been recovered from the less volatile oil products of hydrogenation. Even the sulfur content can be converted to sulfates or sulfuric acid.

The steady state flow processing of liquid-phase hydrogenation as practiced commercially involves the injection of a coal-paste and hydrogen into the reactor and the separation of the products for further processing and/or recycle. The products can roughly be divided into gases, a middle oil cut which has an endpoint of around 320°C or 600°F, and a flash residue phase which contains the unreacted coal, ash, and catalyst mixed with retained oil. This latter product is called "heavy oil let down" or simply HOLD (1, 15, 16, 17).

The middle oil cut is used in the vapor-phase hydrogenation to produce gasoline and other fuels. The middle oil already contains a gasoline fraction which may (15) or may not (17) be separated before processing by vapor phase hydrogenation.

The heavy oil is recycled and also used as the paste oil (1, 15). The HOLD may be processed to remove the solid material for coking, and the retained liquids for recycle or paste. As mentioned, the use of a paste permits the introduction of the feed as a slurry and also recycles the higher molecular weight fraction. The catalyst is also introduced with the paste.

If the liquid-phase hydrogenation is conducted such that all of the heavy oil can be recycled, then a steady state can be reached whereby no net production of heavy oil occurs (17). Otherwise, makeup must be added from some other source or else a net yield of heavy oil product is obtained.

The processing conditions and operations can be varied to change the nature of the products from liquid-phase hydrogenation.

An application of catalytic hydrogenation in the presence of a carrier is the H-Coal process of Hydrocarbon Research, Inc., which is similar to an earlier H-Oil process. An "ebullating" catalyst bed is formed by the suspension of catalyst pellets in a coal-oil slurry along with hydrogen. A heavy-oil recycle is used, and a net yield of heavy-oil product is produced for further processing. The unreacted residue is filtered. The flow diagram is included in the Supplement to Chapter 1. Cobalt molybdate catalysts have been used.

5.3 VAPOR-PHASE HYDROGENATION

Secondary or vapor-phase hydrogenation ordinarily follows the primary or liquid-phase hydrogenation. Its function is to convert higher-boiling oil fractions to gasoline or other naptha fractions, and diesel oil.

Vapor-phase hydrogenation is a catalytic flow process through a fixed bed of catalyst. Reactor conditions which have been used are:

950° F	and	10,000 psi	(16)
850° F	and	3000 psi	(1)
750° F	and	4500 psi	(1)
400-700°F	and	500-2000 psi	(1)
900-950° F	and	3700-10,000 psi	(15)
680-860° F	and	3000-4500 psi	(17)

The conditions used will depend upon the product desired, the feed stock, and the catalyst used. Often a preliminary hyrogenation or supplementary stage is used (1, 17). The preliminary hydrogenation serves to hydrogenate unsaturates and oxygenated compounds which would decrease the life of the vapor phase catalyst and is an upgrading step. Also, recycle of unconverted middle oil is used (15).

The main vapor-phase reaction is called splitting, and is preferably conducted to retain a high aromaticity in the products for a better grade gasoline (1). Higher temperatures (930-1000°F) yield higher aromatic content (17).

Reactions with Hydrogen

Among the catalysts employed for vapor-phase hydrogenation are (1):

MoO_3 - ZnO - MgO mixture
WS_2
WS_2 - NiS
Al_2O_3 - WS_2 - NiS
Fuller's earth

The splitting activity of WS_2 is suppressed by NiS (or Ni_3S_2).

Other catalysts used include activated clay and iron sulfide (15), and aluminosilicates (17).

Both liquid- and vapor-phase hydrogenations recover and recycle hydrogen. Oil absorption is a standard method for separation of the hydrocarbon gases from the hydrogen. A feature used in vapor-phase hydrogenation is the introduction of cold hydrogen throughout the bed for temperature control of the exothermic reaction (1).

Based on a throughput of 25 to 30 pounds of feed per hour per volume of catalyst bed, the retention time is around one hour (1). Conversion is 80 to 95 percent of the middle oil feed, or 50 to 90 percent of the raw coal on an maf basis. The higher the grade of gasoline produced the less the conversion. Typical overall yields mentioned for the hydrogenation of coal are 55 percent gasoline, 30 percent gaseous hydrocarbons, 10 percent H_2O, and 5 percent unreacted residue (17). The composition of the gasoline product will vary considerably with processing procedures (1). The products are characteristically high in naphthenes. The gasoline product from vapor-phase hydrogenation may require still other refining operations for upgrading the hydrogenated product, such as aromatization.

5.4 MILD HYDROGENATION

As perhaps the lower limit in hyrogenation conditions, a mild pretreatment with hydrogen improves the coking qualities of a coal (2).

Hydrogenation in the liquid phase at mild conditions leads to the formation of solid or tar-like products. It has been noted that asphaltenes possibly are intermediates in hydrogenation processes (7). The precursor to this type of operation is the Pott-Broche process (1, 15, 19, 20, 21). Initially, in the Pott-Broche process mixtures of solvents such as tetralin and cresol were used. These were superseded by a middle oil obtained from the hydrogenation of coal tar. Extraction conditions are set forth at 775-800°F and 1500-2200 psi pressure (19), with a residence time of one hour (1). The operation was purely a solution process though an attempt was made to hydrogenate the extract. It led to operating difficulties due to solid formation (19, 21). The extract was filtered from the residue, and the liquids removed from the extract to yield an ash-free solid fuel with a melting point of ~ 450°F (19). The product had a higher hydrogen content and lower oxygen and sulfur content than the coal. Conversion was ~ 70 percent to 80 percent on a dry basis. The residual liquid was used for vapor-phase conversion.

The Consol process of the Consolidation Coal Company involves the relatively mild hydrogenation of a coal-carrier oil slurry, followed by filtration (12). The solubleized portion is subjected to hydrocracking and other processing. The carrier oil is a recycle stream. Originally referred to as Project Gasoline, work continues at the government-owned facilities at Cresap, W. Va. for modification to other processing.

Another development is the Spencer or Pittsburg-Midway solvent-refined or de-ashed process (13, 22). In the Spencer process a coal-solvent slurry is hydrogenated, then filtered to yield an ash-free solid material. Hydrogen is added to a coal-solvent slurry which is heated to 800°F and allowed to react. The mixture is separated by rotary filter or other means and the soluble portion recovered. Solvent to coal ratios are from 2/1 to 4/1. A flow diagram is shown in the Supplement to Chapter 1.

The char remaining from hydrogenation (or from pyrolysis, etc.) can be used as a solid fuel provided the requisite specifications are met. Char product was once even proposed as an additive to petroleum refinery feedstocks (Project Seacoke). Alternately, the char may be reacted with steam to produce hydrogen.

Asphaltic materials may also be obtained by the mild hydrogenation of a coal in the presence of a solvent (14). Reactions carried out in a batch autoclave were at 725-825°F and ~ 2000 psi (cold), for retention times at reactor conditions of one-half to three hours. Coal tar and product fractions were used as the carrier, and runs were also made with a benzene carrier. The asphaltic product was recovered from the +600°F fraction, based on penetration. The product is referred to as a bitumen, and has potential application as a road binder.

In the above type of operation, processing would require a solvent or carrier boiling below 600°F in order to minimize product loss by subsequent reaction to recycle. A recycle oil of less than 600°F would merely be reacted further to gaseous and naphtha (gasoline) fractions, also recoverable as a product.

BITUMENS

The mild hydrogenation of coal under appropriate conditions to produce bitumen-like materials requires comparison and evaluation of those bitumen products with petroleum asphalts.

Petroleum-derived asphalts are generally considered composed of asphaltenes, resins, and oil constituents. The asphaltenes impart the plastic properties generally associated with asphalts. These constituents represent various stages in an oxidation or blowing process. For instance, air-blowing at elevated temperatures will convert oils to resins to asphaltenes. Further reaction leads to decomposition and the formation of coke. Asphalt processing may require blowing in varying degrees depending on the raw material and the final specifications.

Asphalts are also regarded as colloidal suspensions, in which the oily constituents are the dispersant and the asphaltenes, principally, the dispersoid. Oily constituents of an aromatic nature lead to sol-like stable dispersions of lower temperature susceptibility and non-Newtonian behavior. That is, viscosity changes less with temperature but is affected by rate of shear (or the velocity gradient).

While it is conceded that a synthetic bitumen obtained from the hydrogenation of coal is not the same as a petroleum asphalt, still there are similarities which may permit characterization in similar terms. The same element of immiscibility exists between oils and asphaltenes or tars (oxygenated materials).

It has been found that these raw synthetic bitumens contain volatile oils and hard asphaltene-like constituents. The oils are predominately aromatic, giving stable sol-like dispersions with high temperature susceptibilities and exhibiting Newtonian behavior. A high percentage of free carbon (or unreacted coal particles) is apparently present, causing a high specific gravity, as do the asphaltenes. Catalyst present in the product may cause further reaction under test and mixing conditions.

Such a synthetic bitumen product lends itself to the following modifications and uses. Steam distillation under vacuum, a universal practice in asphalt processing, would remove volatiles, which could be air-blown to an asphalt. The residue from distillation, if pitchlike, could be fluxed with a non-volatile oil. Excessive asphaltene-like constituents can be precipitated by solvent refining. Further conversion of an oily product could be produced by air-blowing to an asphalt. The sol-type dispersions should be admirably suited to blending with gel-type asphalts. Washing with water would remove catalyst from the product — as would washing with an acid, for instance, to yield an acidic product with superior adhesive properties for basic mineral aggregates.

REACTION CONDITIONS

Hydrogenation conditions for the production of synthetic bitumens from coal are of the order of 800°F and cold hydrogen pressures of 2000 psi (14). The product obtained will also relate to the residence time and to the degree of contact. An iron catalyst such as the sulfate is apparently necessary in order to yield bitumens with asphalt-like properties. These are relatively mild conditions for hydrogenation.

The mild hydrogenation of coal under pressure in the presence of a solvent carrier or vehicle can be used also to produce a solid pitch-like product (22), as well as a semi-solid asphalt-like bitumen, as previously noted. The former hydrogenation is for the purposes of producing an ash-free solid fuel, the latter for purposes of producing a synthetic bitumen or asphalt for use as a road binder. The degree of hydrogenation and severity of conditions determine the nature and range of the product. Hydrocarbon gases and liquids are produced as byproducts — and will occur increasingly with severity of reaction. In the limit, methane is the product (hydrogasification).

The subject interfaces with the Pittsburg and Midway Solvent Refined Coal process, in which a de-ashed solid fuel is the product. There are also similarities to the older Pott-Broche process and the Consol process.

Ideally the solvent is a recycle liquid byproduct. For small scale batch and continuous reactions, an anthracene oil was used, and also reclaimed liquid products (22). The boiling range of the solvent was about 122°F to 482°F (50-250°C).

In general, the lower the solvent to feed ratio, the higher the viscosity of the exit product stream. However, at least a 2/1 ratio was required for successful solution action and conversion of the coal. Ratios to 4/1 were employed.

Low solution temperature and low H_2 pressure both tended to increase the apparent viscosity of the exit product stream, measured at a reference condition.

The solid final product was obtained after evaporation of liquids present.

These observations delineate the fact that reaction products can be controlled to a considerable extent by reaction conditions. It poses the possibility that the physical properties of any resultant asphaltic product can be somewhat controlled as to penetration, ductility, viscosity, etc. However, to further meet required specifications there is the expectation that the raw synthetic asphalt product will have to be further refined.

THE MAKEUP AND PROPERTIES OF SYNTHETIC BITUMENS

Synthetic asphalts have been made variously by separating petroleum asphalts into asphaltenes, resin, and oils, and recombining the fractions in various proportions (23, 24, 25). Others have proposed separating the constituents into petrolenes (oils plus resins) and asphaltenes, then recombining. Synthetic asphalts have also been made up from the point of view of the dispersant (24). The oil fraction has been separated into cyclics and saturates and recombined with the asphaltics (asphaltenes and resins).

It has been recommended that asphalts be separated into the constituent fractions asphaltenes, resins, and oil constituents, followed by removal of resins to improve weather resistance (26). Other combinations may potentially improve properties, and it may be possible to modify or otherwise reconstitute coal-derived bitumens.

It is therefore conceded that the chemical nature of asphalt is different than that of the synthetic bitumens derived from coal hydrogenation. At the same time, there may be enough similarity to consider that asphaltene-like materials *may* be dispersed in oily constituents of varying aromaticity, subject to the conditions of performance — and may be further treated to meet specifications.

A difficulty encountered in evaluating and comparing asphalt properties lies in devising and correlating elemental laboratory tests to actual road performance for these non-Newtonian substances. The range of usual tests such as shear viscosity and viscosity-temperature-susceptibility (VTS) have not been entirely adequate. The concept of elongational or Trouton viscosity has been proposed, and the subject is discussed in References 27 and 28. Compositional makeup may be the most significant qualification of all.

In general, asphaltenes provide the major element of viscosity, and viscosity-temperature-susceptibility increases with viscosity. These effects are contradictory as some statements are found that asphaltenes decrease temperature susceptibility and give high fluidity (low viscosity). In fact, it seems to be a predominating opinion that asphaltenes lower the temperature susceptibility and that by increasing the asphaltene content (at least up to a point), the temperature susceptibility will be improved. This depends upon a lot of "ifs" and such qualifying phrases as "other things being equal".

The fact that coal-derived products are highly aromatic in character indicates that a sol-type dispersion will result. Such dispersions tend to be Newtonian, with higher values (toward unity) of the coefficient of complex flow. This has been more or less born out from experiment (14).

Aromatics *per se* possess higher viscosity temperature susceptibilities than paraffinics, which tend to produce gel-type dispersions. Additionally, well-dispersed asphalts of the sol-type have been characterized as having higher temperature susceptibilities — though this is not the sole criterion.

Accordingly, there is some reason to believe that the more Newtonian character of the synthetic bitumen, along with its higher temperature susceptibility, is due principally to the highly aromatic character of the dispersant — which would correspond to the oily constituents of an asphalt.

The presence of free carbon solids has been noted in both asphalts and coal-tars, particularly the latter. There are undoubtedly unreacted coal particles present in the products from hydrogenation, though possibly free carbon is also formed in the course of reaction. In any event, the presence of solid particles increases viscosity, though there is supposedly little accompanying effect upon the viscosity temperature susceptibility.

The specific gravity of the hydrogenated product bitumen is very nearly the same as that of asphaltenes, as are some coal-tars. This cannot be inferred to mean that these products *are* asphaltenes, but still they may be somewhat similar in character. If so they should behave somewhat similarly in properties.

It has been noted that the free carbon content of coal-tars may be up to 15 percent. This in itself is enough to cause an increase in specific gravity. And it has been correlated that higher gravity coal-tars do contain more free carbon.

The previous observations have been principally related to the obtaining of a raw bitumen product and its nature. The next step remains that of examining what further modifications should be made to insure that the synthetic bitumen will be usable as a road-binder.

5.5 DEHYDROGENATION

Catalytic dehydrogenation of coal has been accomplished on a lab scale, yielding large amounts of hydrogen (10). About 30 percent of the total hydrogen of a coal (6000 cu. ft./ton) was obtained. Treatment was with a palladium-on-calcium carbonate catalyst in boiling phenanthroline (660°F).

It is of significant note that many catalysts which can be used for hydrogenation, can also be used for dehydrogenation (29).

REFERENCES

1. Donath, E. E., "Hydrogenation of Coal and Tar," in *Chemistry of Coal Utilization,* Supplementary Volume, H. H. Lowry Ed., Wiley, New York (1963), pp. 1041-80.
2. Pinchin, F. J., "The Mechanism of Coal Hydrogenation," British Coal Utilization Research Association, Review No. 193, Dec., 1959, pp. 465-476.
3. von Fredersdorf, C. G. and M. A. Elliott, "Coal Gasification," in *Chemistry of Coal Utilization,* Supplementary Volume, H. H. Lowry Ed., Wiley, New York (1963), pp. 892-1022.
4. Linden, H. R., "Conversion of Solid Fossil Fuels to High-Heating Value Pipeline Gas," in *Hydrocarbons From Oil Shale, Oil Sands, and Coal,* Chemical Engineering Progress Symposium Series Number 54, Volume 61, 1965.
5. Benson, H. E. and C. L. Tsaros, "Conversion of Fossil Fuels to Utility Gas," presented at 149th National Meeting, American Chemical Society, Division of Fuel Chemistry, Detroit, April, 1965 (Vol. 9, No. 2). Also in *Hydrocarbon Processing and Petroleum Refining,* September, 1965.
6. Hawk, C. O. and R. W. Hiteshue, "Hydrogenation of Coal in the Batch Autoclave," *U.S. Bureau of Mines Bulletin 622* (1965).
7. Weller, S. W., "Catalysis in the Liquid-Phase Hydrogenation of Coal and Tar," in *Catalysis,* Vol. IV, P H. Emmett Ed., Reinhold, New York (1956), pp. 513-528.
8. Schlesinger, M. D., "Chemicals from Coal Hydrogenation Products," presented at 143rd National Meeting, American Chemical Society, Division of Fuel Chemistry, Cincinnati, Ohio, January, 1963 (Vol. 7, No. 1).
9. Feldkirchner, H. L. and H. R. Linden, "Reactivity of Coals in High-Pressure Gasification with Hydrogen and Steam," presented before the Division of Fuel Chemistry, American Chemical Society, Atlantic City, N.J., September, 1962.
10. Igoe, J. W. and R. A. Glenn, "Coal Research," *Mining Congress Journal,* February (1962); also Bituminous Coal Research, Inc., 350 Hochberg Road, Monroeville, Pa.
11. Schlesinger, M. D., G. U. Dinneen and S. Katell, "Conversion of Fossil Fuels to Liquid Fuels," presented at 149th National Meeting, American Chemical Society, Division of Fuel Chemistry, Detroit, Michigan, April, 1965 (Vol. 9, No. 2).

12. Gorin, E., "Process for Making Liquid Fuels from Coal," U.S. Patent 3, 143, 489.
13. Bull, W. C., "The Spencer De-Ashed Coal Process," presented to the Coal Technology Group, American Chemical Society, Pittsburg, May 27, 1964.
14. Silver, H. F., B. E. Davis and W. E. Duncan, "Coal Hydrogenation Study," Natural Resources Research Institute, University of Wyoming, June, 1965.
15. Storch, H. H., "Hydrogenation of Coal and Tar," in *Chemistry of Coal Utilization,* Vol. II, H. H. Lowry Ed., Wiley, New York (1945), pp. 1750-96.
16. Donath, E. E., "Chemicals from Coal Hydrogenation," from *Symposium on Gasification and Liquefaction of Coal,* American Institute of Mining and Metallurgical Engineers, New York (1953), pp. 15-27.
17. Rapoport, I. B., *Chemistry and Technology of Synthetic Liquid Fuels,* Second Edition, Translated from Russian, Published for the National Science Foundation, Washington, D. C. by the Israel Program for Scientific Translations, Jerusalem (1962).
18. Powell, A. R., "Relation of Coal Gasification to the Production of Chemicals," from *Symposium on Gasification and Liquefaction of Coal,* American Institute of Mining and Metallurgical Engineers, New York (1953), pp. 196-206.
19. Dryden, I. G. C., "Chemical Constitution and Reactions of Coal," in *Chemistry of Coal Utilization,* Supplementary Volume, H. H. Lowry Ed., Wiley, New York (1963), pp. 232-295.
20. Kiebler, M. W., "The Action of Solvents on Coal," in *Chemistry of Coal Utilization,* Vol. I, H. H. Lowry Ed., Wiley, New York (1945), pp. 677-760.
21. Lowry, H. H. and H. J. Rose, "Pott-Broche Coal Extraction Process and Plants of Ruhrol G.M.bH., Bottrop-Welheim, Germany," U.S. Bureau of Mines Information Circular 7420, October 1947.
22. Kloepper, D. L., T. F. Rogers, C. H. Wright and W. C. Bull, "Solvent Processing of Coal to Produce a De-Ashed Product," Spencer Chemical Division of Gulf Oil Corporation, Report No. 9 prepared for the Office of Coal Research, 1965.
23. Barth, E. J., *Asphalt,* Gordon and Breach, New York (1962).

24. Traxler, R. N., *Asphalt: Its Composition, Properties and Uses,* Reinhold, New York (1961).
25. Zapata, J., "Notes on the Composition and Testing of Asphalts of the Slow Curing Type," *Association of Asphalt Paving Technologists,* Vol. 10, 160-194 (Jan., 1939).
26. Abraham, H., *Asphalts and Allied Substances,* Vol. I and II, 5th Edition, D. Van Nostrand, New York (1945).
27. Hoffman, E. J., "Elongational Flow," *Journal of Materials, 4* (1), 28-30 (March, 1969).
28. Hoffman, E. J., "Viscoelastic Transition," *Journal of Engineering for Industry, 94 B* (2), 732-737 (May, 1972).
29. Berkman, S., J. C. Morrell and G. Egloff, *Catalysis,* Reinhold, New York (1940).

Chapter 6

REACTIONS WITH STEAM

The reaction of coal with steam is, in the broader view, only one of the many reactions occurring during coal conversion. Thus reaction with steam occurs during both combustion and partial combustion. It is all merely a matter of degree — that is, how much intrinsic moisture is present or formed from the coal, and how much excess steam or water is added. Furthermore, temperature, reaction rate and equilibrium are interrelated through the energetics of reaction and the residence times of conversion.

Since the topic has already been examined at some length within the discussions on combustion and partial combustion, a more limited tack is taken here with the emphasis on the so-called water-gas reactions.

6.1 WATER-GAS REACTIONS

The reaction of carbonaceous materials with steam to yield mixtures of carbon monoxide and hydrogen is sometimes referred to as the water-gas reaction. This term is also used for the further reaction of carbon monoxide and steam to yield hydrogen and carbon dioxide, though the latter reaction is more usually referred to as the CO-shift or water-gas shift.

Apart from reacting with coal, steam produces a distillation effect (1, 2). In fact steam distillation has been advanced as a means for the *in situ* extraction of coal deposits. Such a process would act preferentially toward the more volatile portions.

Two independent equations may be written for the reactions of carbon and steam:

		cal/g-mole	Btu/lb-mole
1)	$C + H_2O \rightleftharpoons CO + H_2$	31,340	56,490
2)	$CO + H_2O \rightleftharpoons H_2 + CO_2$	−10,070	−17,710

Further algebraicly dependent equations may be derived from these two such as

$$CO_2 + C \rightleftarrows 2CO \qquad 41,410 \qquad 74,200$$

or the overall expression

$$C + 2H_2O \rightleftarrows CO_2 + 2H_2 \quad 21,270 \qquad 38,780$$

The first reaction is strongly endothermic, and the second somewhat exothermic.

In addition, depending upon conditions, there may be further reaction between CO and H_2 to form hydrocarbons and other products.

The approximately equimolal mixture of CO and H_2 is referred to as water gas or blue gas. It has limited value as a heating medium, but is improved by the simultaneous use of oxygen in pressurized operations (e.g., the Lurgi gasifier) so as to yield significant concentrations of methane. Removal of CO_2 produced also enhances the Btu-rating.

Mixtures of CO and H_2 have further value as a reaction intermediate to produce hydrocarbons or other compounds. As such the mixture is referred to as synthesis gas and usually, but not always, the CO-shift is preferentially promoted to achieve higher concentrations of hydrogen. In the limit, pure hydrogen only may be obtained upon removal of CO_2. An excess of steam favors the shift as does the presence of catalysts.

An inspection of uncatalyzed reaction equilibrium data (3) indicates that reaction (1) is favored by higher temperatures while reaction (2) is favored by lower temperatures. Since the temperature effect of (1) is greater than (2), at first glance it would appear that the overall equilibrium is favored by increased temperature. Such is not necessarily true, however, for both reactions must be figured simultaneously, taking into account the equilibrium composition of intermediate products. Presumably, then, there will be an optimum temperature for the overall conversion to H_2, depending upon initial concentrations.

In other words, the intermediate product is not all reacted as would be implied if we assumed the overall reaction equilibrium could be denoted by multiplying the equilibrium constants together and cancelling out the intermediate product.

At higher steam decomposition, with sufficient CO_2 present, it is possible to exceed the CO-shift equilibrium (3). At temperatures above 2000°F, equilibrium may be assumed.

At levels below 2000°F, however, the CO and H_2 react to form methane, at rates depending upon the operating conditions. Lower space velocities favor C_1 formation, and a fluidized bed yields more C_1 than a fixed bed. Higher pressure also favors C_1 formation as noted. It is also considered that C_1 formation from carbon-steam mixtures occurs by direct reaction of the hydrogen with the carbon surface rather than with CO (4). Apparently a controversy exists.

The pressure range which has been of the most interest in the study and utilization of steam-carbon reactions is from low to moderate pressures, say from atmospheric to 750 psi. Temperature studies have mostly been in the range 900-2000°F. The reaction is considered slight below 900°F. However, there are important exceptions for catalyzed reactions which we will wish to consider.

The comparative roles and mechanisms of the reactions involved apparently is a question not readily settled (4, 5). However, it seems to be that the reaction (1) is the predominately important reaction. The rate is pronouncedly affected by the type or form of carbonaceous material involved and is affected by pretreatment (4).

The carbon-steam and carbon-CO_2 reactions compete at temperatures above the transition range of 1300-2200°F, wherein the rates become about the same (6). Below the transition range, the carbon-steam reaction is much the faster.

While reaction (1) definitely is a reaction involving the surface of a solid, it has also been advanced that the CO-shift, reaction (2), also occurs at the surface (4). However, the CO-shift is also regarded as occurring in the gas phase.

The addition of O_2 to the steam-carbon reaction (to supply the endothermic heat) will yield less H_2 and thus give lower H_2/C ratios. For the same O_2/C ratio, lower rank coals are more reactive (7). The reaction is a function of particle size.

The addition of hydrogen tends to inhibit the carbon-steam reaction (8). This could be explained by virtue of the fact that hydrogen is a reaction product. On the other hand, pretreatment with N_2 at reactor conditions has been found to enhance coal activity twofold.

For the reactions of coal, it has been stated that, ordinarily, direct gasification of coal is preferable to first carbonizing the coal and then reacting the coke (9).

The proportions of steam used not only affect the composition of the product (via the CO-shift, at lower temperatures), but an excess of steam is also required to drive the reaction. Thus it is characteristic to use about 50% excess steam to drive the reaction, as can be observed from the operating data for coal-steam gasifiers (e.g., as cited in the Appendix). This excess is condensed along with any tars or oils produced. The effect, energy-wise, is to reduce the overall thermal efficiency of the operation due to the waste-heat losses that will ensue (see Appendix).

6.2 CATALYTIC EFFECTS

There have been studies made upon the effect of catalysts on the carbon-steam reaction, notably by Taylor and Neville, White and Cobb and coworkers. Earlier literature references are given in References 10 and 11. The most interesting effect of catalytic action is that it permits a lowering of the reaction temperature.

Taylor and Neville (12) reviewed the early patent literature prior to 1920 and noted that some of the catalytic substances advocated to lower the reaction temperature are lime or caustic soda, hydrates or carbonates, chlorides, sulfates, sulfides, silicates, metallic oxides, alkalies, and metallic couples such as copper-iron. Taylor and Neville found potassium and sodium carbonates to be the best, particularly potassium carbonate. Catalysts which worked for the steam-carbon reaction would also work for the carbon dioxide-carbon reaction.

Using impregnation from solution, Taylor and Neville studied the effect of alkali carbonates between 490 and 570°C (915 and 1060°F). The optimum concentration of catalyst was found to be 20 percent. At 915°F reaction was only about one-fifth as fast as at 1060°F. The reactor was a fixed bed of charcoal through which steam was passed.

Ferric oxide was found to be only about one-tenth as effective as the alkali carbonates for the steam-carbon reaction. However, it is very effective for the CO-shift. It was not mentioned in what form the ferric oxide was applied.

White and coworkers (13, 14, 15) have examined the role played by sodium carbonate. It was hypothesized that the carbonate was reduced to yield the oxide or the metal,

$$Na_2CO_3 + C \rightarrow Na_2O + CO$$

followed by regeneration with CO_2 present,

$$Na_2O + CO_2 \rightarrow Na_2CO_3$$

Evidently the assumption is made that a near equilibrium simultaneously exists for the CO-shift,

$$CO + H_2O \rightarrow CO_2 + H_2$$

which is catalyzed by the Na_2CO_3 and is the source of H_2 and CO_2 (5). This train of reasoning has apparently not been pursued in any subsequent investigations.

Experimental data was obtained in the range of 1400 to 1800°F. With five percent Na_2CO_3 conversion was about two percent at 1400°F and 95 percent at 1730°F based upon the steam (14). With one percent Na_2CO_3, there was little catalytic activity (15). The same result could be

Reactions with Steam

obtained at a markedly lower temperature by using the sodium carbonate, which in these experiments was added by solution impregnation. Sodium carbonate, incidentally, is a solid below 1560°F. A product composition obtained at 1730°F was (14):

    36.8% CO
    49.7% H_2
    6.5% CO_2
    nil CH_4

Lewis, Gilliland, and Hipkin (10) have studied the steam-carbon reaction in the presence of K_2CO_3 at pressures up to 750 psi. A common reaction temperature of 1200°F was used for all of the reported data.

The reactor was a fixed-bed flow system. About ten percent K_2CO_3 was used and mixed dry with the wood charcoal charge. High conversions were obtained with the water-gas shift going essentially to completion. After an initial change, the outlet product composition remained constant. A typical analysis was as follows:

    65 mole % H_2
    30   CO_2
    5   CO
    1   CH_4

It was noted that at low steam rates, higher pressures increased the C_1 formation at the expense of the H_2. At higher rates pressure had little effect. Adding additional H_2 increased C_1 formation. The water-gas shift was considered always at equilibrium, while the steam-carbon reaction was below equilibrium. The gas-phase reactions to produce C_1 were above equilibrium.

The following commentary was made by the above authors

"The reaction of carbon with steam at low temperatures offers the possibility of producing hydrocarbons directly. At temperatures of 700°F to 900°F the equilibrium favors the formation of methane and CO_2. At lower temperatures, higher molecular weight hydrocarbons should be formed."

It was also commented that the overall heat of reaction for conversion on to hydrocarbons would be small, but that there was the problem of low reaction rate.

No data was given at temperatures below 1200°F, nor were any appreciable amounts of hydrocarbons obtained, possibly because no additional catalyst was used to promote hydrocarbon formation.

As an added note, K_2CO_3 or Na_2CO_3 is an effective additive or promotor catalysts used in the Fischer-Tropsch reaction between CO and H_2. This will be discussed in the appropriate sections of Chapter 7.

Tuddenham and Hill (11) have studied the catalytic effects of cobalt, iron, nickel and vanadium oxides on the steam-carbon reaction. The reactions were studied from 1030-2000°F. The charge consisted of a carbon rod which had been soaked in solutions of the catalyst agent, and was heated by resistance. The reaction was analyzed by the CO/CO_2 ratio produced.

The impregnating solutions were formed from $Fe(NO_3)_3$, $Co(NO_3)_2$, $Ni(NO_3)_2$, and NH_4VO_3. It was observed that reaction rates were obtained for the impregnated charges which were up to 30 times the rates for the untreated charges. Minute amounts of catalyst had a pronounced effect, pointing out the difficulty to performing and evaluating supposedly uncatalyzed systems. The activity of iron, cobalt, and nickel impregnated systems was observed to fall off above 1400°F.

Results indicate that more CO_2 is formed with charcoal than with carbon (5). This increase is thought to come from the CO-shift being catalyzed by the ash in the charcoal, particularly the iron oxides. Other workers have also suggested this catalytic effect (16). Accordingly, various inorganic substances have been investigated for their effect. It is very possible that the presence of these substances also can have a catalytic effect upon the primary coal or charcoal reaction with steam.

Among the various inorganic substances found not to have any activity are silica, alumina, and fireclay (5). However, calcium oxide or carbonate, iron oxide, and sodium carbonate showed considerable activity in the carbon-steam reaction, having an increasing effect in the order given. The steam was supplied at 1000°C (1830°F) for the tests.

While iron oxides are notable for catalyzing the CO-shift, it was found that cobalt oxide reduced in hydrogen would also catalyze the CO-shift, even at temperatures as low as 283°C (540°F) and space velocities as high as 1800 (apparently reciprocal hours).

The Fischer-Tropsch reaction between CO and H_2 to produce hydrocarbons is catalyzed by iron and cobalt compounds (which will be discussed in the appropriate section.) However, in the case of the CO-shift, using cobalt oxide, the formation of hydrocarbons could be suppressed by adding various promoters. Among these are iron and copper. The CO-shift will be considered further under the reactions of carbon monoxide and hydrogen.

Long and Sykes (17, 18) carried out studies of specific metal catalysts (Fe, Na, Ca, and Al) at 750°C using cocoanut charcoal. The

Reactions with Steam

results indicate that the steam-carbon reaction is catalyzed chiefly by Na, the CO_2-carbon reaction by Fe and Na, and the oxidation of CO by steam is enhanced by all except aluminum. Other reports (19, 20) point out the effectiveness of the alkali carbonates as catalysts for carbon-oxygen-steam reactions.

Taylor and Neville (12) indicated that potassium carbonate is superior, principally because of the acceleration of the carbon-carbon dioxide reaction yielding carbon monoxide:

$$C + CO_2 = 2CO$$

Cobb et.al., on the other hand, indicate a superiority for sodium carbonate (3).

Alkali carbonates with cobalt or copper oxide have a tendency to form metal carbonates thereby reducing the volatility and increasing the effectiveness.

The subject of catalysis of carbon-steam reactions is but a facet of the catalysis of carbon gasification reactions in general. Further research and discussion are provided in References 21 and 22. Bibliographies are presented in References 22 and 23.

Interestingly, alkali salts are regarded as semiconductors (J. L. Dye, *Scientific American,* July, 1977). See also Chapter 8, wherein all catalysts are viewed as semiconductors (e.g., F. F. Vol'kenshtein).

Inorganic acids (H_2SO_4, HF, H_3PO_4) are used in petroleum refining as reactants or catalysts (treating, alkylation, polymerization), and such acids or their anhydrides may be expected also to have an effect on the conversion of coal — the acidic counterpart of the action of alkalies.

PROCESS APPLICATIONS

The Exxon catalytic coal gasification process (Figure 6.1) uses an alkali carbonate in steam gasification. Supported by ERDA (DOE), the work embodies a fluidized bed at ~ 500 psi and ~ 1300°F. The reactor product is composed of $CO/H_2/CH_4$ in varying proportions as indicated by Figure 6.2. In the commercial embodiment the CO and H_2 are to be cryogenically separated from the CH_4.

It is stated that the reactor product gases are at equilibrium, indicating that the alkali serves more to enhance the reactivity than to produce a selective conversion. The conversion is endothermic since appreciable amounts of CO and H_2 are produced.

Alkali and alkaline-earth impregnation of coal is the basis for the Battelle process. Not only is part of the sulfur captured, but combustibility and steam gasification are enhanced.

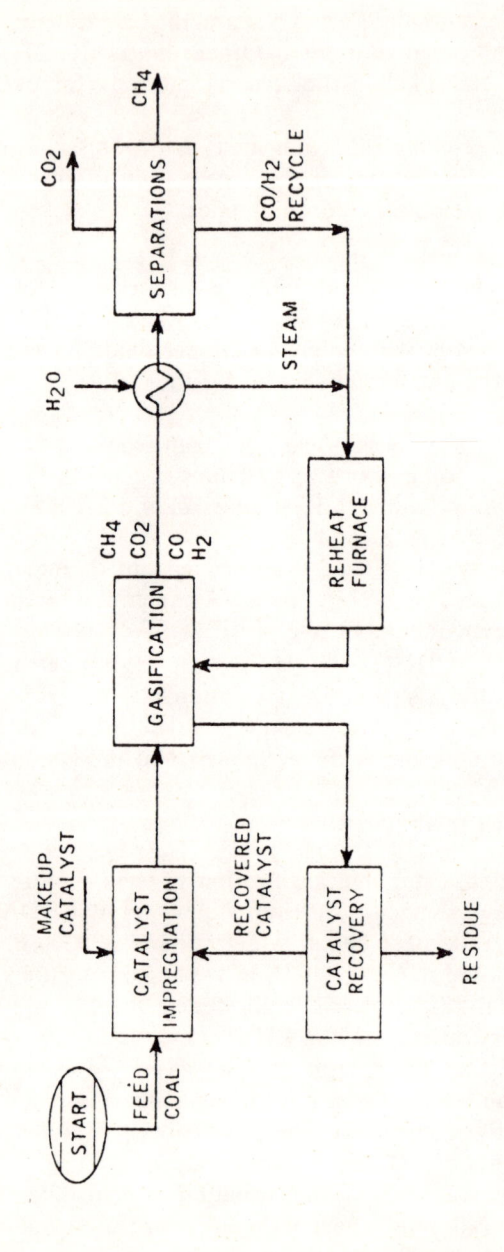

Fig. 6.1. Exxon catalytic coal gasification. [From W. R. Epperly and H. M. Siegel, "Catalytic Coal Gasification for SNG Production," Proceedings of the Eleventh Intersociety Energy Conversion Engineering Conference, State Line, Nevada, Sept. 12-17, 1976.]

Reactions with Steam

Fig. 6.2. Effect of temperature on CO + H$_2$ yields. (From Epperly and Siegel, *op. cit.*)

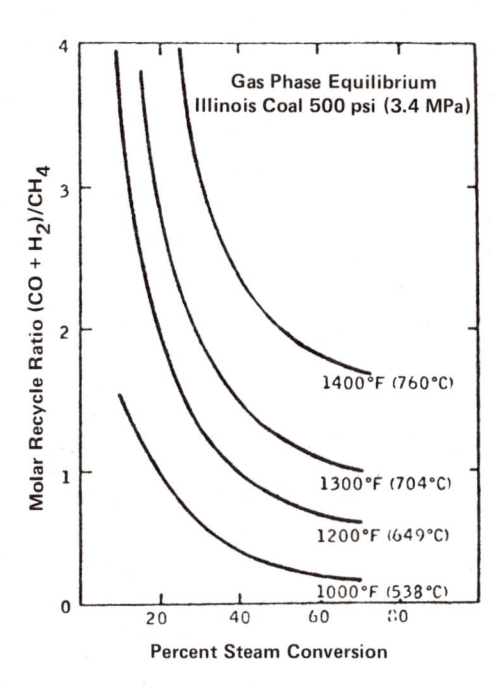

[Graph: Molar Recycle Ratio (CO + H$_2$)/CH$_4$ vs Percent Steam Conversion. Gas Phase Equilibrium, Illinois Coal 500 psi (3.4 MPa). Curves at 1400°F (760°C), 1300°F (704°C), 1200°F (649°C), 1000°F (538°C).]

Work by TRW is reportedly aimed at conversion all the way to hydrogen. Such would be expected to depend upon steam/coal ratios as well as the use of catalytic materials. The conversion is endothermic.

6.3 ENDOTHERMICITY

Schemes for reacting carbon and steam are all concerned in one way or another with means for supplying the endothermic heat of reaction. The heat may be supplied directly from the system *per se* or indirectly from some other source than the reactants.

Of the direct means, the use of the steam itself is a possibility. Steam has been successfully used in the low temperature carbonization of coal, albeit an expensive source of energy for this purpose (4). In a variation of the Rochdale process, steam yields simultaneously a coke product and a CO-H$_2$ gaseous product. The steam first reacts with the hot coke to produce a water gas which then carbonizes the coal charge higher up in the reactor.

The preheating of the coal or coke charge is another direct means of supplying heat to the system. This may be done continuously or by cyclic operations, depending upon whether a moving or fixed bed is employed.

214   Coal Conversion

The other direct means of interest is autothermic gasification. In an autothermic process, the heat required is generated within the system. For the steam-carbon reaction, the addition of oxygen or an oxygen-bearing medium will supply heat by the exothermicity of the carbon-oxygen reaction.

As a special case, heat may be generated within the system by the passage of electric current through the charge. This is used in electrothermal (or electrofluidic) gasification (24).

The most obvious indirect means is by heat transfer through the walls containing the medium, and considerable documentation exists for this approach (5). Nuclear energy is a potential heat source which is advanced from time to time.

Other means consist of using an added non-reacting medium to supply heat (4). The medium may be a gas, liquid, or solid. Media suggested or tried include pebbles and slag. Whether such means should be called direct or indirect is a semantics question. Ash or char provides such a medium, as embodied by the COGAS and Agglomerating Ash processes shown in the Supplement to Chapter 1.

In the Kellogg coal gasification process steam and powdered coal are injected into a molten salt (Na_2CO_3). The unreacted residue is separated by dissolving the Na_2CO_3, which in turn is reprecipitated as the bicarbonate and regenerated.

In the CO_2-Acceptor process, dolomite is regenerated as the oxide and then reacts with the CO_2 produced by the steam-carbon reaction. The hot calcined dolomite also supplies the heat for reaction. About ten percent C_1 is produced along with the synthesis gas.

The TOSCOAL process uses ceramic spheres as the heat transfer medium. Still other processes may use other media, including sand, as noted. Appropriate flow diagrams are shown in the Supplement to Chapter 1 and elsewhere.

A comprehensive treatment of the balancing of the reaction heat requires a thermodynamic analysis for each particular reactor system under consideration. Techniques for making such an analysis, including losses in efficiency, are demonstrated in the Appendix.

6.4 STEAM-GASIFICATION SYSTEMS

It is not so much a purpose here to report the various processes, but to examine the operating conditions and their effect.

In commercial practice, the production of water gas or synthesis gas has been predominately by cyclic operations. In the one phase, air is

Reactions with Steam

reacted with coal or coke charge to heat the charge to reaction conditions. In the other phase of the cycle, steam is added to the charge to yield the CO-H_2 product. The endothermicity of the reaction cools the charge, whereby the cycle is repeated.

Steady-state or continuous operations are at present of more interest. These are exemplified by the Lurgi and Winkler generators. The former is essentially a fixed-bed operation, while the latter is a fluidized system. Oxygen, preferably, is introduced with the steam and reacts to counteract the endothermicity of the steam-carbon reaction. There are also other processes such as a fully entrained bed, etc., which will be described further in Chapter 9.

A Lurgi gasifier operates at around 300-450 psi and perhaps 1600°F as a minimum. A higher pressure favors C_1 formation and saves on compression costs. The C_1 formation may be suppressed by adding CO_2, which also increases the proportion of CO (4). A Lurgi gasifier may run as high as 2200°F with steam introduced at 1000°F at system pressure (4). With greater O_2 input, temperatures mentioned are ~ 2600°F at steam/oxygen ratios of 2/1 (3). Typical product gas compositions are (3):

- 30% CO_2
- 17-25% CO
- 37-42% H_2
- 8-12% CH_4

Steam/oxygen molar ratios were around 8/1; conversions around 90 percent. In addition, liquids (tars and oils) are produced which may amount to ~ 10% of the feed. The excess steam produces an aqueous condensate.

The means of operation of a Lurgi gasifier determine to a large extent the product composition. Lower temperatures toward the top of the bed favor C_1 in the products (25). Increasing steam rates gave higher H_2/CO ratios (3, 26). For instance, for a molar steam/oxygen ratio of 1.0, the H_2/CO ratio is 25.5/72.7; for a ratio of 7.0 the H_2/CO ratio is 44.3/18.2. At the lower steam/oxygen ratios, the gasification process would yield what would be characterized as a producer gas. Higher steam/O_2 ratios insure that temperatures remain below the ash fusion point (6).

On increasing the reaction temperature by decreasing the steam/oxygen ratio, the unreacted steam is decreased with a decrease in steam consumption (3). However, the temperatures for steam decomposition and C_1 formation are not compatible (27). There results an increase in gas heating value due mainly to an increase in CO content (4). Such may also be accomplished by adding CO_2 as noted.

A means recommended for increasing the C_1 production in a Lurgi gasifier involves removing gases from the lower part of the bed, and adding more steam to the upper part of the bed (26).

Steam temperatures of 930°F are recommended, though lower temperatures reduce steam consumption (26). It is mentioned, however, that preheating the steam-oxygen mixture to 1300°F lowers the oxygen consumption (3).

At temperatures above 2200°F, slagging conditions occur. The removal of the slag poses special problems of its own, and the operation of slagging gasifiers is still in the experimental stage.

In experimental studies, lower rank fuels gave higher heating value gas with lower oxygen consumption (4). Increasing the presssure increased C_1 formation, which also increases heating value.

The Winkler generator has been characterized as insensitive to the type of fuel used (5). Fuel may vary from coke to lignite. This is a characteristic of fluidized bed systems (4). The process operates at atmospheric pressure, and with the fluidized bed at 1470-1830°F. A typical product stream is as follows (28):

13-20%	CO_2
47-36%	CO
39-41%	H_2
0.6-0.4%	CH_4
0.4-0.5%	N_2

The use of air instead of oxygen of course will yield a different composition. One volume of O_2 per two volumes of steam is characteristically used (5).

Advantages of fluidization which have been listed are (4):

(1) The ability to tolerate variations in fuel quality.
(2) High gasification rates.
(3) Variable output.
(4) Safe and reliable for gasification reactions due to the large charge of unreacted material.
(5) Localized hot spots avoided.

Disadvantages of fluidized beds listed are (4, 6):

(1) Loss of sensible heat from the reaction.
(2) Loss of unreacted solids.

Reactions with Steam

(3) Buildup of ash in the fluidized beds.
(4) Slagging.
(5) Fluidization characteristics of the fuel are limiting.

In gasification, fully entrained systems inherently yield a product with little tar because of the complete conversion of volatile matter (4). Carbon losses are reduced by recycle and steam decomposition is lower (6). Fully entrained systems have the usual disadvantages of concurrent flow and a low concentration of solid particles.

Though the reaction of coal or coke and steam has been used commercially for many years the reaction is still being studied towards more efficient means of gasification. A few of the more interesting of past investigations are described in the following paragraphs.

In pilot plant scale studies using a fluidized bed, it was found that lignites and brown coal react readily with steam at temperatures as low as 1400°F, without adding any catalyst (29). For bituminous coals, 1650°F seemed to be the minimum useful reaction temperature (for about 50 percent decomposition). In this instance, the reactor was heated indirectly by combustion in an annular space around the reaction chamber. The heat transfer coefficient was about 40 Btu/hr-°F-ft².

In experiments with a small fluidized reactor at 800-2000°F, steam or nitrogen was used to effect a pyrolysis of coal (12). No catalyst was used. Conversion was 25 percent at 800°F and 50 percent at 2000°F. The tar produced decreased with temperature, and the gases produced increased with temperature. No mention of the type of reactions involved was given.

A small-scale fluidized reactor reacted steam and coal at 900°F and yielded gases, liquid distillate, tar, pitch, and char (30). No catalyst was used. The conversions calculate to about five percent gas, two percent distillate, 19 percent tar, up to 14 percent pitch and the remainder char. Appreciable concentrations of methane occurred in the gaseous phase. The liquid distillate was predominately divided between acids, paraffin and aromatics.

Steam itself has been used to supply the heat for the pyrolysis of coal (31). Superheated steam at 1000-1200°C was used. Evidently there was little reaction, and no catalyst was present. It was commented that steam was an expensive source of heat.

A falling particle technique has been used to study the reaction of steam with coal (32). This is a suspended system whereby the solid particles are introduced at the top of the bed and removed at the bottom. The steam feed is also introduced at the top. Motion of the coal particles is about six times as fast as the steam. The coal mesh size was −60 to +65. Conditions were 1700°F at atmospheric pressure, with steam introduced at 800-1000°F. Up to 70 percent conversion was obtained. The reactivity decreased with increased rank.

In gasification studies in a fluidized pilot plant unit using both steam and oxygen, 90 percent conversion was obtained at 100 psi under slagging conditions (33). Superheated steam was introduced at 1000°F. A distinguishing feature was that the pulverized coal feed was introduced in a fluidized state using an inert gas.

In studies with a fluidized system at 1600°F at one atm pressure, it was found that pretreatment with N_2 at reactor conditions for up to 24 hours would give gasification rates up to twice the untreated rate (8). The steam conversion was from 45 to 65 percent. It was noted that the reaction between carbon and steam was controlling, the CO-shift reaction being much faster. The effect of added hydrogen was studied in H_2/steam ratios from 0.1 to 1.0. The H_2 was found to have an inhibiting effect (possibly as a reaction product). Methane was in as high a concentration as the CO_2 in the exit gas.

Studies at the Eyring Research Institute are aimed at high gasification rates using pulverized coal with preheated oxygen and superheated steam (34).

Laboratory, bench-scale and pilot-plant studies are being succeeded by demonstration and implementation. Commercial gasifiers, both past and present, are discussed in Chapter 9.

REFERENCES

1. Lyon, L. B., "Primary Extraction, Conversion and Upgrading Hydrogen-Deficient Fossil Hydrocarbons," from Proceedings of the First Intermountain Symposium on Fossil Hydrocarbons, Salt Lake City, October, 1964. Sponsored by Brigham Young University.
2. Lyon, L. B., "Liquid Fuels from Coal, Other Fossil Hydrocarbons," *Coal Age, 69,* No. 12, 70-2, 75 (Dec., 1964).
3. van der Hoeven, J. C., "Producers and Producer Gas," in *Chemistry of Coal Utilization,* Vol. II, H. H. Lowry Ed., Wiley, New York (1945).

4. von Fredersdorf, C. G. and M. A. Elliott, "Coal Gasification," in *Chemistry of Coal Utilization,* Supplementary Volume, H. H. Lowry Ed., Wiley, New York (1963), pp. 892-1022.
5. Morgan, J. J., "Water Gas," in *Chemistry of Coal Utilization,* Vol. II, H. H. Lowry Ed., Wiley, New York (1945), pp. 1673-1749.
6. Hoy, H. R. and D. M. Wilkins, "Total Gasification of Coal," British Coal Utilization Research Association, Review No. 174, Feb./March, 1958, pp. 57-110.
7. "Synthetic Liquid Fuels. Part I. Oil from Coal," *U.S. Bureau of Mines Report of Investigations 5236,* July, 1956.
8. Goring, G. E., G. P. Curran, R. P. Tarbox and E. Gorin, "Kinetics of Carbon Gasification by Steam," *Ind. Eng. Chem., 44,* 1051-65 (1952).
9. Powell, A. R., "Relation of Coal Gasification to the Production of Chemicals," from *Symposium on Gasification and Liquefaction of Coal,* American Institute of Mining and Metallurgical Engineers, New York (1953), pp. 196-206.
10. Lewis, W. K., E. R. Gilliland and H. Hipkin, "Carbon-Steam Reaction at Low Temperatures," *Ind. Eng. Chem., 45,* 1697-1703 (1953).
11. Tuddenham, W. M. and G. R. Hill, "Catalytic Effects of Cobalt, Iron, Nickel, and Vanadium Oxides on Steam Carbon Reaction," *Ind. Eng. Chem., 47,* 2129-33 (1955).
12. Taylor, H. S. and H. A. Neville, "Catalysis in the Interaction of Carbon with Steam and with Carbon Dioxide," *J. Am. Chem. Soc., 43,* 2055-71 (1921).
13. Fleer, A. W. and A. H. White, "Catalytic Reactions of Carbon with Steam-Oxygen Mixtures, *Ind. Eng. Chem., 28,* 1301-9 (1936).
14. Fox, D. H. and A. H. White, "Effect of Sodium Carbonate Upon Gasification of Carbon and Production of Producer Gas," *Ind. Eng. Chem., 23,* 259-66 (1931).
15. Weiss, C. B. and A. H. White, "Influence of Carbonate upon the Producer Gas Reaction," *Ind. Eng. Chem., 26,* 8307 (1934).
16. Wilson, P. J. and J. D. Clendenin, "Low-Temperature Carbonization," in *Chemistry of Coal Utilization,* Vol. III, H. H. Lowry Ed., Wiley, New York (1963), pp. 395-460.
17. Long, F. J. and K. W. Sykes, "The Effect of Specific Catalysts on the Steam-Carbon System," *Proc. Roy. Soc. (London), 215A,* 100-110 (1952).

18. Long, F. J. and K. W. Sykes, "The Mechanism of the Steam-Carbon Reaction," *Proc. Roy. Soc. (London), A 193,* 377-99 (1948).
19. Marson, C. B. and J. W. Cobb, "Influence of the Ash Constituents in the Carbonization and Gasification of Coal, with Special Reference to N and S (I) preparation and Preliminary Examination of Special Cokes," *Gas Journal, 175,* 882-9 (1926). CA 20:490.
20. Kroger, C., "The Gasification of Carbon in Air, Carbon Dioxide, and Water Vapor, and the Effect of Inorganic Catalysts," *Angew Chem., 52,* 129-39 (1939). CA 33:3563.
21. Long, F. J. and K. W. Sykes, "The Catalysis of the Oxidation of Carbon," *J. Chim. Phys., 47,* 361-78 (1950). CA 44:7132c.
22. Walker, P. L., M. Self and R. A. Anderson, "Catalysis of Carbon Gasification," in *Chemistry and Physics of Carbon,* Vol. I, P. L. Walker Jr., Ed., Dekker, New York (1965), pp. 287-383.
23. Ergun, S. and M. Mentser, "Reactions of Crabon with Carbon Dioxide and Steam," in *Physics and Chemistry of Carbon,* Vol. I, P. L. Walker Ed., Dekker, New York (1965), pp. 203-261.
24. Pulsifer, A. H. and T. D. Wheelock, "Observations of Coal Processing in an Electrofluid Reactor," Dept. of Chem. Eng., Iowa State University, Ames (undated).
25. Linden, H. R., "Conversion of Solid Fossil Fuels to High-Heating Value Pipeline Gas," in *Hydrocarbons From Oil Shale, Oil Sands, and Coal,* Chemical Engineering Progress Symposium Series Number 54, Volume 61, 1965.
26. Igoe, J. W. and R. A. Glenn, "Coal Research," *Mining Congress Journal,* February (1962).
27. Rapoport, I. B., *Chemistry and Technology of Synthetic Liquid Fuels,"* Second Edition, Translated from Russian, Published for the National Science Foundation, Washington, D.C. by the Israel Program for Scientific Translations, Jerusalem (1962).
28. Montgomery, R. S., "Coal Chemicals from Coal Oxidation Products," presented at 143rd National Meeting, American Chemical Society, Division of Fuel Chemistry, Cincinnati, Ohio, January, 1963 (Vol. 7, No. 1).
29. Jolley, L. J., A. Poll and J. E. Stantan, "Fluidized Gasification of Non-Coking Coals with Steam in a Small Pilot Plant," from *Symposium on Gasification and Liquefaction of Coal,* American Institute of Mining and Metallurgical Engineers, New York (1953), pp. 60-72.

30. Naugle, B. W., C. Ortuglio, L. Mafrica and D. E. Wolfson, "Steam-Fluidized Low-Temperature Carbonization of High Splint Bed Coal and Thermal Cracking of the Tar Vapors in a Fluidized Bed," U.S. Bureau of Mines Report of Investigations 6624 (1965).
31. Wilson, P. J. and J. D. Clendenin, "Low-Temperature Carbonization," in *Chemistry of Coal Utilization,* Supplementary Volume, H. H. Lowry Ed., Wiley, New York (1963), pp. 395-460.
32. Sebastion, J. S. and R. L. Day, "The Reactivities of Suspensions of Coals," presented before the Division of Fuel Chemistry, American Chemical Society, Atlantic City, N.J., September, 1962.
33. McGee, J. P., L. D. Schmidt, J. A. Danko and C. D. Pears, "Pressure Gasification Pilot Plant Designed for Coal and Oxygen at 30 Atmospheres," from *Symposium on Gasification and Liquefaction of Coal,* American Institute of Mining and Metallurgical Engineers, New York (1953), pp. 80-107.
34. Coates, R. L., "High Rate Coal Gasification," in *Coal Processing Technology,* Vol. 3, American Institute of Chemical Engineers, New York, 1977.

Chapter 7

REACTIONS OF CARBON MONOXIDE AND HYDROGEN

Depending upon the reaction conditions and catalyst, mixtures of carbon monoxide and hydrogen react to form a wide variety of hydrocarbons and oxygenated compounds. Under certain conditions elemental carbon is also formed, and, with certain catalysts not commonly employed, some unusual compounds result.

The proportions of CO and H_2 present have a great deal to do with the resulting products. With the catalysts commonly used, the effect is more one of activity rather than selectivity. An exception is in the production of methanol. Higher concentrations of CO favor oxygenated compounds, as would be expected.

The reaction conditions generally are much milder than for hydrogenation or the formation of synthesis gas. As a rule, severity of reaction conditions favors oxygenated compounds. Higher temperatures tend to produce more gaseous and lower molecular weight compounds, while higher pressures have an opposing effect. The residence times or space velocities will also have an effect.

The overall reactions may be generalized as

$$xCO + yH_2 \rightarrow C_xH_{2(y-z)}O_{x-y} + zH_2O$$

and

$$xCO + yH_2 \rightarrow C_xH_{2(y-x)} + xH_2O \quad (z = x)$$

The reactions are further complicated by the CO-shift reaction which occurs simultaneously:

$$CO + H_2O \rightarrow CO_2 + H_2$$

This parallel and consecutive reaction is also markedly affected by the nature of the catalyst used. Thus, in general, both H_2O and CO_2 would be expected as reaction products. Since the oxygenated compounds are at a higher level of oxidation, it has been advanced that they are intermediates to the formation of the hydrocarbons (1, 2, 3). Olefins may also be intermediates (3).

Oxygenated reaction products which have been identified include alcohols, aldehydes, ketones, acids and esters. Hydrocarbons produced are both paraffinic and olefinic, varying from methane to waxes. Isomers are also produced, and aromatic and cyclic compounds as well.

Reactions forming both hydrocarbons and oxygenated compounds from mixtures of CO and H_2 are all exothermic (negative) in varying degree. The heats of reactions only vary slightly with temperature (4). For the formation of paraffins, the heat of reaction per carbon atom becomes more endothermic (less negative) with carbon number. For olefins, the heat of reaction becomes more exothermic with carbon number (In the above-mentioned reference, negative values are used for exothermicity, positive for endothermicity in common with the more usual convention.)

In comparison, the heats of reaction to produce CO_2 will always be more exothermic than to produce H_2O, since the CO-shift is exothermic. The equilibrium formation of CO_2 is favored at lower temperatures, as examination of the behavior of the equilibrium constant for the CO-shift will show at Fischer-Tropsch synthesis temperatures (see Table 4.2).

The equilibrium constants for the formation of at least most of the reaction products decrease with increasing temperature (4). Equilibrium is such that to maintain a given conversion an increase in temperature requires an increase in pressure.

Since for any appreciable reaction to occur requires the presence of a solid catalyst and the reactions are thus heterogeneous, all thermodynamic considerations involving the free energy and equilibrium concepts of homogeneous reaction analysis should be viewed as suspect in the extreme (see Chapter 8).

7.1 PRODUCTS AND CONDITIONS

The origins and developments of the study of catalysis parallel the investigations of the reactions of carbon monoxide and hydrogen (5). Sabatier and Senderens, in 1902, reported the reaction of CO and CO_2 with H_2 to form CH_4 over reduced nickel and cobalt catalysts. This, and

Reactions of Carbon Monoxide and Hydrogen

subsequent developments, preceded the discovery of catalytic processes for the synthesis of ammonia, hydrogenation of olefins, and the catalytic refining of petroleum (5, 6).

Other significant developments include German patents issued in 1913 and 1914 for the formation of liquid products containing alcohols, aldehydes, ketones, fatty acids, and aliphatic hydrocarbons using a catalyst of cobalt and osmium oxide.

In 1923, Fischer and Tropsch reported the successful use of alkalized iron turnings to yield an oily liquid containing chiefly oxygenated compounds. Conditions were at 750 to 850°F and 1500 to 2200 psi. By reducing the pressure to about 100 psi, the chief products were paraffinic and olefinic hydrocarbons. In 1925-26 Fischer and Tropsch used iron or cobalt catalyst at atmospheric pressure and 480-570°F to yield a hydrocarbon product from C_2 to solids which was virtually free of oxygenated compounds (7).

The high-pressure process of Fischer and Tropsch yielding the oxygenated compounds, particularly alcohols, was called the "Synthol" process and the product "Synthin" (5). The hydrocarbon product of the lower pressure process was originally called "Kogasin" (6, 7). This latter process is now uniformly called the Fischer-Tropsch process. It has also been called the "Synthine" process — a contraction for synthetic gasoline. If the reaction is to produce predominately methane, it is termed "methanation." Other designators are "Synthoil" for hydrocarbon liquids and "Synthol" for alcohols. as noted.

There is the possibility that uncatalyzed mixtures of CO and H_2 may react. For instance, water gas passed through an empty steel tube at 400 psi and 850°F gives C, CO_2 and low molecular weight hydrocarbons, especially methane (1). It can be argued, however, that the steel surface serves as a catalyst. The argument can be extended to include any surface as having some sort of catalytic effect, and that the reactions are in fact heterogeneous.

Be as it may, for any conversions of an important enough magnitude, significant quantities of catalyst must be employed. Catalysts containing Fe, Co, Ni, or Ru have been given the most extensive study. Of these more commonly used catalysts, Ni or Co and their compounds are the most active (5). Rubidium is specific toward solid hydrocarbons (1, 5, 7, 8). However, Ni and Co catalysts are expensive, making Fe catalysts the most desirable for large scale commercial operations (7). Ni tends to form volatile carbonyls, Co less so (1). These decompose at reaction conditions, however.

The more active catalysts Ni and Co favor the formation of methane or other light hydrocarbons (8). A less active catalyst such as iron will favor higher molecular weight hyrocarbons and oxygenated compounds (5). The product olefinic content will increase as Ni, Co, or Fe catalysts are used, in that order (7). The unsaturated content decreases with product molecular weight (5). The Fischer-Tropsch reaction yields primarily straight chain hydrocarbons, though the proportion of isomers increases with increasing carbon number (9). It is considered that straight chain compounds are produced by a primary reaction, since they are less stable thermodynamically than the branched chains.

Due to the water-gas shift, reactions in the presence of iron catalysts, for instance, tend to produce CO_2 as a final product; in the presence of Ni or Co, H_2O is produced (1, 3). The iron acts as a catalyst for the CO-shift, while the Ni and Co do not. Thus the water-gas shift is a slower reaction than the synthesis reaction in the presence of Ni or Co, but is comparable or faster in the presence of Fe (1, 4). With higher concentrations of CO (and H_2O vapor), however, the CO-shift occurs appreciably even with Co catalysts (2). Thus the effect of Co as a catalyst for the water-gas shift apparently is somewhat controversial. It has been found that catalysts made by the fusion of cobalt oxide, and reduced in H_2, would catalyze the water-gas shift at temperatures as low as 540°F (10). The added presence of Fe, on the other hand, would suppress the formation of C_1.

It has been noted that at 450°F, the water-gas shift is 13 times the rate of the synthesis gas reaction in the presence of an iron catalyst (2). As the temperature decreases, the rate of the CO-shift decreases faster than the rate of the synthesis reaction. At temperatures below 350°F the CO-shift is negligible (9). Above 2000°F, the shift may be assumed at equilibrium (11).

The CO- or water-gas shift, as well as the formation of water gas, apparently is even catalyzed by the ash present in coal, particularly the iron oxides (10). Both Fe_2O_3 and Na_2CO_3 have been shown to have a catalytic effect, but not silica, alumina, or fireclay.

Methanation, the reaction of CO and H_2 to produce predominately methane, is conducted at higher temperatures than the usual Fischer-Tropsch synthesis reactions. For instance, at 1750°F and 1 atm using an Fe catalyst, methane was produced, but not at 750°F (12). However, on increasing the pressure to 600 psi at 750°F, methane was obtained. Other conditions used for methanation are 800°F and 300 psi pressure over a nickel-alumina catalyst (13), and 660-750°F at 150-450 psi using a Ni catalyst (14). H_2/CO ratios vary from 2/1 to over 3/1. A Raney-nickel catalyst produced methanation at 570-660°F and 300-450 psi (15).

Usually methane yields are highest at lower temperatures and higher H_2/CO ratios (12). Pressure does not effect C_2 formation until

Reactions of Carbon Monoxide and Hydrogen

temperatures of 800°F or higher. Then an increase in pressure favors C_1 formation.

As a rule nickel catalysts work better than iron. In practice, excessive reaction temperatures lead to carbon deposition and catalyst disintegration for iron catalysts (4, 15).

A variant of the Synthol process, called the Synol process, is directed toward the production of a wide range of straight chain alcohols (1, 6, 9). Lower temperatures and pressures are used, around 400°F and 350 psi, with an iron catalyst. Another variation is the Schwarzheide synthesis for alcohols. In practice, the only important difference between the Synol process and the usual Fischer-Tropsch synthesis is that higher space velocities and lower temperatures are used in the Synol process (9).

A process which yields organic oxygenated products indirectly is the Oxo process in which an olefin is reacted with CO and H_2 over a cobalt catalyst (1, 6). The products tend to be branched. This type of reaction is considered to be intimately involved in the Fischer-Tropsch reactions (1).

Alcohols decompose and react in the presence of Ni and Co catalysts, and also in the presence of Fe catalysts (2). In some respects, the reactions are the reverse of the synthesis reaction.

The reforming of methane or natural gas to produce hydrogen may be looked upon as the reverse of methanation, the latter a Fischer-Tropsch reaction. Thus methane may be reacted with steam and/or CO_2 to yield mixtures varying from $H_2/CO = 1/5$ to pure H_2 (5, 16). The course of the reaction with steam is as follows (13):

$$CH_4 + H_2O \rightarrow CO + 3H_2$$

By using the CO-shift, there is a further conversion to H_2:

$$CO + H_2O \rightarrow CO_2 + H_2$$

Temperatures for steam reforming to CO are $\sim$ 1560°F; all the way to CO_2 with excess steam, $\sim$ 1380°F (7, 16). A Ni catalyst serves. Above 2400°F no catalyst is needed. (See Chapter 8 for furhter commentary.)

A partial combustion will also yield a mixture of CO and H_2 (1):

$$CH_4 + O_2 \rightarrow CO + 2H_2$$

No catalyst is needed at 2500°F (7). A Ni catalyst causes reaction at 1560°F. The CO can also be used to generate further hydrogen by the CO-shift. At 2200°F, CH_4 is oxidized directly by Fe_2O_3.

In addition to reacting methane with steam or oxygen, the reaction with CO_2 can be used (7):

$$CH_4 + CO_2 \rightarrow 2CO + 2H_2$$

Ni or Co catalysts work. The reactions with CO_2 and steam are both endothermic.

All of the aforementioned reforming reactions are, in general, reactions within the system CH_4, H_2O, CO, CO_2, H_2 (and O_2). The direction in which the reactions go depends upon the respective concentrations and conditions. At lower temperatures (<570°F), CH_4 and H_2O are the stable components, at higher temperatures (>1470°F), CO and H_2 tend to be the stable components.

A reaction apparently dissimilar in nature to the previously discussed reactions is the catalytic formation of methanol. Here ZnO acts specifically toward the formation of only the one compound.

Though the reaction of CO and H_2 to methanol is thermodynamically one of the least favorable (17), the reaction is specifically catalyzed by ZnO catalysts, with Cr_2O_3 as a promotor (17, 18). Copper oxide is also a promotor (14). Iron oxide has been used as a catlayst at 400°F and 375 psi (14), though Fe catalysts favor C_2 formation also (17). Pressures of 1500 to 2200 psi and temperatures of 480 to 750°F have been used, with the ranges 3000-3700 psi and 570-660°F apparently the most common.

The reaction to methanol is probably a reduction reaction passing through the aldehyde (17). Other side reactions occur such as the reaction of methanol to methyl ether. Unusual products have been noted using ThO_2 catalysts (1).

On adding alkali to methanol catalysts, higher alcohols will be produced (19).

In isosynthesis, a mixture of CO and H_2 is passed over certain difficulty reducible oxide catalysts, chief of which is thoria (1, 20). Conditions are 750-850°F and 4500-9000 psi. At these conditions, thorium oxide provides an intense dehydrating action (1). At lower temperatures, high yields of alcohols are obtained. As the temperature is raised, iso-olefins are produced, then isoparaffins.

At temperatures above this range, naphthenes and then aromatics are produced (1).

Higher alcohols may thus be produced in a number of ways (21):

(1) Fischer-Tropsch derived catalysts and conditions — the Synthol and Synol process.
(2) Synthesis using methanol derived catalysts, including isosynthesis at lower temperatures.
(3) Oxysynthesis.

Aromatics may be synthesized directly from CO and H_2 using catalysts of chromium, molybdenum and thorium oxides which contain 5-10 percent potash (8). Conditions are 930°F and 450 psi. The synthesis was still at the laboratory stage.

Reactions of Carbon Monoxide and Hydrogen

In the presence of Co or Ni, hydrocracking occurs at higher H_2 concentrations (2), but not in the presence of Fe_2O_3 catalyst. The presence of CO inhibits hydrocracking. It is remarked that all hydrocarbons are thermodynamically unstable above 212°F, except C_1, and C_2, which are unstable above 1000°F and 400°F, respectively (4). Thus the need for product quenching or cooling.

On more active catalysts of Ni, Co and Fe, alcohols decompose (2).

Other subsequent reactions may be catalyzed to change the nature of the reaction products (8). Isomerization and aromatization have already been mentioned, whereby ThO_2, ZnO and Al_2O_3, or other catalysts may be used (1). The oxygenated compounds may also be further converted (8). Further processing includes the following.

The olefins can be polymerized in the presence of $AlCl_3$ catalyst (8, 16). That is, gaseous olefins polymerize to liquids using $AlCl_3$ catalysts (7, 16). ThO_2 also causes polymerization (7). Low molecular weight olefins may alternately be polymerized in the presence of H_3PO_4 (8, 22). Fatty acids can be produced by oxygenation of paraffin waxes (7, 8, 16).

Hydrocarbons convert to alcohols using a boric acid catalyst, and oxidize all the way to acids with air over $KMnO_4$ solution (8). Alcohols catalytically dehydrate to ketones or aldehydes.

Paraffins dehydrogenate to olefins over chromium-alumina catalyst at 930°F. Waxes can be cracked to diesel fuels (16). Alkenes hydrolyze to alcohols using H_2SO_4.

n-Butane isomerizes for alkylation with olefins. Passing gasoline over bauxite at 530-570°F improves the octane number by 6-10 points (9). Many of the above mentioned are of commercial importance.

Almost any olefin hydrogenates at 200-400°F and 1500 psi H_2 pressure with nickel-kieselguhr or Raney-nickel catalysts (23). The reaction can be carried out without affecting aromatics present. Other catalysts are the platinum group and the sulfides of Ni, Mo, Co, or W. Iron successfully catalyzes the hydrogenation of low molecular weight olefins even as low as −150°F, but its activity is reduced by K_2O. Iron will not catalyze the hydrogenation of benzene.

The oxygenated compounds (synthol) as a group are convertible hydrocarbons by hydrogenation at 800°F (8).

Aromatic compounds hydrogenate to yield cyclic saturated compounds (24). The most useful catalysts are Raney-nickel and Adams platinum oxide. Ni requires a higher pressure. Platinum will cause reaction at room temperature. Fe catalysts are unsatisfactory. Below 650°F benzene forms cyclohexane; above, cracking occurs with product variety. Alkyl substituted benzenes react as do phenols.

Catalytic cyclization and aromatization converts paraffins, olefins and 5-ring naphthenes into aromatics without degradation to lower molecular weight compounds (22). The catalysts are oxides of subgroups 5 and 6. Chromium and molybdenum oxides are examples. Pt also works, Conditions are 840-1020°F and H_2 pressure.

Isomerization takes place as well. Applications are the upgrading of petroleum fuels, and the production of toluene from petroleum fractions.

In this connection it should be mentioned that the presence of alcohols in a gasoline upgraded the product (8, 25). Also, passing gasoline over bauxite at 540-570°F improves the octane number 6-10 points (9).

In addition to hydrogenation reactions, aromatics, unsaturates, aldehydes, ketones, ethers, alcohols and even saturates will react with CO in the presence of suitable catalysts to yield a diversity of compounds (26).

The chief commercial developments of the Fischer-Tropsch process were made by Germany to produce liquid fuels in addition to that obtained by hydrogenation and coal carbonization (1, 7). These operations were of a fixed bed type, of a partitioned construction to permit removal of exothermic heat.

A commercial venture of significance carried out in the United States was the Hydrocol process (1, 9). This application of the Fischer-Tropsch reaction was directed toward the production of gasoline fractions, but also yielded up to 25 percent oxygenated compounds. Reaction was in a fluidized bed using an iron catalyst. Natural gas was reformed to provide the synthesis gas. For various reasons, including the increasing value of natural gas, the project was discontinued. The most recent large-scale venture is the SASOL plant in South Africa which produces gasoline and other products, with new efforts directed at methane.

The experimental and operating results for the Hydrocol process have been reported extensively in the literature (9). Experiments were carried out in a fluidized bed using a −100 to +325 mesh iron catalyst prepared from mill scale. Conditions were 600°F and 250 psi. In experimental runs averaging 86 percent conversion, 8 percent of the CO was converted to CO_2, 26 percent to C_1 and C_2, 58 percent to higher hydrocarbons, and 8 percent to oxygenated compounds. The hydrocarbons, C_3+, were from 65 to 80 percent olefinic, and sometimes higher. The fractions tended to be 25 to 45 percent isomeric. Aromatic and isomeric percentages increased with carbon number.

The oxygenated fractions were about 50 percent alcohols, 10 percent aldehydes, 10 percent ketones, and 27 percent acids.

TABLE 7.1 Comparison of Fischer-Tropsch Processes.

Cat.	Temp. °F	Press. psi	Gasoline % of Prod.	Diesel Percent	Heavies Percent	Oxys Percent	Oct No.	Cetane No.
Granular Catalyst, Externally Cooled, No Gas Recycle								
Co	350-400	1 atm	56	33	11	<1	50	100
Co	350-400	150	35	35	30	<1	50	100
Fe	400-450	150	32	18	35	15		
Granular Catalyst, Externally Cooled, Gas Recycle								
Co	190-224	150	50	22	22	6		
Fe	230	300	19	19	56	6		
Fe	275	300	68	19	8	5		
Powered Catalyst, Externally Cooled, Gas Recycle								
Fe	480-530	300	25	30	51	4		
Granular Catalyst, Internally Cooled, Gas Recycle								
Fe	460-540	300	58	10	24			
Granular Catalyst, Hot Gas Recycle								
Fe	570-600	300	70	17	1	12	75	50
Fluidized Catalyst, Gas Recycle								
Fe	570-600	300	73	7	3	17	76	

	1 atm	150 psi	1500 psi	15,000 psi

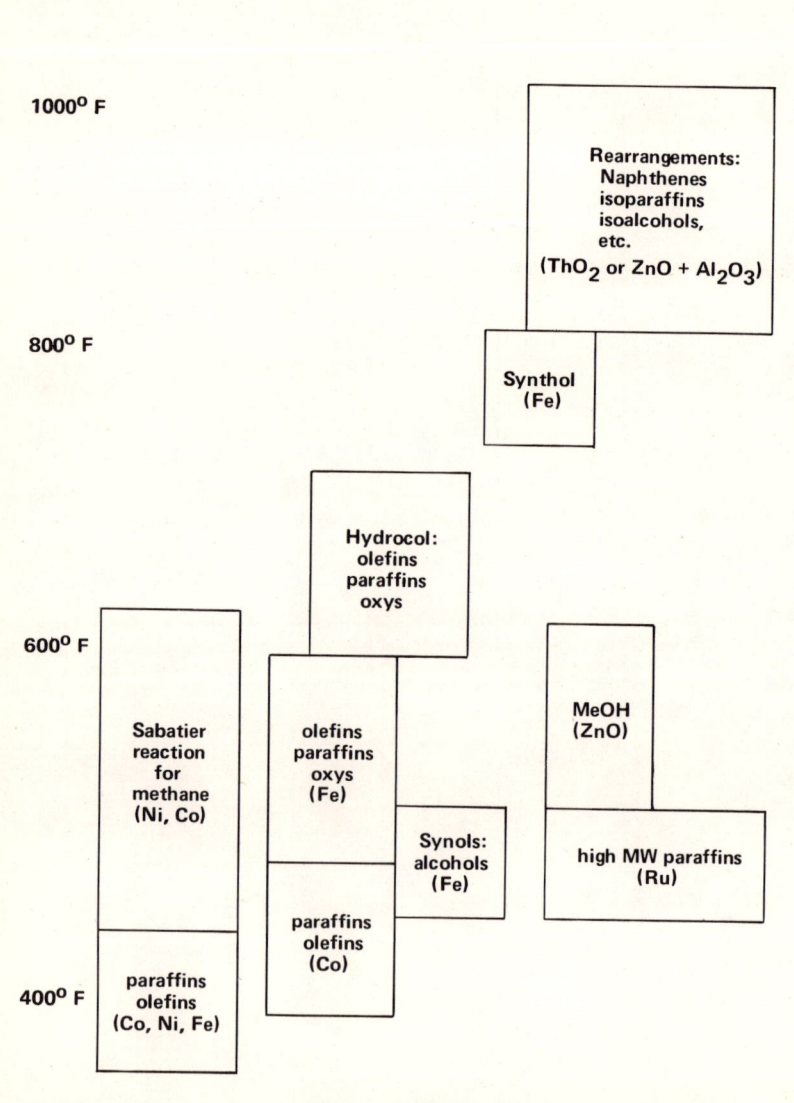

1000° F

Rearrangements:
Naphthenes
isoparaffins
isoalcohols,
etc.

(ThO_2 or $ZnO + Al_2O_3$)

800° F

Synthol
(Fe)

Hydrocol:
olefins
paraffins
oxys

600° F

Sabatier
reaction
for
methane
(Ni, Co)

olefins
paraffins
oxys
(Fe)

MeOH
(ZnO)

Synols:
alcohols
(Fe)

high MW paraffins
(Ru)

paraffins
olefins
(Co)

400° F

paraffins
olefins
(Co, Ni, Fe)

Fig. 7.1. Operating ranges for Fischer-Tropsch type reactions.

A comparison of conditions and products for various catalysts and processes for the Fischer-Tropsch reaction is shown in Table 7.1, based on Reference 9.

In general, the less active the catalyst, the more severe the reaction conditions necessary (1). Thus, from left to right and from bottom to top in the accompanying diagram, there is a change from Ni and Co catalyst at the less severe conditions, to Fe at the more severe. The other catalysts tend to be specific, though at atm. pressure Ru has given liquids (7), and at all pressures may yield 25 percent gases.

In summary, representative reaction conditions are also diagrammed in Figure 7.1, after Pichler (1).

7.2 CATALYST COMPOSITION AND PROPERTIES

As mentioned, the Fischer-Tropsch type of reaction has been intimately involved with the development of the study of catalysts, and is itself almost wholly dependent upon catalytic activity for the reactions to occur. By Fischer-Tropsch type of reactions, we mean the reaction of CO and H_2 to form hydrocarbons (and oxygenated compounds) of more than one C atom (27) — though methanation may be considered a special case.

The Fischer-Tropsch catalysts have as their basic constituents the metals of Group VIII of the periodic table (8). These transition elements in their order of increasing atomic number are Fe, Co, Ni, Ru, Rh, Pd, Os, Ir, and Pt, and are found in almost all catalysts showing any trace of F-T catalytic activity (27).

These elements have been arranged in the order of activity based on the formation of methane (8). On this basis, in the order of decreasing activity, the arrangement is: Ru, Ir, Rh, Ni, Co, Os, Pt, Fe, Mo and Pd. This order does not necessarily hold for other reactions. Another authority considers Pt, Pd and Ir not to be very active, but Rh and Os to be fair (5). However, all catalyst mixtures of any importance which have been used may be classed into the following four groups, based upon the base metal (8): nickel catalysts, cobalt catalysts, iron catalysts, and ruthenium catalysts.

Nickel and cobalt to a lesser extent have greater hydrogenating action than Fe catalysts, and promote the formation of methane (1). Thus at 750°F and atmospheric conditions, methane is formed according to Reference 1. However, at lower temperatures, 400°F, nickel catalysts are suitable for forming higher hydrocarbons. The higher hydrogenating power orients Ni, Co, and Ru catalysts toward hydrocarbons, while less active Fe catalysts can be used to produce oxygenated compounds as well.

Characteristics of this group of elements and their use are (8, 27):

1) All are transition elements; i.e., the "d" shell is incomplete.
2) The basic element is used in the form of various oxides.
3) The oxides are easily reduced by H_2 and/or CO at temperatures in the range of the F-T synthesis reactions.

The catalysts are considered easily susceptible to sulfur poisoning (8). In comment this is also a matter of temperature, however. (See Chapter 8.)

Further characteristics assigned to these catalysts are (9):

1) Only metals that form carbides are active in synthesis.
2) These catalysts are active at temperatures where carbides are stable.
3) The rates of carbide formation and synthesis are somewhat alike. Ruthenium is listed as an exception.

The catalysts are commonly of a mixed type containing promotors and other additives (8). The catalysts may be precipitated from solution onto supports, prepared from alloys or prepared from oxidized metal particles, and sintered, cemented, or made up in various other ways (3, 5, 8, 9, 16). Ceramic materials successfully perform as supports; they are also resistant to abrasion in fluidized beds.

Catalysts of three or more active components evidently possess no advantage (16). Alkalies are sometimes added to Fischer-Tropsch catalysts, however, notably to iron and nickel. With iron, for instance, the oxide and alkali carbonate are heated to sintering temperatures, and ferrites may form. Alkali is a component of nickel catalysts used for the naphtha-steam conversion to produce methane. For more on the conversion of petroleum-derived liquids to methane, consult "Substitute Natural Gas from Hydrocarbon Liquids," SNG Symposium I, Institute of Gas Technology, Chicago, March, 1973.

Catalyst constituents are divided into phases of four classes (27):

1) Active phases at which the reaction is known to proceed at a reasonably fast rate: α-iron, α-cobalt, nickel, platinum, magnetite, ϵ-iron nitride, cementite, Hägg iron carbide, and others.
2) Inactive phases at which the reaction proceeds at lower rates: Examples are cobalt carbide and nickel carbide.

3) Phases from which active phases are formed by reduction with H_2, CO, etc. Such phases may be active themselves. Examples are α-$Fe_2O_3.H_2O$, β-$Fe_2O_3.H_2O$, α-Fe_2O_3, Fe_3O_4, FeO, NiO, etc.

4) Phases which may be formed due to reaction, though not initially present. Examples might be ϵ-iron carbonitride, ϵ-iron carbide, Hägg carbide, cementite, magnetite, siderite, etc. These phases may be active or inactive.

Active catalysts are ferromagnetic; inactive, paramagnetic (16). Magnetic susceptibility increases with activity (8, 16). An uncompleted "d" shell is necessary but not sufficient condition for ferromagnetism (27). Thus α-Fe_4O_3, hematite, though of this classification, is anti-ferromagnetic. On the other hand, maghemite (magnetic ferric oxide). γ-Fe_2O_3. is magnetic. Iron, cobalt, and nickel are particularly prone to develop ferromagnetic properties both as elements and as compounds. This, then, is a potential test for activity.

As a yardstick for comparison, ferromagnetic substances, which are highly magnetic, will have magnetic susceptibilities of $\sim 10^{+6}$. The examples are Fe, Ni, and Co. Paramagnetic substances, which are only weakly magnetic, may be divided into weakly paramagnetic with magnetic susceptibilities of 100×10^{-6} to 1000×10^{-6}. Examples of paramagnetic substances are Pd, Pt, and the alkali metals. Diamagnetic substances, which are anti-magnetic, may have negative values of magnetic susceptibilities of the order of -10×10^{-6}. Examples of diamagnetic substances are Bi, Sb, Zn, P, H_2O and H_2. The distinction is not always clear-cut.

Active iron catalysts contain iron carbides (Fe_2C), α-Fe_2O_3, α-Fe, sometimes γ-Fe_2O_3, and Fe_3O_4 (8). It is distinguished that α-Fe_2O_3, or α-$Fe_2O_3.H_2O$ is active, while β-$Fe_2O_3.H_2O$ is inactive (16). Both are paramagnetic (27). Neither α or β structures are stable at synthesis conditions (16).

The degree of incompleteness of the "d" shell is an indication of catalyst activity (27). If the "d" shell is completed by the donation of electrons, then the material is no longer ferromagnetic. Fe needs three electrons for completion of the "d" shell.

The iron system has a greater number of catalytically active phases than either Ni or Co. These include

    α-Fe
  Carbides: ϵ-Fe_2C
      χ-Fe_2C (Hägg carbide)
      Fe_3C
      FeC

Nitrides:	ϵ-Fe$_2$N
	γ-Fe$_4$N
Carbonitrides:	ϵ-Fe$_2$X
	γ-Fe$_4$X
Magnetite:	Fe$_3$O$_4$

The carbonitrides are formed by the reaction of CO with nitrides. The ratios of carbon to iron and nitrogen to iron will vary.

Interstitial compounds form important Fischer-Tropsch catalysts. These are nonionic, intermetallic compounds having electrical conductivities similar to metals. All are formed with transition metals, in which it is believed that the metalloids donate electrons to complete the "d" shell. These compounds are "interstitial" in that the metalloid element — carbon, nitrogen, hydrogen, or (boron) — is located in the interstices of the metal lattice.

Interstitial compounds may be stable or unstable. The stable compounds tend to be refractory, hard, acid resistant, with two crystal structures. The unstable are relatively soft, decompose at ~ 1850°F, form hydrides with acids, and have many crystal structures. The unstability is assumed to be significantly related to catalytic activity.

The carbides, nitrides, (borides), hydrides and oxides of the transition elements do not obey Dalton's law of constant composition. Thus ϵ-Fe$_2$N may have a composition ranging from Fe$_3$N to Fe$_2$N. The iron oxides are well known for this behavior.

Cobalt catalysts do not noticeably react during synthesis, but iron catalysts do (1, 3). An active iron catalyst reacts with CO to form an accumulation of carbides (1). Specifically, an iron catalyst carburized at 615°F was transformed to a ferromagnetic higher iron carbide of formula Fe$_2$C. A reduced iron catalyst was found to contain Fe$_2$C. FeC has also been identified in catalysts. This conversion of iron to carbide is acclaimed to form an active and durable catalyst (6, 25). Fischer-Tropsch reaction conditions are also close enough to those for the formation of carbonyls.

Carbides and oxides of iron exist in sizable amounts in used iron catalysts (3, 9). It is considered that carbiding increases the catalyst life and activity for iron catalysts (25) but not for Co catalysts. In fact, carbiding reduces the activity of Co catalysts (7). Nitriding tends to increase the yield of oxygenated compounds.

Iron catalysts have been observed to be composed of a mixture of compounds which changes during synthesis. No large change of activity is noticed with variation of composition. For instance, two catalysts of comparable activity were compared (5): one was predominately magnetite, the other Hägg carbide. In another instance, a sequence of catalyst changes observed during a reaction was: reduced catalyst → Hägg carbide → magnetite → FeC. All existed simultaneously in varying proportions, with the changes caused by simultaneous carburization and oxidation.

An analysis of an iron Fischer-Tropsch catalyst showed a constant nonmagnetic iron composition of 30 percent with use (9). The α-Fe change from 70 percent down to about 10 percent. At the same time the Hägg carbide (Fe_2C) went from nil to 25 percent, and the Fe_3O_4 content also went from nil up to 40 percent.

After synthesis, a precipitated iron catalyst was shown to be predominately magnetite (2).

In a fixed bed, the fraction of reduced iron has been observed to increase, then decrease through the bed (5).

In another instance, the proportions of Hägg carbide and Fe_3O_4 varied from 29 to 12 percent, respectively, to 21 percent and 38 percent (8). The α-iron present decreased from 69 to 8 percent. It was stated the activity depends upon the percentage of iron present as Fe_2C, Fe_3O_4 and nonmagnetic iron. The higher the Fe_2C and the lower the Fe_3O_4, the more active the catalyst. Fe_2C has itself been characterized as a highly active and durable catalyst (5, 6). The carbides of Ni and Co are relatively inactive, however, compared to the reduced catalyst (5).

The presence of carbides at the catalyst has been linked off and on to the Fischer-Tropsch synthesis reactions. It was first advanced by Fischer and co-workers as entering into the reaction (7), and has had its ups and downs theory-wise (1). The carbide is regarded as reacting with hydrogen to deliver a methylene group which polymerizes and reacts to form hydrocarbons. Reaction with CO leads to oxygenated compounds. Complicating matters, cobalt carbide reacts with H_2 only to yield CH_4; iron carbides react not at all. Cobalt forms carbides above 140°F, Fe above 375°F. Carbides of Ru are unknown at synthesis conditions.

The disproportionately low yield of C_2 in synthesis products indicates that ethylene is a building block (3, 7, 9). Ethylene enters into reaction with CO and H_2 over Co catalyst, but not over Fe catalyst (1, 2, 7). Nonetheless, ethylene will not react alone over the same catalyst, nor will it react with CO and H_2 without the catalyst.

Methane will react directly with Fe_2O_3 at 2200°F and be oxidized (7).

We have previously examined instances where catalyst activity was attributed in part to the presence of carbides. There are other instances where catalyst composition and activity are related to iron and its oxides, and where the activity of iron catalysts is considered to depend upon the Fe_3O_4, Fe_2O_3, FeO and Fe content (8).

There is evidently an optimum ratio between the metal and oxide (8). In one instance it was observed that the activity of a catalyst (presumably based on the oxide) was increased with reduction until 40 percent was reduced (25). The oxide form is regarded as being less active than the metal and favors alcohols or oxygenated products (1). The presence of metallic iron causes a reduction in product molecular weight (5), which is an indication of increased activity.

A reduced and activated iron catalyst was observed to contain Fe_2O_3, Fe_3O_4, FeO and Fe (16). Only about 8 percent was Fe.

X-ray studies have indicated that all active iron-copper catalysts contain maghemite, the magnetic iron oxide γ-Fe_2O_3, and that the presence of Fe_3O_4 lowers activity (7). All fused iron oxide catalysts revert to magnetite at the fusion temperature in the presence of O_2 (5). However, reduction occurs if H_2 activation is used.

Besides appearing as oxides, catalysts are precipitated or otherwise prepared as the carbonate. Components include oxides, oxide-hydrates, carbonates or oxide carbonates (16). Catalysts are prepared by impregnation from solutions of the nitrate or sulfate. The nitrate or sulfate may also be ignited.

Evidence of two or more oxidation levels in iron catalysts is repeatedly set forth. A ferrous-ferric mixture was classed as a very good precipitated catalyst (1). A catalyst prepared from either ferrous or ferric chloride was inactive, but a catalyst prepared from a mixture of the two worked very well (5, 9). Ferric salts are prepared so that some ferrous salt is present (9). A catalyst prepared from Fe^{++} was not active; however, the addition of either Cu or Fe^{+++} activated the catalyst (8).

In conclusion, it may be remarked that catalyst composition is characterized by heterogeneity. This may be rendered by the addition of a promoter or carrier, by a difference in oxidation level, possibly by interstitial differences, or even by differences in the electron distribution within a given crystal structure.

7.3 CATALYST PREPARATION AND ACTIVITY

Various promoters and carriers are used for the catalyst and may play a part in the reaction. Alkalies or alkali carbonates are effective promoters (5, 8, 9, 16). Carriers or supports used include kiesulguhr or diatomaceous earth, silica gel, alumina, silica, dolomite, etc. (5, 8, 9).

Catalysts may also be prepared by impregnation from solution (5), by using binding agents (3, 9), and by sintering or fusion (3, 9, 16). Catalysts are sprayed on surfaces (e.g., Raney nickel) and are sometimes produced by spray drying (in the micron particle size range). The process or preparation may affect the interstitial pores and hence have an effect on the activity.

The carrier itself has an effect on catalyst activity (1). The effects are both structural and chemical (2). Alumina, for instance, lowers the amount of oxygenated compounds produced (1). Kieselguhr (diatomaceous earth) from different origins affected the activity of a cobalt catalyst. Even catalysts of quite different compositions can have the same activity (9).

While Co catalysts are prepared only by precipitation on a carrier, iron catalysts can be prepared in a number of ways, as previously enumerated. Ni catalysts have been sprayed on surfaces as Raney-Ni (28).

Alkali-derived promoters are the only significant promoters for iron catalysts (2). They serve to increase chain growth and secondary reactions such as the water-gas shift (1, 2). An alkali alone will have no effect (2). It should be noted, however, that the effect of alkali is considered controversial (3). Some say it helps, some say not. No promoter has been found for Ru catalysts (1). Alkali promoters are also used for NH_3 synthesis (5), and have an effect *per se* upon the carbon-steam reaction as previously noted.

The presence of Na_2CO_3 has been observed not to increase activity but to favor oil formation (8, 9). K_2CO_3 has a similar effect (8). Alkali was observed not to increase activity but to favor paraffins and oxygenated products (16) and to increase molecular weight (5).

On the other hand, alkali carbonates have been reported to enhance activity (7). While some have declined to explain the effect, others consider it to stabilize cubic Fe_2O_3 and to prevent transition to Fe_3O_4 which is less active.

The effectiveness of alkali promoters decreases with increased pressure (5). Alkalies affect the type of compound produced (1), favor higher molecular weight compounds (1, 8), and favor liquids over gases. The effect of K_2O increases with the molecular weight of the catalyst (5).

$CaCO_3$ and $BaCO_3$ are relatively poor promoters (8).

Fe-Cu catalysts have been successfully used (7, 8). Copper is itself not a catalyst, but serves as a promoter for iron catalysts (8). The copper increases the yield of gasoline (9). Too high a proportion of copper, however, has a deleterious effect. Copper is also considered to give better catalyst reduction and activation for iron catalysts (1, 8). It is unsatisfactory for Ni, forming an alloy, and helps neither Ni or Co. Cu, as well as alkalies, is ascribed as preventing the formation of elemental carbon and aiding in the formation of carbides.

Ni and Co catalysts, in general, have greater surface area than an Fe catalyst (9). It is observed that fused iron catalysts, while mechanically strong, have low pore volume and surface area (3). Cemented iron catalysts, on the other hand, have a larger pore volume and area. The smaller the particle size (and the larger the surface area), the more active the catalyst. A fused iron catalyst was observed to increase activity with decreasing particle size until 28 mesh was reached (25). In another instance activity was observed to increase as particle size decreased until 40 mesh was reached (9). On the other hand, a granular catalyst is reported to have produced less C_1-C_4 (and less CO_2) than a pelleted catalyst (16).

Grain or crystal size itself is noted to have an effect on activity.

The appearance of pure iron compounds is radically altered in mixtures. For instance, pure ferrous hydroxide is white but a small amount of oxidation converts it to green (27) — an intermediate in the oxidation of steel. Ferrous oxide is black as is Fe_3O_3. Ferric hydroxide is reddish-brown while Fe_2O_3 may be red to black depending on crystalline structure. The hydrate geothite may vary from yellow through red to black. Appearances noted for Fe catalysts are that catalysts precipitated from ferric salts were brown, and from ferrous salts were black (16). As a general statement precipitated iron catalysts are either brown or black (3). Precipitated ferric catalysts are also characterized as being hard and glassy, while ferrous catalysts are of an earthy consistency (1).

Pure reduced magnetite is malleable (3), but a reduced, promoted catalyst is brittle.

It has been noted that catalysts prepared by precipitation of ferric chloride are inactive (16), but those from ferric nitrate are active. The inactivity may be due to chloride ion. Sodium ion possibly may also render a catalyst inactive. On the other hand, sodium compounds are used as promoters, sometimes favoring liquid production and sometimes gas (9). The effect of alkalies has been previously discussed. It is further discussed in Chapter 8.

However, it has also been observed that a pure ferrous salt is no good as a catalyst, and that ferric salts are prepared as a catalyst so that some ferrous salt is present. Furthermore, when a mixture of ferric and ferrous catalyst was prepared from the chlorides, it worked very well (5, 9), though either alone were ineffective. The point is regarded by some as questionable (3).

The catalysts are usually regarded as easily poisoned by sulfur and certain aromatics (1, 5, 7, 8). There are exceptions, however, that make this statement controversial (7, 16, 25). Thus the addition of H_2S has actually been shown to increase yields up to a point, particularly liquid yields. The presence of sulfur is also reputed to retard C_1 formation, but to aid the CO-shift (12).

Tolerances have been set up for the sulfur concentration of the synthesis gas. Representative figures are from 0.1 to 0.2 grams sulfur per 100 cu. m. of gas (1, 8).

However, consider the following remarks concerning the addition of sulfur compounds (7). Catalysts are much more tolerant of organic sulfur than of H_2S. Small additions of H_2S tended to increase the activity of a Ni-Mn catalyst and of a Co-Th-kieselguhr catalyst. The first addition of sulfur caused a marked increase of liquid hydrocarbons and a falling off of gaseous hydrocarbons. The yield increased till 8 milligrams/gm of catalyst had been added, then fell off. Even when 33.5 milligrams/gm had been added, the yield of liquids was still 20 percent greater. Further addition caused a decline finally to complete poisoning.

This increase in liquid hydrocarbons caused by sulfur is noted elsewhere (16), and the upper poisoning limit is higher than thought (25).

Direct reactions can occur between the sulfur compounds and catalyst. An increase in steam concentration decreases the conversion of Ni to inactive NiS in the presence of sulfur compounds (1). A decrease in activity, fortunately, can be rectified by raising the reaction temperature. Catalyst beds, however, should operate at the lowest temperatures consistent with catalyst activity (12).

Molybdenum is a possibility as an additine to improve the sulfur tolerance to nickel catalysts. Moreover, scavengers can be added to preferentially react with sulfur. An interesting alternative is the potential of tungsten sulfide to serve as a methanation catalyst. The problem of sulfur merits continued investigation.

Various ways exist for removing H_2S and organic sulfur from the synthesis gas. A comprehensive list is given in the cited Reference 11 and elsewhere. Of special mention are hot carbonate solution (25), amine solution (8), charcoal (5, 8), iron oxides (6, 7). Some of these methods also apply to the removal of CO_2 from the product gas. For further details consult Chapter 10.

Organic sulfur can be removed by catalytic conversion to H_2S followed by absorption (7). Hydrorefining is a version of this route (8). A MoS_2 or Cr_2O_3 catalyst at 680-790°F and 300 psi will react the sulfur with H_2. A highly alkalized iron catalyst also converts organic sulfur (1).

Iron oxide at 540°F will remove organic sulfur (6).

Not all catalysts are affected by sulfur. For instance ThO_2 is affected by carbon deposition but not by sulfur (1). Tungsten sulfide is resistant to sulfur.

Catalyst particles are rendered inactive by the deposition of carbon and of solid hydrocarbons from the reaction (1, 9). The formation of carbon is aggravated by higher temperatures (5), making heat control a serious problem (1). Carbon formation can be lowered by using higher H_2 concentrations in the synthesis gas. Regeneration by reduction or solvent washing will re-activate the catalyst under certain circumstances. It has been mentioned that, for a fluidized bed, carbon deposits up to 50 percent are not harmful (8). Also, increasing the space velocity decreases carbon formation (9). High hydrogen content lessens carbon formation (2). Carbon will not form for synthesis gas H_2/CO ratios greater than 3/1 (12).

Sintering is another cause of loss of catalyst activity (1). It can be reduced by the addition of high melting oxides. Crystal growth is another factor, however, and may reduce activity.

During synthesis, iron catalysts are oxidized by water vapor and reduced by CO and H_2 (5). Iron carbides and iron (and iron oxides) are regarded as oxidized by the water present in synthesis gas (4). The final universal product tends to be magnetite (Fe_3O_4). The presence of CO inhibits oxidation and increases catalyst life (8).

It is also observed that the presence of water vapor or steam affects the catalyst activity. (This would be the effect of a reaction product in part.) Thus, the activity of a Co catalyst was reduced by the presence of steam (1). However, Co is not affected nearly as much as Fe catalyst.

If the partial pressures of steam and CO_2 exceed a certain level during synthesis, depending upon catalyst and conditions, catalyst oxidation begins with a loss of activity (1). Thus, the less the concentration of water produced, the lower the rate of formation of metal oxides (3). Accordingly, higher space velocities are in order.

An iron catalyst once completely oxidized cannot be reactivated, possibly due to the formation of nonreducible iron oxides (8).

Pressure tends to make an iron catalyst more durable (9, 18). Pressure also tends to make a catalyst stay active longer (1, 9, 16). As an instance, a catalyst would have a useful life of a few weeks at one atm, but several months at 75 to 100 psi. Another figure cited for total catalyst life was 8 months. An optimum pressure cited was 150-450 psi. At atmospheric pressure, Fe catalysts are more durable than Co, and with repeated hydrogen activation may last 6-12 months (6). The SASOL operation uses iron catalysts for 45-60 days.

In general, the effect of pressure on catalyst life appears to be controversial, for others have found that as pressure increased, catalyst life decreased (7). Tests made with a precipitated iron oxide catalyst at ~ 450°F showed no reaction at one atm but near 90 percent conversion at 150 psi (9). Moreover, the activity held up at 150 psi whereas it decreased at 450 and 900 psi.

The method of pretreatment or activation of the catalyst can cause drastic changes in the nature and amount of product (1, 5). It is preferable to reduce a catalyst as rapidly and at as low a temperature as possible to avoid sintering. Low-temperature treatment also limits the crystal size (1). The use of CO increases the surface area by producing carbon which breaks up the structure. The presence of CO_2 or CO, however, reduces the rate of catalyst reduction, while H_2 pressure increases the rate (3).

Catalysts may be activated by reducing with H_2 at 680°F (16). Other temperatures mentioned are 660-930°F (1), and 700-775°F (9). A routine pretreatment recommends water-gas at 500°F (3). Pretreatment with CO and H_2 has been observed to give Hägg carbide, magnetite and carbon (5). Iron catalysts, carburized at 615°F, were changed to a ferromagnetic iron carbide with the formula Fe_2C (1). Precipitated catalyst can be treated with synthesis gas; fused catalyst must be treated with H_2 (5). Treatment with water-gas is less effective.

The presence of H_2O or CO_2 inhibits catalyst reduction (1, 5). For this reason it is best to keep the partial pressure of H_2O very low during catalyst reduction (1, 8). Hence temperatures as high as 900-1550° F have been recommended for reduction (7, 8), presumably to counteract moisture effects.

Reduction times recommended are on the order of 50 hours (6).

For maximum activity it has been stated that Ni catalysts should be completely reduced, and Co 60-70 percent reduced, while the activity of Fe catalysts depends upon the Fe_3O_4, Fe_2O_3, and Fe content (8). The activity of a catalyst was observed to increase with reduction until 40 percent reduction had occurred (25).

Iron oxide in its raw, unreduced state exists mostly as an amorphous gel-like material (27). The composition is related to hematite (α-Fe_2O_3), magnetite (Fe_3O_4), or maghemite (α-Fe_2O_3, magnetic ferric oxide). The ferrous form wustite (FeO), is unstable and rarely found. Geothite (α-$Fe_2O_3.H_2O$) is found, however.

Hydrous oxides have the gel-like appearance associated with raw iron oxide. They revert to the oxide on losing water: thus $Fe(OH)_3 \rightarrow \alpha$-

Fe_2O_3. A hydrous oxide is the oxide in a finely divided state holding absorbed water as contrasted with a hydrate which has a definite crystalline structure.

On the reduction with H_2 at 480-930°F, the hydrous oxides lose absorbed water and the hydrates also dehydrate. The hematite is reduced to magnetite at an increasingly slower rate. Still further reduction will produce metallic α-Fe. If the H_2 is saturated with water, the reaction will stop with magnetite.

For NH_3 synthesis catalyst, Fe_3O_4 is regarded the stable oxide, and hydrogen-steam mixtures are used for activation by oxidation using not less than 16 percent steam (1). Reforming catalysts are also activated by steam oxidation (see Chapter 8). The method of activation (reduction or oxidation) depends upon the course of reaction desired.

The treatment of iron with CO will yield free carbon as well as carbides above 615°F, but only carbides below 440° F (27). Magnetite is also a product. The oxides of iron will also yield these products with CO. Hydrocarbons over a reduced iron catalyst produce carbides. Additionally, CO reacts with iron to form CO_2, carbon, and carbides.

Iron catalysts may be made from readily available iron ores and from industrial by-products (3). Catalyst materials which have been tried and found effective include the following:

1. Nitrided steel wool (25).
2. Steel turnings, oxidized with steam, alkali added, and reduced with H_2 (25).
3. Iron ore, alkalized with, K_2O, alumina added (1, 5).
4. Roasted pyrites, the S removed by oxidation (5).
5. Pretreated iron ores (1).
6. Fused iron catalyst from NH_3 synthesis (1).
7. Mill scale, K_2O added (5, 9).
8. Iron alloys, extracted with NaOH (8).
9. Steel shot and lathe turnings (5).
10. Magnetite, K_2O added and fused (9).
11. Hematite, K_2CO_3 added and sintered (9).
12. Luxmasse, a technical grade iron oxide, appearing as an alkaline by-product of Al industry (9, 29).
13. Lautamasse, iron oxide residue from alumnina recovery from bauxite (9, 30).
14. Siderite, $FeCO_3$ (3).

Reactions of Carbon Monoxide and Hydrogen

A prepared list of catalysts from commercially available sources is as follows (3):

Active:
1. Alkalized geothite ore (α-$Fe_2O_3.H_2O$).
2. Alan Wood magnetite, cemented with aluminum, and alkalized with K_2O.
3. Pigment grade iron oxide bonded with alumina and alkalized.
4. Mill scale bonded and alkalized.

Less Active:
1. Magnetite.
2. St. Peter's sandstone (with a small percent of Fe).
3. Limonite (a mixture of geothite and hematite) with high S content.
4. Alkalized wrought iron.
5. Common ferrosilicon (43 percent silica).

A fluidized catalyst employed consisted of 97 percent Fe_2O_4, 2.5 percent Al_2O_3 and 0.5 percent K_2O (6). The fused catalyst is ground and reduced with H_2.

Another catalyst was prepared by burning pure iron in oxygen to form a molten mass of oxide. Aluminum nitrate and potassium nitrate were then added to the melt and the fused mixture crushed to one to three mm. The resulting composition was 97 percent Fe_3O_4, 2.5 percent Al_2O_3, 0.2 to 0.6 percent K_2O, 0.16 percent S and 0.03 percent C. The catalyst was activated by reducing for 50 hours with pure dry H_2 at 840°F. 2000 volumes H_2/hour were passed through one volume of catalyst at a linear velocity of 20 m/sec. The reduced catalyst was transferred to the reactor under a CO_2 blanket.

7.4 EFFECT OF OPERATING VARIABLES

The operating variables temperature, pressure, and the feed rate and reactor size all affect the course of the synthesis reaction. The latter two are usually incorporated as residence time or space time, which is reported as

$$\frac{\text{Vol reactor}}{\text{Vol feed/time}} = \text{residence time}$$

or space velocity, which is

$$\frac{\text{Vol feed/time}}{\text{Vol reactor}} = \text{time}^{-1}$$

Thus the one is the reciprocal of the other.

Additionally, the course of reaction is affected by the composition of the feed, which may be reported as the synthesis gas ratio: the moles of hydrogen present divided by the moles of CO: H_2/CO. The presence of inerts, recycle or reaction products also affects the reaction, as do such matters as the type of reactor. Catalysts have already been discussed, and affect not only equilibrium but the direction of reaction (see Chapter 8).

All of the individual effects act in parallel and must be taken into account simultaneously to predict the overall course. Statements concerning the action of the operating variables are categorized under the appropriate headings.

TEMPERATURE

Lower temperatures favor hydrocarbons (1, 8). Decreasing the temperature favors C_1 homologs of higher molecular weight (1). Below 350°F, reaction is meager (16). However, 350-410°F is recommended as the optimum for a Co catalyst (1).

Lower operating temperatures increase alcohol production but decrease olefins (5, 9). Lower temperatures and higher space velocities are used to produce alcohols instead of hydrocarbons (9). In general, it is reported lower temperatures favor oxygenated compounds (5).

Increasing the temperature favors methane (8) and other light hydrocarbons (7). Above 570°F, the C_1 yield becomes increasingly predominate (7, 16). More isomers are produced at increased temperatures (8). In general, increasing the temperature increases the yield, but decreases the molecular weight (5). Increasing the temperature increases the space-time yield (9).

The RMProcess (Ralph M. Parsons Co.) methanates CO and H_2 over nickel at up to 1400°F and is tolerant to sulfur and other impurities. (Beyond this, excessive surface temperatures may destroy activity.)

Although increasing the temperature affects the CO-shift, no variation in CO_2 content was observed (9). Methane yields are highest at lower temperatures, but require higher H_2/CO ratios (12) — a possible contradiction to the above.

The production of methanol is favored by lower temperatures (and higher pressure) (17). The optimum temperature depends upon space velocity but is in the range 660 to 750°F.

PRESSURE

Lower pressures yield more hydrocarbons (8), but require more catalyst for the same conversion (7, 16). The overall yield is increased by lowering the pressure (6). This last mentioned effect is contradictory since reaction velocity is said to be reduced with decreasing pressure (16), and space time yield is also reduced (9).

There is also the classic effect of the change in volume of the system and how affected by pressure. Other factors may be overriding, however.

Increasing the pressure is said to increase conversion (1, 3). Increasing the pressure increases the boiling range of the products (16). Increasing the pressure increases the H_2/CO usage ratio, though more low molecular weight compounds are produced at low pressures(8). More isomers are produced by increasing the pressure (8).

Increasing the pressure above ~ 200 psi increases the yield of oxygenated compounds (5, 7, 8, 9, 16) but decreases the olefins (1, 9). The effect on olefins, however, may be slight (5). It increases the yield of high molecular weight, solid hydrocarbons, but decreases the yield of liquid hydrocarbons (1, 7, 9, 16). This increase in average molecular weight has been found to occur up to ~ 400 psi (5). There is a possible contradiction here.

More C_1-C_4 hydrocarbons are produced under pressure (16). Pressure is said not to appreciably affect C_1 formation until temperatures of 800°F or higher are used (12). Above this an increase in pressure increases the C_1 yield.

In experiments with an iron catalyst, increasing the pressure did not affect the conversion, but increased the fraction of lower molecular weight compounds and decreased the higher molecular weight compounds (9). The percent olefins decreased but the percent alcohols increased.

With an iron catalyst, the average molecular weight increased with pressure to 280 psi, then decreased. The gasoline fraction decreased, while the high molecular weight materials increased in going from 0 to 280 psi. The percent olefins stayed relatively constant.

In yield studies, it was found that conversion increased up to 70 psi and then fell off (7). An intermediate pressure operating range is considered the most desirable.

It is said that nickel cannot be used successfully above atmospheric pressure due to the formation of volatile carbonyls (1). For the same reason the optimum upper range for Co is 75-300 psi, Fe is 150-450 psi, and Ru is 1500-7500 psi. This, however, depends on temperature — and reaction temperatures are generally beyond the decomposition temperature of the carbonyls.

The reaction rate is more dependent upon pressure for Fe catalysts than for Ni or Co catalysts (9). More explicitly, while with Fe catalysts the rate of reaction is dependent on pressure, with Co catalysts the reaction rate is independent of pressure in the range of 1-20 atm (300 psi) (3). In general, the rate of reaction increases with pressure.

REACTION TIME

Shorter residence times favor C_1 production (8). Higher space velocities or shorter residence times increase the percent olefins (7, 9). Shorter times have also been ascribed as lowering total yield and lowering the production of higher molecular weight compounds (6, 7) or increasing the production of low molecular weight compounds (9).

Higher space velocities at lower temperatures yield alcohols in preference to hydrocarbons (9). The space time yield increases with space velocity, but the specific yield decreases.

In experiments with an iron catalyst, increasing the space velocity did not change the total conversion but increased the production of low molecular weight compounds and decreased the production of higher molecular weight compounds. The percent olefins went up slightly, but the percent alcohols decreased sharply.

Longer residence times favor CO_2 production over H_2O, but favors hydrocracking (1, 2).

It has been ventured that there may be a maximum space time (residence time) for a given unit to operate (6). A longer converter (longer residence time) has been noted to give less C_1 (16). A longer residence time also favors CO_2 as a reaction product (1).

For an initial period at the start of a Fischer-Tropsch reaction, only C_1 or other light hydrocarbons are produced (7, 16). Presumably, this initial period is at a high activity due to using fresh, active catalyst which has greater hydrogenating power.

After this initial period, a catalyst declining in activity would be expected to yield more liquid products. One reference, however, cited more C_1 production with time (9). Another mentioned that increasing time and temperature caused lower molecular weight products (1).

It is mentioned that many times erroneous conclusions have been drawn because the process never reached steady state (9). The attainment and duration of a run is tied in with the durability and life of the catalyst. Some representative figures given are:

Ruhrchemie	150-200 days
Fluid-iron	20 days
Oil circulation	100-200 days

COMPOSITION

The methanation reaction, for example, as well as the other reactions, is one of reversibility. Other simultaneous and reverse reactions can occur depending on conditions (1):

$$CO + 2H_2 \rightleftharpoons CH_4 + CO_2$$

$$CO_2 + 4H_2 \rightleftharpoons CH_4 + 2H_2O$$

$$2CO \rightleftharpoons C + CO_2$$

Etc.

CH_4 and H_2O are considered the stable forms below 570°F; CO_2 and H_2 above 1470°F. The net effect is that the presence of both reactants and products influences the course of the reaction, depending upon conditions.

C_2 yield increases with the H_2/CO ratio (3). Thus increasing the H_2/CO ratio to 3/1 or 5/1 gives C_1 at 450-660°F (8).

Though lowering the temperature may decrease the C_1 (8), lower temperatures and higher H_2/CO ratios are recommended for C_1 production (12). Excess H_2 produces a saturated product and higher C_1 formation (7, 16). It also reduces carbon formation (2).

The rate of synthesis supposedly increases with H_2 content up to an H_2/CO ratio of 2/1 (3, 9).

Decreasing the H_2/CO ratio increases the product molecular weight (1) and reduces C_1 formation (8). As such, oils and waxes are favored (9). Lower H_2/CO ratios favor alkenes and will produce a gasoline fraction of higher octane (8). More CO_2 is also produced (7).

Decreasing the H_2/CO ratio not only increases the percent olefins produced (1, 9, 16) but also the percent alcohols or other oxygenated compounds (9). For instance, with a Co catalyst, results obtained are (1):

$H_2/CO = 2/1$	20 percent olefin
$H_2/CO = 1.2/1$	40 percent olefin
$H_2/CO = 1/2$	70 percent olefin

An optimum H_2/CO ratio cited for the Fischer-Tropsch reaction is 1.5/1 (8), though the ratios used have varied from 1/2 to 2/1 (14). Other figures mentioned are H_2/CO ratios of 2/1 to 1/1, and usually 1/1 (8). Above 3/1 the yields fall off. For hydrocarbon production an H_2/CO of 2/1 is cited (7, 9). For methanol production the best H_2/CO ratio is greater than 2/1 (14, 17).

The optimum yield is said to be expected when the H_2/CO usage ratio is equal to the synthesis gas ratio (9). Figures given are 1.185 at 200 psi and 1.2 at 1 atm.

H_2/CO ratios lower than 1/2 should not be used due to the formation of carbon by the reaction (1):

$$2CO \rightarrow CO_2 + C$$

Higher CO content increases catalyst life, however. The effect may be one of activation.

The effect of the H_2/CO ratio has been found to vary with catalyst (9). For a Co catalyst, increasing the H_2/CO ratio increases the yield of solid paraffins. But for an Fe catalyst the yield of light hydrocarbons increases.

Hydrocracking occurs at high H_2/CO ratios using Ni and Co catalysts (2).

The ratio of hydrogen reacted to CO reacted, H_2/CO, is termed the usage ratio. It may vary in value from 0.5 to 3, depending upon conditions and the H_2/CO ratio present in the synthesis gas (3). Usually the usage ratio is lower than the synthesis gas ratio, which means more CO reacts than H_2. However, optimum yields are said to occur when the usage ratio equals the synthesis gas ratio (9), and figures cited are 1.185 at 15 atm and 1.2 at 1 atm as mentioned.

The H_2/CO usage ratio increases as the H_2/CO synthesis gas ratio increases (9). Pressure may increase the usage ratio (5). The H_2/CO usage ratio decreases with space velocity to a minimum, then increases (9).

The removal of water vapor increases the usage ratio H_2/CO and the yield. Thus the presence of water vapor increases the CO-shift and reduces the synthesis reaction rate.

Dilution with inerts up to 15-20 percent has no effect on yield (6, 7, 8). In fact, in some instances, inerts have caused conversion to go up (9). Dilution decreased the product molecular weight. The presence of the reaction products CO_2 and CH_4 was reported to have no effect up to 20 percent concentration over a Co catalyst (7). Above this there was an effect, particularly for CO_2.

Excess H_2 favors C_1 production (1). As a further consequence, H_2O will tend to be produced rather than CO_2, even with an iron catalyst.

The presence of appreciable concentrations of CO retards the formation of CH_4.

High moisture content reduces the yield, though the CO-shift is favored (8, 9).

The presence of olefins retards the synthesis reaction (9). This may be merely the effect of a reaction product.

The use of recycle increases the tendency to form H_2O as a product rather than CO_2, for an iron catalyst (1).

The use of recycle ups the H_2/CO usage ratio (2, 9). Ordinarily this would lead to production of lower molecular weight materials. By lowering conversion at the same time, however, a lower C_1 yield was obtained. The yields of alcohols and olefins increased with recycle rate (9).

The uses and effects of recycle should be further distinguished, however: whether of main product or of byproduct.

7.5 EXPERIMENTAL STUDIES

The two principle types of reaction systems which have been studied for the Fischer-Tropsch process are the fixed-bed systems utilized principally in Germany and the fluidized-bed systems as developed in the United States. Other arrangements or variations of these systems have been tried, chiefly for the purpose of removing the exothermic heat of reaction. Each particular reaction system will require different reaction conditions, space velocities, etc., and will give different products.

Fluidized systems are considered advantageous over fixed beds (1, 8). The advantages include:

    No heat transfer problem

    Higher temperatures and cheaper catalyst can be used

    Catalyst can be regenerated

The fluidized bed is also more efficient. Yields may be 10-30 times that of a fixed bed (9). Installation is cheaper, with heat removal more readily accomplished.

A survey of various Fischer-Tropsch processes and reaction conditions indicates that Fe catalysts require higher reaction temperatures, as does recycle over no recycle, and fluidized systems over fixed-bed systems (9). As an illustration a fluidized system may react in the neighborhood of 570-600°F, whereas a fixed bed system may be as low as 400-440°F with no gas recycle.

The Synol process to produce alcohols in a fixed bed requires higher space velocities (and lower temperatures). Thus, values of 2500 hr^{-1} are used, at 365-435°F and 300 psi, with a 20-25 recycle ratio (6).

However, space velocities of only 100 hr^{-1} have been used in fixed beds for the Fischer-Tropsch process. Other figures cited include 25 and 200 hr^{-1} depending upon temperatures (1). Fluidized beds may operate at space velocities calculated at from 200 to 400 hr^{-1} and give much higher conversion rates (1, 9).

To better describe the sort of experimental conditions and arrangements which have been used, some of the representative information, data and results reviewed in the literature are set forth.

A representative sequence of operation for German plants employing the Fischer-Tropsch process was as follows (1):

A. Water gas manufacture.
B. Gas purification, including sulfur removal.
C. CO-shift to yield $H_2/CO = 2/1$, $Fe-Cr_2O_3$ catalyst.
D. Synthesis over Co catalyst at 1-7 atm. Catalyst between heat transfer surfaces, cooled by boiling circulating water.
E. Condensation of liquid products, recovery of gasoline and LPG.
F. Fractionation and refining of synthesis products.

A representative hydrocarbon product is as follows (6, 7):

LPG	8%	55% olefins
Gasoline to 300°F	46%	45% olefins
300-400°F	14%	25% olefins
Liquid 400°F+	22%	10% olefins
Wax	10%	

Other results from an iron catalyst are (6):

$C_1 + C_2$	17%
C_3	7%
C_4	7%
Olefinic gasoline	44%
Diesel	11%
Wax	1%
Alcohols	7%

At atmospheric pressure, Fischer-Tropsch reactors can use beds of partitioned sheets of catalyst separated by coolant; at higher pressures, the catalyst is placed in tubes (6).

Reactions of Carbon Monoxide and Hydrogen

If a two-stage reactor is contemplated, it is preferable to remove liquid products intermediately (1). There is an advantage to using the older or less active catalyst in the first stage and the fresher or more active catalyst in the second stage.

In addition to the removal of heat indirectly by vaporizing a liquid medium, the heat of reaction may be removed by direct liquid-vapor contact (7). Thus the heat removal problem for the synthesis reaction has been approached by the use of an oil recycle process (1, 9). The heat of reaction is absorbed by the oil. The system may be used in a fixed catalyst bed, or the catalyst may be suspended in the oil as a slurry. Synthetic diesel oil can be used. Conditions are 330 psi and 480°F.

(The latter system is somewhat analogous to the H-Oil or H-Coal hydrogenation process in which the oil itself is also hydrogenated in the presence of a suspended catalyst. The bed is described as ebullating, and is mentioned elsewhere in this study.)

Difficulties have been encountered with catalyst erosion and catalyst particles cementing together for a fixed bed catalyst. The moving bed is apparently much more satisfactory.

Using oil-catalyst suspensions of 12 percent catalyst, 70 percent conversion was obtained at 390-570°F (5, 9). End-gas recycle was used. Space velocities were 300 volumes of catalyst per hour per volume of slurry. The product distribution was:

CO_2	45%
$C_1 + C_2$	6%
$C_3 + C_4$	7%
C_5	37%
Water solubles	5%

The evaporation of an oil phase in the synthesis gas has also been tried as a means of removing heat (5, 6).

Another system tried for heat removal by direct heat absorption is hot gas recycle (9). The fresh inlet synthesis gas is diluted with recycle gas at recycle ratios of as much as 100/1. Conditions used were 600°F at 300 psi with an iron catalyst. The conversion per cycle is calculated as around ~ 1 percent, but the overall net conversion obtained was around 80 percent.

In methanation experiments over Raney-nickel catalyst sprayed on plates, 95 percent conversion was obtained at 460-545°F and 570-650°F at 300-450 psi (28). The process was conducted with hot gas recycle at ratios of 8/1 to 26/1 were used. The inlet synthesis gas ratios were stoichiometric at $H_2/CO = 3/1$. At the lower temperature the product was

    22 percent H_2
    65 percent C_1
    4.6 percent C_2
    2.6 percent C_3

At the higher temperature the product was

    7.5 percent H_2
    87 percent C_1

The synthesis reaction products, upon condensation, yield a water phase as a part of the product. A major portin of the oxygenated compounds, being water soluble, will appear in this phase (8). And oily layer will also appear with the water layer upon condensation. This will contain hydrocarbons as well as some distributed oxygenated compounds.

An iron catalyst in a fluidized reactor is noted as producing large amounts of oxygenated compounds at conditions of 570°F and 300 psi or higher (9). In fact, 25 percent at the product exclusive of H_2O, CO_2, and gases may be oxygenated compounds. Of these, 75 percent were distributed in the water layer, and 25 percent in the oil layer, as follows:

	Water layer (75% of total)	Oil layer (25% of total)
Alcohols	52%	33%
Aldehydes & Ketones	21%	34%
Acids	27%	33%

In other experiments, 72 percent of the product was composed of hydrocarbons and 28 percent oxygenated compounds, exclusive of CO_2 and H_2O (5). Of the 28 percent that was composed of oxygenated compounds, there was distribution between the hydrocarbon and water phases such that 11 percent was in the hydrocarbon phase and 17 percent in the water phase. The total oxygenated compounds were broken down as:

    50 percent alcohols
    25 percent acids
    8 percent aldehydes
    2 percent esters

Over a nitrided iron catalyst the proportion of oxygenated products was found to be even higher, being 40 percent of the C_3+ product (25). Of this, the breakdown was:

    85 percent alcohols

    3-8 percent aldehydes & ketones

    1-6 percent esters

    1 percent acids

Most pilot-plant work on fluidized systems using iron catalyst has been done in reactors of 2-8 inch diameter and 10-24 feet high (9). In particular instances, temperature control has been by internal heat transfer to keep a 45°F gradient. Recycle ratios of 2/1 have been practiced. The catalyst ws prepared by adding K_2O to magnetite or mill scale and reducing in a fluid bed with H_2 at 660-860°F. Reaction conditions used include 550-570°F with gas velocities of 0.7 feet/second.

The reactors have for the most part been of the fixed fluidized type with little carryover of catalyst (9). Conversions have been 90 percent giving 150 g of C_3+ per cubic m. of fresh synthesis gas. The space time yields are of the order of 180 kg/hour per cubic m. of fluidized catalyst. These yields are 10-30 times that of the fixed bed German processes, and the equipment cheaper to install. Waste-heat is convertible to 300-400 psi steam.

The product is chiefly motor gasoline, with a secondary product of water soluble oxygenated compounds (6). These organics were chiefly ethanol and acetic acid with smaller amount of acetone, acetaldehyde, and higher alcohols.

Operating difficulties were encountered due to carbon formation and the breaking down of the catalyst (9). The iron catalyst particles break down rapidly to smaller size and become difficultly fluidizable, the mass of catalyst assuming a fluffy nature. Breakdown is attributed to the reaction of iron with CO. The addition of inert solids for the purposes of extending the bed life of the catalyst only resulted in classification.

Catalyst agglomeration and the resultant defluidization was attributed to the formation of waxy carbonaceous material which coats the catalyst and causes the particles to stick together.

Instead of a fixed fluidized bed, an entrained or fully suspended bed has been proposed wherein the catalyst is circulated to a stripping or regenerating vessel for treatment with H_2 or synthesis gas. Carryover and recirculation of 500-2000 pounds of catalyst per 1000 cubic feet of gas has been proposed.

In fluidized operations, recycle has been used at ratios of 3 to 4 times the feed rate, giving conversions of 90-95 percent (6).

The initial catalyst density was 60-80 pounds/cubic foot, decreasing to 10-20 pounds/cubic foot. Superficial gas velocities of 1.5-2.5 feet/second were employed.

For fluidization, a catalyst particle size of 2-100 microns is recommended (9). This will result in a bulk density of 100-150 pounds/cubic foot and a fluidized density of 50 pounds/cubic foot.

A full scale reactor would be a cylindrical vessel 25 feet in diameter and about 45 feet high containing around 120 tons of catalyst (9). At a dispersed density of 50 pounds/cubic foot, the fluidized catalyst occupies only about one-fourth of the reactor volume. Estimated yields are 130 g of C_4+ per cubic m. of fresh feed gas, or 156 g of C_4+ per hour per cubic m. of catalyst volume. Such a plant would yield about 1,400 gallons of 34 bbl. of C_4+ product per day per MMSCF/D of synthesis gas.

In fluidized experiments carried out by the M. W. Kellogg Co., a cobalt catalyst was employed (9). The reactor was 2 inches by 13 feet, of which the bed height was 7.3 feet. Gas velocity was 0.6 feet second, and the density of the fluidized catalyst was 46 pounds/cubic feet. 76 percent of the CO reacted, of which 21.7 percent formed C_1 and C_2.

A summary of fluidized experiments carried out by Stanolind are as described in the following paragraphs (9).

Catalyst prepared from mill steel had the following screen analysis:

+100 mesh	23%
−100+140 mesh	32.8%
−140+200 mesh	19.8%
−200+325 mesh	19.8%
−325 mesh	4.6%

The catalyst was impregnated to 0.5-2.0 percent K_2O from aqueous solutions of K_2CO_3 or KNO_3. Activation was in the fluidized state at 700-775°F for about 70 hours using H_2 at 50 psi. The activated catalyst contained 95.5 percent elemental iron and 97.1 percent total iron.

Two different reactors were employed. One, 2 inches by 20 feet, was used with 15 pounds of catalyst; the other, 8 inches by 30 feet, was used with 214 pounds of catalyst. Initial bed densities were 100 pounds/cubic foot. After 400 hours of operation, the densities were only 30 pounds/cubic feet. Stone fillers at the top caught entrained reactor dust.

Operating conditions averaged 600°F at 250 psi yielding about 86 percent conversion of the CO in the feed. The linear gas velocity in the bed was 0.6-0.65 feet/second. Recycle ratios of 1.7/1 were used. The

H_2/CO ratio was 2.6/1 for the fresh synthesis gas. The total stream analysis including recycle was:

       30 percent H_2
       10 percent CO
       17 percent CO_2
       23 percent N_2
       20 percent C_1+

The average percentage of carbon converted to products was:

CO_2	8.0%
$C_1 + C_2$	26.6%
$C_3 + C_4$ olefins	19.6%
$C_3 + C_4$	4.4%
C_5	33.0%
Oxys	8.4%

Carbon deposition amounted to 0.05 pounds of carbon per hour per 100 pound catalyst. The rate of increase of -325 mesh particles was 0.10 pounds/hour.

The Carthage Hydrocol, Inc. plant using the Hydrocol process for the conversion of synthesis gas in a fluidized reactor was sponsored by the following companies (1):

       Texas Company
       Standard Oil Development Company
       Socony-Vacuum
       La Gloria Company
       I. S. Abercrombie Company
       Hydrocarbon Research, Incorporated

Construction was by the Arthur G. McKee Company. The capacity of the plant erected at Brownsville, Texas was 8000 bbl/day (6, 9).

The main features of the Hydrocol process were (1):

A. Recovery and separation of C_4+ and C_4- from natural gas.
B. Separation of O_2 from air.
C. Partial combustion of C_4- from natural gas to CO and H_2 (no catalyst used).
D. Conversion of the $CO-H_2$ synthesis gas to gasoline and other products using a fluidized iron catalyst.

E. Separation of products.

F. Further separation and treatment of gasoline fractions to remove oxygenated compounds, C_3 and C_4 polymers, blending, etc.

G. Manufacture of the catalyst.

The Carthage plant used a fixed fluidized bed of sintered iron (16). In the synthesis converter, the fluidized system flowed around tubes of boiling water under pressure. Synthesis was at 570-660°F, with recycle (1). Other figures cited were 550-650°F at 400 psi (9). Gasoline was the chief product. From 10-20 percent oxygenated compounds, mostly alcohols, were simultaneously produced (1). The plant utilized 64 MMSCF/D of natural gas and 40 MMSCF/D of O_2 at 300 psi or higher (1).

Equipment sizes were approximately as noted (9). Two refractory lined vessels of 200 cubic feet each reacted the natural as at 400-500 psi with O_2 and steam. Synthesis was carried out in six converters, each of 3,500 cubic feet volume, giving a total of 21,000 cubic feet per 8000 bbl/day rating. This rating corresponds to a space time yield of 80 kg/hour per cubic m. of reactor volume. The fluid catalyst volume was 80/180 = 0.445 of the total reactor volume.

An elementary hydrocarbon product list was as follows (9):

$C_3 + C_4$	32%
naphtha	56%
diesel	8%
residue	4%

The C_2+C_4 contained 80 percent olefins, the naphtha 70 percent and the diesel 65 percent. Much more detailed analyses are given in the reference, particularly with respect to isomers, which were characterized by a high ratio of mono- to-dimethyl isomers.

The 400°F end product naphtha had an octane number of 62 which could be increased to 76 by treating, and to 82 by adding tetraethyl lead.

The water phase solutions produced amounted to about 2.4 gallons per gallon of oil produced. The water phase solutions in turn contained from 5-10 percent oxygenated compounds. Separation of these oxygenated materials posed formidable difficulties due to the formation of constant-boiling mixtures (azeotropes). Solvent extraction and extractive distillation procedures were employed to effect separation. About 50 percent of the oxygenated compounds were alcohols with ethanol predominating. Ketones were about 11 percent with acetone

Reactions of Carbon Monoxide and Hydrogen

predominating. Aldehydes were about 10 percent with acetaldehyde predominating. Acids were about 27 percent, mostly acetic. A detailed breakdown is given in Reference 9.

The latest and most successful commercial attempt to produce liquids via the Fischer-Tropsch synthesis is the SASOL operation in South Africa, previously mentioned. Based on design and technology developed by the M. W. Kellogg Company and Arge-Arbeit Geminschaft Lurgi and Ruhrchemie, both fluidized and fixed-bed reactors are used. Catalyst lifetimes for fluidized catalysts are 45-60 days as mentioned. Lurgi pressure gasifiers provide the synthesis gas. Operations are to be expanded to include methanation.

A diagram of the Fischer-Tropsch synthesis process is provided in Figure 7.2. Typical product yields (31) are given in Table 7.2.

With regard to methanation, work in Scotland at a Lurgi facility has successfully shown that sustained runs can be maintained over nickel catalyst with temperature control of the highly exothermic reaction.

TABLE 7.2. Typical range of products from Fischer-Tropsch synthesis.

	Fixed Bed Process	Fluid Bed Process
	— Liquid Product Composition —	
Liquid Petroleum Gas ($C_3 - C_4$)	5.6	7.7
Petrol ($C_5 - C_{11}$)	33.4	72.3
Middle Oils (diesel, furnace, etc.)	16.6	3.4
Waxy Oil or Gatsch	10.3	3.0
Medium Wax, mp 203-206°F	11.8	
Hard Wax, mp 203-206°F	18.0	
Alcohols and Ketones	4.3	12.6
Organic Acids	traces	1.0

	Fixed Bed Process		Fixed Bed Process	
	———— Liquid Product Composition ————			
	C_5-C_{10}	C_{11}-C_{18}	C_5-C_{10}	C_{11}-C_{14}
Parafins, Vol %	45	55	13	15
Olefins	50	40	70	60
Aromatics	0	0	5	15
Alcohols	5	5		
Carbonyls	traces	traces	6	5

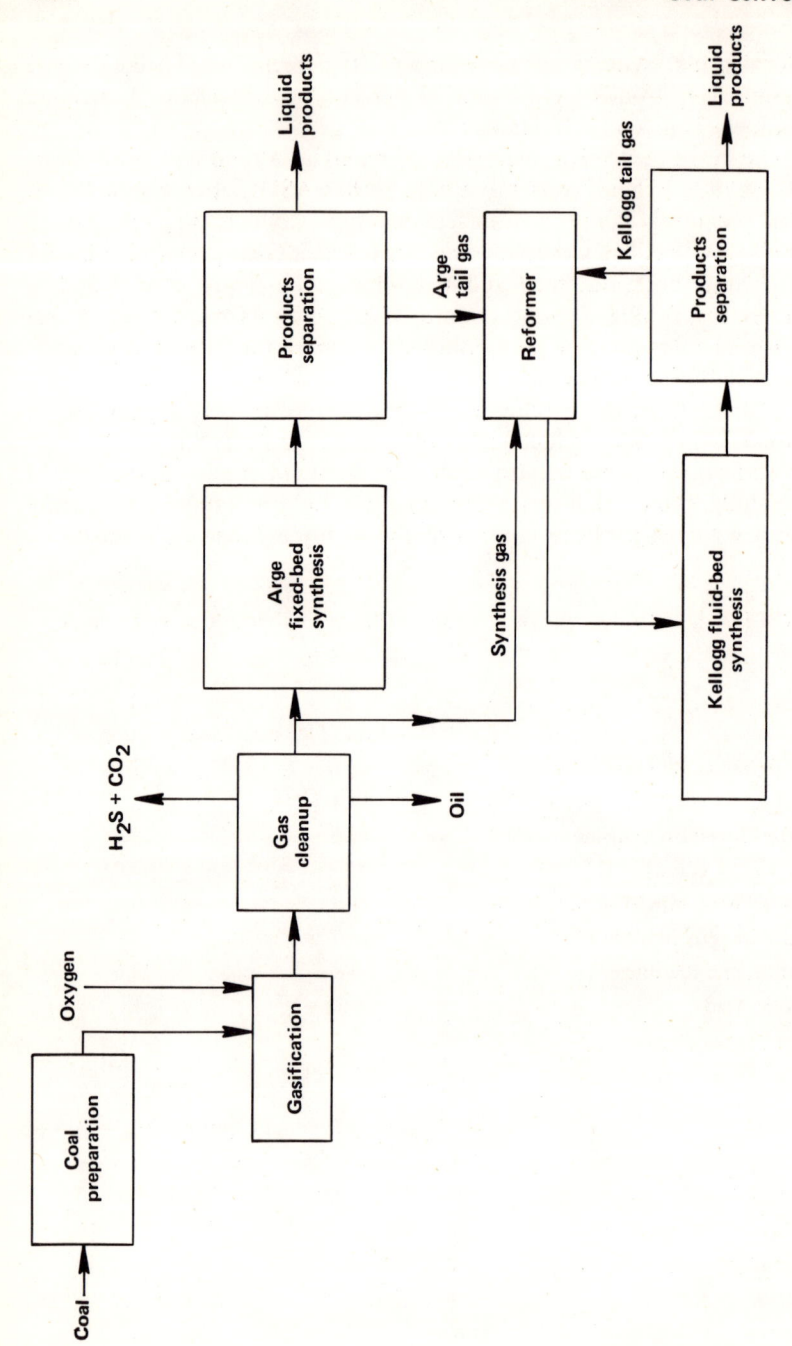

Fig. 7.2 Fischer-Tropsch synthesis process.

REFERENCES

1. Pichler, H., "Twenty-Five Years of Synthesis of Gasoline by Catalytic Conversion of Carbon Monoxide and Hydrogen," from *Advances in Catalysis,* Vol. IV, Academic Press, New York (1952), pp. 271-341.
2. Anderson, R. B., "Kinetics and Reaction Mechanism of the Fischer-Tropsch Synthesis," in *Catalysis,* Vol. IV, P. H. Emmett Ed., Reinhold, New York (1956), pp. 257-371.
3. Schultz, J. F., L. J. E. Hofer, E. M. Cohn, K. C. Stein and R. B. Anderson, "Synthetic Liquid Fuels from Hydrogenation of Carbon Monoxide," Part II, *U. S. Bureau of Mines Bulletin 578* (1959).
4. Anderson, R. B., "The Thermodynamics of the Hydrogenation of Carbon Monoxide and Related Reactions," in *Catalysis,* Vol. Iv, P. H. Emmett Ed., Reinhold, New York (1956), pp. 1-27.
5. Anderson, R. B., "Catalysts for the Fischer-Tropsch Synthesis," in *Catalysis,* Vol. IV, P. H. Emmett Ed., New York (1956), pp. 29-255.
6. Storch, H. H., "The Fischer-Tropsch and Related Processes for Synthesis of Hydrocarbons by Hydrogenation of Carbon Monoxide," in *Advances in Catalysis,* Vol. I, Academic Press, New York (1948), pp. 115-56.
7. Storch, H. H., "Synthesis of Hydrocarbons from Water Gas," from *Chemistry of Coal Utilization,* Vol. II, H. H. Lowry Ed., Wiley, New York (1945).
8. Rapoport, I. B., *Chemistry and Technology of Synthetic Liquid Fuels,* Second Edition, Translated from Russian, Published for the National Science Foundation, Washington, D.C. by the Israel Program for Scientific Translations, Jerusalem (1962).
9. Storch, H. H., N. Golumbic and R. B. Anderson, *The Fischer-Tropsch and Related Syntheses,* Wiley, New York (1951).
10. Morgan, J. J., "Water Gas," in *Chemistry of Coal Utilization,* Vol. II, H. H. Lowry Ed., Wiley, New York (1945), pp. 1673-1749.
11. Von Fredersdorff, C. E. and M. A. Elliott, "Coal Gasification," in *Chemistry of Coal Utilization,* Supplementary Volume, Wiley, New York (1963).
12. Greyson, M., "Methanation," in *Catalysis,* Vol. IV, P. H. Emmett Ed., Reinhold, New York (1956), pp. 473-511.

13. Linden, H. R., "Conversion of Solid Fossil Fuels to High-Heating Value Pipeline Gas," in *Hydrocarbons From Oil Shale, Oil Sands, and Coal,* Chemical Engineering Progress Symposium Series Number 54, Volume 61, 1965.
14. Hoy, H. R. and D. M. Wilkins, "Total Gasification of Coal," British Coal Utilization Research Association, Review No. 174, Feb./March, 1958, pp. 57-110.
15. Field, J. H., J. J. Demiter, A. J. Forney and D. Bienstock, "Development of Catalysts and Reactor Systems for Methanation," *Ind. Eng. Chem. Prod. Res. Develop., 3* (2), 150-153 (1964).
16. Storch, H. H., R. B. Anderson, L. J. E. Hofer, C. O. Hawk, H. C. Anderson and N. Golumbic, "Synthetic Liquid Fuels from Hydrogenation of Carbon Monoxide," Part I, U. S. Bureau of Mines Technical Paper 709 (1948).
17. Natta, G., "Synthesis of Methanol," in *Catalysis,* Vol. III, P. H. Emmett Ed., Reinhold, New York (1955), pp. 349-411.
18. Hirst, L. L., "Methanol Synthesis from Water Gas," in *Chemistry of Coal Utilization,* Vol. II, H. H. Lowry Ed., Wiley, New York (1945), pp, 1846-68.
19. Powell, A. R., "Relation of Coal Gasification to the Production of Chemicals," from *Symposium on Gasification and Liquefaction of Coal,* American Institute of Mining and Metallurgical Engineers, New York (1953), pp. 196-206.
20. Cohn, E. M., "The Isosynthesis," in *Catalysis,* Vol. IV, P. H. Emmett Ed., Reinhold, New York (1956), pp. 443-72.
21. Natta, G., U. Colombo and I. Pasquon, "Direct Catalytic Synthesis of Higher Alcohols from Carbon Monoxide and Hydrogen," in *Catalysis,* Vol. V, P. H. Emmett Ed., Reinhold, New York (1957), pp. 131-174.
22. Steiner, H., "Catalytic Cyclization and Aromatization of Hydrocarbons," in *Catalysis,* Vol. V, P. H. Emmett Ed., Reinhold, New York (1956), pp. 529-560.
23. Corson, B. B., "Catalytic Hydrogenation of Olefinic Hydrocarbons," in *Catalysis,* Vol. III, P. H. Emmett Ed., Reinhold, New York (1955), pp. 79-108.
24. Smith, H. A., "The Catalytic Hydrogenation of Aromatic Compounds," in *Catalysis,* Vol. IV, P. H. Emmett Ed., Reinhold, New York (1957), pp. 175-256.

25. "Synthetic Liquid Fuels. Part I. Oil from Coal," U.S. Bureau of Mines Report of Investigations 5236, July, 1956.
26. Orchin, M. and I. Wender, "Reactions of Carbon Monoxide," in *Catalysis,* Vol. V, P. H. Emmett Ed., Reinhold, New York (1957), pp. 1-72.
27. Hofer, L. J. E., "Crystalline Phases and Their Relation to Fischer-Tropsch Catalysts," in *Catalysis,* Vol. IV, P. H. Emmett Ed., Reinhold, New York (1956), pp. 373-441.
28. Field, J. H., J. J. Demiter, A. J. Forney and D. Bienstock, "Development of Catalysts and Reactor Systems for Methanation," *Ind. Eng. Chem. Prod. Res. Develop., 3* (2), 150-153 (1964).
29. Weller, S. W., "Catalysis in the Liquid-Phase Hydrogenation of Coal and Tar," in *Catalysis,* Vol. IV, P. H. Emmett Ed., Reinhold, New York (1956), pp. 513-528.
30. Donath, E. E., "Hydrogenation of Coal and Tar," in *Chemistry of Coal Utilization,* Supplementary Volume, H. H. Lowry Ed., Wiley, New York (1963), pp. 1041-80.
31. Bodle, W. M. and K. C. Vyas, "Clean Fuels from Coal — Introduction to Modern Processes," *Clean Fuels from Coal,* Symposium II, Institute of Gas Technology, June, 1975.

Chapter 8

CATALYSIS AND HETEROGENEOUS REACTIONS

Preceding descriptions of catalysts point up some observations and generalizations that can be made concerning composition heterogeneity. Additionally, the differences between homogeneous and heterogeneous or catalytic reaction theory require comment on the role and purpose of catalysis.

A feature repeatedly apparent in the listings of catalytic activity is heterogeneity: differences in composition and structure. Even with a pure element — e.g., metal — the "hole theory" of structure evidences departures from uniformity of structure. This is further evidenced by the fact that compounds of the transition elements do not obey Dalton's law of constant composition. Additionally, mixtures of elements and compounds, by their very constitution, are heterogeneous.

The last mentioned situation is similar to the well known electrochemical theory for corrosion, in which anodic and cathodic areas form cells due to differences in composition. Broadly speaking, corrosion can be regarded as a chemical reaction whereby the corroding surface is the catalyst. Commonly, the heterogeneity is produced by different degrees of oxidation. Contact of dissimilar metals will also provide heterogeneity with resultant corrosion.

Thus it is noticeable that in all cases of catalytic activity there is the feature of heterogeneity. Whether or not there is a direct relation between this heterogeneity and the formation of active "centers" may not always be resolvable. It may be a semantics problem, analyzable only by the criterion of mathematical description for the mechanism or models advanced.

By inference, at this point, it is reasonable to assume that a particular set of reactions may be catalyzed by some special group of substances, and that any means of enhancing the heterogeneity within the group of substances will promote catalytic activity.

Whereas the rate descriptions of homogeneous reactions are straightforward and well-recognized, such is not the case for heterogeneous reactions. This difficulty in part arises from the fact that two or more phases are involved. One of the phases may actually enter into the reactions and be used up, as coal reacts, or a phase may act as a catalyst and not be depleted, though the composition or structure — i.e., activity — be changed.

The phases involved, moreover, may be systems of gas-solid, liquid-solid, liquid-liquid, or some other multiple combination. Usually the first two combinations mentioned are of most interest, particularly if the solid phase acts as a catalyst.

As would be expected, the more complex a phenomenon is, the more ways that can be devised for explaining or representing it. With heterogeneous reactions, the mechanisms or models range from using the relations of homogeneous reactions to concepts involving the transfer of reactants, surface reactions, and transfer of products. A well-accepted mathematical version for the latter is presented by Hougen and Watson (1), which has also been generalized (2). When the added effects of heat of reaction and rate of heat transfer are included as well, the mathematics of the problem becomes formidable.

It is not the purpose here to present all the various theories of catalysis, though the electron theory will be dealt with subsequently. Prior to this, it is in order to make some candid comments as to how preconceived notions may go awry in attempting to generalize from homogeneous reaction theory to heterogeneous reactions. In homogeneous reaction theory we are accustomed to the point of view that the rate of reaction is more or less proportional to the concentrations of the reactants, and the reverse reaction to the products. However, with catalytic or heterogeneous reactions, such is not necessarily true. The effects of mass transport between phases, for instance, may throw an unsuspected effect upon behavior. That this is so can be seen by the much more complicated expressions used to represent catalytic reactions (1, 2).

The previous paragraphs bring up an interesting point. In the kinetic analysis of heterogeneous reactions, equations are fit to conceptual models which to our minds appear reasonable. Reaction data is merely fit to the model to obtain values for the undetermined coefficients (e.g., rate constants) introduced via the mathematical mechanism, by the method of least squares or some other method of fit. The model or mechanism which best fits the data is regarded as the correct, or most nearly correct, description. It follows, however, that innumerable mathematical relations could be devised which would fit the data with varying degrees

Catalysis and Heterogeneous Reactions

of precision. These expressions could be quite arbitrary — for instance, the linear forms of statistical analysis involving variates made up of arbitrary combinations of the variables. With enough variates, it is possible to obtain quite precise reproduction of a given set of data.

Hence we can argue that a mathematical form does not really tell us anything — it is just a way to reproduce a set of data, and interpolation or extrapolation to other conditions should be viewed with suspicion. It is, however, for these very reasons that models or mechanisms are conceived — to provide a plausible basis for interpolating known data to other conditions. Success cannot be fully demonstrated till the mechanism is used over all ranges of conditions — a requirement usually never met.

The effects of temperature and pressure are usually quite predictable for any simple homogeneous reaction: the rate increases with temperature, and shifts toward the more dense system with pressure. For a catalytic reaction, other subtleties may come into play. Temperature and pressure may have unsuspected effects upon the adsorption and desorption of reactants and products. However, until these effects are known, the use of these simpler concepts are a very useful rule of thumb.

It is in the area of equilibrium conversions that the concepts of homogeneous equilibria can be grossly misleading. In homogeneous reaction theory, the catalyst is regarded as an intermediate product and reactant which cancels out overall. Whereas, in the more detailed mechanisms of heterogeneous reaction theory, the overall equilibrium depends upon other equilibria: the transport of reactants and products and the surface reaction (1, 2). Moreover, the reaction — if considered as occurring homogeneously — is completely diverse from the reaction regarded as occurring in the stages of adsorption, reaction at the surface, and desorption of the products. Even further, there is no absolute way to relate the absolute activities of the components in the two systems of description, except arbitrarily. Two phases are involved; though an expression for activity can be assigned to the one phase, there remains the problem of assigning an expression for activity at the other: the catalyst or the solid reactant (2).

In applying reaction equilibrium constants to successive homogeneous reactions involving intermediates, the overall equilibrium constant becomes the product of the equilibrium constants of the individual reactions. The magnitude of this overall product can be misleading. For to solve the component distribution requires a simultaneous solution of the equilibrium relations for each reaction, and involves all components — reactants, intermediates, and products. Only from this simultaneous solution can the quantity of product be determined.

In consequence, if there is evidence that a reaction will not go according to homogeneous concepts, this viewpoint should not necessarily follow if a heterogeneous catalyst is used. Indeed, this is the purpose of the catalyst: not only to affect rate but equilibrium and the course of reaction as well. This is not meant to mean, however, that the reacting system will not revert to true homogeneous equilibrium once the confines of the catalyst are left. This in fact is the purpose of using a quench-cooler following a catalytic reactor: to insure that the reacting system does not revert to homogeneous equilibrium. There is thus a distinction between the chemical equilibrium of homogeneous and heterogeneous (catalyzed) reacting systems. Furthermore, different catalysts may produce different product species, and conversions may exceed homogenous equilibrium.

Apropos of the reactions to produce hydrocarbons and oxygenated compounds, the following paragraphs contain pertinent statements found in the literature.

Comments made by Anderson (3) are as follows:

"The synthesis of hydrocarbons .... which approximate the over-all reactions involved in the production of synthetic liquid fuels from coal is accompanied by positive $\Delta F^0/n$ values at all temperatures ... . However, coal is much more reactive than graphite, and with some coals these reactions may be thermodynamically possible."

Statements of Lewis, Gilliland and Hipkin (4) regarding the reaction of steam and carbon are:

"The reaction of carbon with steam at low tempertures offers the possibility of producing hydrocarbons directly. At temperatures of 700°F to 900°F the equilibrium favors the formation of methane and CO_2. At lower temperatures, higher molecular weight hydrocarbons should be formed."

The article goes on to say that the overall heat of reaction for conversion on to hydrocarbons would be small, but there remained the problem of low reaction rate. It should be emphasized that the problem of low reaction rate is a problem of proper catalysis.

A statement made by Ergun (5), concerning the reactions of CO_2 and steam with coke, reflects upon the diversity between heterogeneous reactions and homogeneous reactions:

"Most of the confusion about the role of physical factors such as diffusion and gas velocity resulted from attempts to interpret data on the heterogeneous gasification reactions in a manner analogous to homogeneous reactions."

For a heterogeneous catalytic reaction, the overall equilibrium constant, expressed in terms of gas composition, contains the equilibrium constant for the surface reaction and also the adsorption and desorption mass transfer constants for the reactants and products (1, 2). The particular expression will depend upon the mechanism adopted. There is no *a priori* constraint that this overall constant be equal and equivalent to the equilibrium constant of any uncatalyzed homogeneous reaction for the same system. Moreover, the rate expressions are dissimilar. In short, these heterogeneous and homogeneous processes are diverse from one another, both in physical and analytical interpretations. Thus $CO-H_2$ mixtures, at the same conditions, convert differently with catalysts.

The foregoing discussion points up the possibility of proceeding directly from carbon (coal) and steam to hydrocarbons or oxygenated compounds in one stage. If carbon monoxide and hydrogen were to be formed intermediately, the reaction would proceed on to the final products, possibly through a nascent condition.

In any event, to proceed directly from steam and coal to a final hydrocarbon or oxygenated product in one reactor is the simplest and potentially the most economic route.

Considerations of homogeneous reaction theory generally assume that the steam-carbon reaction to form carbon monoxide and hydrogen must proceed at a higher temperature level than the consummating reaction between CO and hydrogen. It has been demonstrated, however, that steam and carbon can react at considerably lower temperatures by the use of catalysts such as Na_2CO_3. This sets forth the favorable possibility of conducting the reaction at conditions whereby the carbon and steam react through carbon monoxide and hydrogen to hydrocarbons and other compounds. This proposition is predicated upon the use of proper catalysts: one catalyst for the first part of reaction, possibly, and another for completion.

Perhaps the most striking example of the direct conversion of carbonaceous materials to a hydrocarbon and carbon dioxide occurs within microbiological systems: the anaeorobic digestion of organic materials to yield methane and carbon dioxide. For the glucose monomer, the reaction reads

$$CH_2O \rightarrow \tfrac{1}{2}CH_4 + \tfrac{1}{2}CO_2$$

or put more simply $C + H_2O \rightarrow \tfrac{1}{2} CH_4 + \tfrac{1}{2} CO_2$. This indicates the thermodynamic feasibility at lower temperatures. A different system yields enthanol at the same conditions, etc.:

$$CH_2O \rightarrow 1/3\ C_2H_6O + 1/3\ CO_2$$

As discussed, the use of homogeneous equilibria data — experimental or calculated — should be viewed with suspicion when applied to heterogeneous catalytic reactions. This includes free energy changes.

However, an indication of the probability of formation of the various possible products is provided to some extent by use of the heat of reaction. Consider the reaction

$$C + n\, H_2O\,(g) \rightarrow x\, product\,(g) + y\, CO_2\,(g)$$

Heats of combustion and latent heats can be used to calculate the heat of reaction per mole of carbon for the formation of a wide variety of compounds from steam and carbon.

A positive ΔH means heat absorbed during reaction, a negative ΔH indicates heat evolved. A positive ΔH provides some measure that a compound is less likely to form unless heat is supplied. A negative ΔH indicates the compound is more likely to be formed especially if heat is removed from the system. Hence, in a manner of speaking, the more negative the number the more the likelihood of formation. Or else the less likelihood of formation on increasing the temperature. While it would be desirable to utilize free energies of formation of the individual components involved to determine the overall free energy of conversion and the equilibrium constant for the reaction, such information is most often not available where needed (see Appendix). Accordingly, an alternative procedure must sometimes suffice, as indicated.

The ratio of moles of product (as a gas) produced to the moles of steam used provides a measure of the effect of pressure. The lower this ratio the more probable the formation under pressure. In addition, the moles of H_2O needed to react per mole of carbon is an indication of the effect of steam concentration on the direction of reaction. The lower this value, the more chance for reaction.

All of these effects enter simultaneously, and would apply to a given catalyst. The effect of increasing catalyst activity is to produce compounds of higher H/C ratios.

A variety of lower molecular weight compounds have been used to calculate the aforementioned entities. These are listed in Table 8.1 in the order of values of ΔH per mole of carbon.

Another facet of heterogeneous catalysis is the fact that not only may the conversion exceed the bounds of homogeneous equilibrium, but the course of reaction may itself be changed. For example, at the same identical conditions of temperature and pressure, carbon monoxide and hydrogen in the presence of nickel catalyst will yield methane, whereas in the presence of iron higher hydrocarbons and oxygenated compounds are

Catalysis and Heterogeneous Reactions

TABLE 8.1. Heats of Reaction and Molar Ratios.

Formula	Name	$\Delta H/C$	Products/H_2O	H_2O/C
CH_4	methane	−0.76	1.0	1.0
$C_5H_{11}OH$	butyric acid	−0.48	0.42	0.8
CH_3COOH	acetic acid	−0.3	0.5	1.0
C_2H_5COOH	n-propionic acid	0.74	0.5	0.85
C_2H_6	ethane	0.885	0.83	0.86
C_3H_7COOH	n-butyric acid	0.94	0.5	0.8
C_3H_8	propane	1.07	0.8	0.8
C_4H_9OH	n-butyl alcohol	2.75	0.6	0.83
$CH_3COC_2H_5$	methyl ethyl ketone	3.00	0.63	0.73
C_3H_7OH	n-propyl alcohol	3.3	0.63	0.89
$CH_3COOC_2H_5$	ethyl acetate	3.81	0.5	0.8
C_4H_8	n-butane	4.05	0.75	0.67
$C_2H_5OC_2H_5$	diethyl ether	4.21	0.60	0.83
CH_3COCH_3	acetone	4.53	0.67	0.75
C_2H_5OH	ethyl alcohol	5.3	0.67	1.0
C_3H_6	propylene	5.69	0.83	0.67
CH_3CHO	acetaldehyde	7.112	0.75	0.80
$CH_3COOOCH_3$	methyl acetate	7.95	0.5	0.86
$HCOOC_2H_5$	ethyl formate	8.31	0.5	0.86
CH_3OCH_3	dimethyl ether	8.70	0.67	1.0
C_2H_4	ethylene	8.75	0.75	0.67
CH_3OH	methyl alcohol	9.2	0.75	1.33
$HCOOCH_3$	methyl formate	12.6	0.5	1.0
H_2	hydrogen	21.6	1.5	2.0
C_2H_2	acetylene	23.4	1.5	0.4
CHO	formaldehyde	26.9	1.0	1.0

Units:
$\Delta H/C$ kcal/g-mole
Products/H_2O moles/mole
H_2O/C moles/mole

the result. Over a suitably sufficient interval of time, at temperature, the products would revert to the true homogeneous equilibrium composition — by definition. Hence the need for a quench in order to preserve the conversion attained.

8.1 CATALYST ACTIVATION AND REGENERATION

The formulation and activation of catalysts — in this case, Fischer-Tropsch type catalysts of Group VIII transition metals — is as much art as science. Usually if not always in the form of oxides, the various states of oxidation, the presence of the metallic state, and the presence of impurities are all of inter-related concern. Semiconductors as a class are now believed to play a prominent role in catalytic activity, and modern theories of catalysis introduce and apply this concept.

In the following discussion there is an implied emphasis on nickel catalysts, though iron or other catalysts are of parallel interest.

ACTIVATION AND REACTION

Fischer-Tropsch catalysts, including those of nickel, are activated by reduction with hydrogen, though a producer gas containing carbon monoxide can also be used. Presumably two or more different states of oxidation occur, including the presence of the metal, since the starting material is an oxide (or oxides). For reasons to be discussed, this heterogeneous combination of states of oxidation induces a set of reactions to take place at the catalyst surface.

The mechanism of reaction for such a heterogeneous catalytic system is implied rather than known. It is generally assumed to involve the adsorption of reactants at active "sites" or "centers" on the catalyst surface, followed by reaction at these active sites to form an activated complex which decomposes into the products, and consequent desorption of the products.

The presence of hydrogen and carbon monoxide in this case — both of which are reducing agents — evidently maintains an equilibrium condition at the catalyst surface such that a continual, partially reduced state is maintained. This is noticed by the high degree of heat evolution that takes place when the active catalyst is exposed to air, and oxidation takes place.

We distinguish the preceding from homogeneous catalytic reactions as embodied by the use of metal carbonyls, of which nickel carbonyl is an example. At lower temperatures (below the decomposition point) nickel carbonyl will exist in the vapor state and is used as a homogeneous vapor-phase catalyst for certain classes of reactions. At higher temperatures, nickel carbonyl decomposes and thus would not exist at the temperatures for methanation. Hence the ruling out of vapor-phase homogeneous catalysis — including the toxicity problem of carbonyls — at Fischer-Tropsch conditions.

THE SURFACE REACTION

As to the mechanism of the surface reaction itself, and the "driving forces" causing the reaction, the phenomenon can be regarded as of electrochemical origin (2, 6).

It has long been known that corrosion cells in aqueous solutions are formed when two levels of oxidation exist at points contiguous on the exposed metal surfaces. It is but the principle of the battery over again. A galvanic action is set up which produces half-cell reactions at the shorted anodic and cathodic points.

The net overall reaction which occurs in solution can be viewed as taking place due to a "catalytic" action of the exposed metal surfaces, which may exist in the metallic and/or oxide form.

The concept allows generalization to any surface of heterogeneous composition such that an emf is generated at points of dissimilar composition, and considering the electrochemical potential difference so generated as the driving force.

If the cell so formed is short-circuited, then a migration of ionic material to or from the electrode surfaces ensues, with dissolution or deposition. We are then in general speaking of heterogeneous reactions, reactions occurring between phases.

If the electrodes (anodic and cathodic points or sites) are joined by a semiconductor rather than a conductor, such that electron flow is inhibited, then a different situation arises.

There will be an induced polarity causing anions to move to the anode and remain, and cations to move to the cathode and remain, with the possibility of limited interaction occurring at the electrodes. It might be said that a condition of metastable equilibrium is approached.

If the anodic and cathodic points are adjacent, then the ions are free to react. Since we are considering ions in solution, there would be at the most only an increase in respective concentrations at the elctrode points. A different situation arises, however, if non-ionized reactant species exist as (or in) a gas or non-conducting (non-ionized) liquid.

Thus if the preceding be extended to reactions between solid and gaseous phases, where there is no change in the solid or catalyst phase due to direct reaction (as there would be for combustion of a coal particle for instance) or to ionic migration (as in galvanic cells in aqueous solutions), a counterpart concept is required. This is furnished by the concept of the ionization potential, the energy required to remove an electron from orbit. In these terms, heterogeneous catalytic activity is provided a driving force.

Again, we are speaking of anodic and cathodic points or sites joined by a semiconductor (which may constitute one or the other or both of the electrodes) such that the flow of electrons between points is impaired.

A difference in ionization potential so set up may be theorized to induce a positive charge in a receptive molecular particle at the anodic point and a negative charge in a different species of receptive molecular particle at the cathodic point (or atomic particles, as the case may be). That is, an electron is transferred from the particle species at the anodic point to the particle species at the cathodic point. The impaired electron transfer through the semiconducting medium is such that a potential difference is maintained.

The positively-charged particle species is then free to react with the negatively-charged species provided the anodic and cathodic points are sufficiently close together.

While no doubt many other theories may be, and have been, developed for explaining heterogeneous catalytic activity, the idea of differences in the ionization potentials of heterogeneous materials involving semiconductors provides a device for introducing and maintaining a driving force for causing reaction to occur between gaseous (or liquid) species.

Thus for the catalyzed reaction of hydrogen and carbon monoxide at the catalyst surface, a symbolic representation is as follows:

$$H_2 \xrightarrow[(-2e)]{\text{anodic transfer}} 2[H^+]$$

$$[2H^+ CO] \rightarrow \tfrac{1}{2}CO_2 + \tfrac{1}{2}CH_4$$

$$CO \xrightarrow[(+2e)]{\text{cathodic transfer}} [CO^-]$$

where the electron migration or transfer is from the anodic point to the cathodic point — the anodic point being where oxidation occurs with the loss of electrons, and the cathodic point being where reduction takes place with the absorption of electrons.

In other words, a difference in ionization potentials of the anodic and cathodic points causes "activation" of the hydrogen and carbon monoxide which then react to form a complex which decomposes into the final products.

Catalysis and Heterogeneous Reactions 275

The conventions above are illustrative and arbitrary only, and if so desired the reactants could be shown to first form a complex with an active site, etc.

Suffice to say, as to the solid catalyst itself, the element of heterogeneity must be present such that an ionization potential difference is generated and a semi-electrical conductance maintained. Furthermore, the heterogeneity of anodic and cathodic points must be distributed uniformly over the surface, with the anodic and cathodic points contiguous. It is implicit that a very large surface to volume ratio must exist for the catalyst to be effective.

A corresponding situation would exist in ionic solutions where an emf is generated due to analogous electrochemical potential differences. In fact, this effect can be used as an indication or measure of catalytic practicability of materials by use of disjoint electrodes of differing composition which are immersed in a common electrolyte.

Anodic and cathodic points are distinguishable by a valency difference or by other differences in composition. The full range of criteria remains to be established.

The potential difference may be destroyed if the anodic and cathodic points are short-circuited by the presence of a conducting medium. On the other hand, the effect is also destroyed if the anodic and cathodic points are joined by a nonconductor. The appearance of either could be caused by a chemical reaction due to a contaminant.

In further comment, it is noted that a nickel catalyst is also used for the steam reforming of methane (natural gas):

$$CH_4 + H_2O(g) \rightarrow CO + 3 H_2$$

This is the exact inverse of the methanation reaction. The surface reaction sequence can only remain speculative. A likely symbolic representation is as follows:

$$2 H_2O \xrightarrow{\text{anodic}} [H_3OH^+]$$
$$2 [H_3OH^+ \cdot CH_4^-] \rightarrow 2 CO + 6 H_2$$
$$2 CH_4 \xrightarrow{\text{cathodic}} 2 [CH_4^-]$$

While the sequence is anyone's guess, the point is that the overall reaction occurs and is catalyzed by nickel catalysts which remain stable over long periods of time.

In reforming, however, the catalyst is activated by steam which serves as an oxidizing agent. Presumably, then, NiO is partially oxidized to a higher state of oxidation (nickel is noted to have a number of valencies or oxidation states: $-1, 0, 1, 2, 3, 4$). This condition is, on the evidence, selective to the reforming reaction. Furthermore and significantly, a condition of steady-state oxidized equilibrium is maintained at the catalyst surface.

For purposes of simplicity the higher state of oxidation can be regarded as NiO_2. On further inspection, however, nickel appears either to be non-stoichiometric or else to form a series of combinations of NiO and NiO_2 or other oxides. The early literature is full of such speculations (e.g., Mellor, *A Treatise on Inorganic and Theoretical Chemistry*). In the other direction, there are lower oxides or mixed oxides.

The subject of oxides of nickel is similar to that for oxides of iron, another of the Group VIII transition metals. More modern theories indicate vacancies in the lattice structure which would represent a host of non-stoichiometric compositions, and in itself could account for the catalytic activity of these materials.

And as presented, heterogeneous catalysis enters the topical subject area of semiconductors. For in fact the oxides of the Group VIII transition metals, as well as other oxides and compounds, are semiconductors — and in this connection, a large body of theory has been built up to explain the behavior of semiconductors, notably in terms of electrons and "holes".

Vol'kenshtein, in particular, has explained catalytic activity in these terms in his *The Electronic Theory of Catalysis on Semiconductors* (6). Vol'kenshtein maintains that all solid catalysts are in fact semiconductors or involve semiconductors. The catalytic activity of platinum even is due to a minute film of oxide at the surface. The degree of electrical conductance therefore may be an indication of activity.

In wrapup, based on the foregoing descriptive development, the selectivity of catalysts to reactants and vice versa would presumably lie in the matching of ionization potentials and/or other characteristics of the catalyst with the receptivity of the reactants. The characterization and measure of parameters for activation and reaction is a field for further study and correlation. Grain growth is one such parameter.

As to de-activation and the effect of sulfur or other contaminants or poisons, it may turn out that the control of catalyst activity is more a matter of controlling oxidation states or heterogeneity than of merely eliminating the contaminant from the system. The issue is clouded by the fact that so-called de-activated catalysts may at times in themselves exhibit activity or selectivity toward other reactions. Thus nickel sulfide may serve as a catalyst (e.g., in hydrogenation). Other sulfides are also catalysts (e.g., WS_2).

8.2 THE PREMISE OF DIRECT CONVERSION

A consideration of the reaction of coal and the patterns of research in coal conversion underscores the fact that the production of liquid and gaseous fuels is at present the most desirable form of conversion. It possesses the potential of a large, ready market if the conversion process can be made economically competitive with petroleum-derived products.

A study of the reactions of coal and steam and the further reaction of carbon monoxide and hydrogen but emphasizes the desirability of a direct conversion of coal with steam to yield a hydrocarbon product in either, but preferably both, the liquid or gaseous forms.

All other variations involve two or more stages: the production of synthesis gas or hydrogen, followed by reaction of the synthesis gas or hydrogenation of the coal. The reduction to the simplest possible denominator would be the one stage conversion of coal to the final consumable hydrocarbon product, an operation previously recognized.

The chief objection to such an ideal has been the opinion that a one-stage conversion is not thermodynamically feasible. This objection has been met by a study of the role of catalysts. The use of catalyst permits conversions at conditions other than dictated by the results of homogeneous reaction rates and equilibria.

For the coal-steam system, reaction is obtained at temperatures well below homogeneous conditions by using Na_2CO_3 and other catalysts. In fact, the temperatures approach the region for the Fischer-Tropsch reaction of carbon monoxide and hydrogen using Group VIII transition metal catalysts. Consequently, the overall reaction of coal and steam to produce hydrocarbons and other oxygenated products is rendered feasible by the use of appropriate catalysts. This is the basis for the Hoffman process (7), also to be discussed in the next section. With nickel catalysts methane is the principal product. With iron catalysts, which are emphasized here, higher-boiling materials are also produced.

It is expected that any alkaline carbonate or oxide would serve to initiate the reaction. Na_2CO_3, however, possesses the advantage of cheapness and availability. Moreover, it is very possible that it can be used in its unrefined forms.

Iron catalysts, too, possess the advantages of cheapness and availability. There is reason to believe that the iron can ultimately be used in a variety of unrefined or unpurified forms, including the ore.

Accordingly, as a first choice of catalyst, a mixture of sodium carbonate and iron oxides are preferable. The exact proportions and the various useable degrees of oxidation of the iron have yet to be fully established but no doubt can be varied. Partly reduced Fe_2O_3 (say around 50 percent) should suffice — or partly oxidized metallic iron. It may be required to calcine the carbonate to soda ash as part of the regeneration cycle. Incidentally, the alkali compounds may also serve as promoters for the iron catalysts.

The catalyst mixture may be mixed with the coal, or as an alternative, impregnated from solution. The coal-catalyst system can be introduced dry or as a slurry — particularly into the steam.

A fluidized-reaction system is one choice. The coal-catalyst mixture would be fluidized by steam and recycle gas. The course of the reaction will also generate more gaseous products at the reactor conditions for continuation and support of fluidization.

The reactor conditions are of the order of one to fifty atmospheres or more, requiring superheated steam on the order of 1400°F. The exact temperature level in the reactor will depend upon several factors as yet unknown. It turns out to be in the neighborhood of 1200-1400°F.

It may be preferable for the coal to be preheated. In fact the coal may be first carbonized and the overhead collected independently or combined with the reactants or products. Other pretreatment to increase coal reactivity is a possibility.

An ebullating bed of uniform-sized Group VIII catalyst particles assures that the coal-char and alkali particles will not become captured by the bed, as in the case of a true fluidized bed where the particles have a size distribution. Particle durability is a condition. Refractory catalysts stabilize after the edges attrit.

The fluidized system should be variable. That is, fresh charge and catalyst can be introduced at the bottom and removed at the top of the bed at any desired rate. Additionally, any entrained catalyst requires separation from the overhead gas. It is not anticipated that the catalyst should be fully entrained. The catalyst from the reactor may be generated or discarded depending upon its form, or used elsewhere. Since unreacted coal may be present, it is recommended that regeneration and recycle be used. The type of regeneration needed will depend upon the oxidation-reduction conditions in the reactor. An independent source of H_2, synthesis gas or CO is required for activation.

The overhead product from the reactor passes through a separator to remove entrained solids. The gases then pass through a partial condenser followed by separation into an off-gas and liquid hydrocarbon and water phases.

With iron catalysts, the water phase will contain oxygenated compounds for disposal or other uses. The study of the separation of the

Catalysis and Heterogeneous Reactions 279

compounds by azeotropic and extractive distillation or other methods constitutes a research program unto itself. At some future time this phase may prove of great value, but at present the production of oxygenated compounds is to be minimized. Ammonia (or its derivatives) may also appear in the water phase.

The condensed water phase from the overhead also provides a potential means of removing solids carryover.

The hydrocarbon phase requires processing for the production primarily of gasoline by conventional refining procedures.

Carbon dioxide and H_2S also, if present, require removal. The net production of off-gas is principally methane and hydrogen with some LPG. If rich enough, the off-gas may be processed through an LPG and natural gasoline recovery system.

The use of gas recycle poses the possibility of suppressing the formation of C_1 by its presence as a reaction product, and at the same time recirculating hydrogen. Additionally, the gas recycle would aid in fluidization and provide temperature control.

The condensed water phase from the reactor overhead product may also be recycled for steam regeneration. Any oxygenated compounds which may be present can be first removed, or recycled with the water-steam to suppress further oxy-forming reactions.

The presence of sulfur is not anticipated as being of trouble in the fluidized reaction. Sulfur tolerance is much higher than previously thought at the reaction conditions, and the presence of alkali preferentially captures the H_2S driven off from the coal. This is affected by the presence of CO_2, however. As an additional precaution, a scavenger can be used to capture the sulfur (H_2S).

Impetus for the direct conversion of coal was given in a 1973 report prepared by the National Research Council, National Academy of Engineers for the Office of Coal Research (*Evaluation of Coal Gasification Technology. Part I. Pipeline Quality Gas,* R & D Report No. 74, Interim Report No. 1, National Research Council, National Academy of Engineering, 1973):

> New catalysts or improved catalysts are always a possibility. In addition, reaction rates under different conditions should be known. Catalyst research should be strongly supported on an essentially continuous basis. A catalyst for the thermally neutral and thermodynamically feasible reaction
>
> $$2C + 2H_2O \rightarrow CH_4 + CO_2$$
>
> could revolutionize coal gasification.

8.3 THE HOFFMAN PROCESS

The possibility of utilizing catalyst-induced reactions of coal as a means for reducing costs and process complexity is the inspiration for several "third generation" efforts in coal conversion.

There is the Exxon process using the catalytic effect of alkali carbonates on the coal-steam reaction. By conducting the reaction at 1200-1400°F, sufficient yields of methane are produced (at 500 psi) to justify the cryogenic separation of the methane from the carbon monoxide and hydrogen also produced, which are then recycled.

Other alkali-related processes under study include that of TRW, which utilizes alkali carbonates to enhance the coal-steam reaction to produce hydrogen (presumably involving the use of higher H_2O/ coal ratios).

Battelle has conducted work on the impregnation of coal with alkali or alkaline earth materials in order to enhance subsequent reactivity, particularly with steam in gasification.

Prior to and beyond the aforementioned work is the so-called direct methanation process, or Hoffman process (7, 8), an embodiment for producing methane directly from coal-steam systems by using alkali and nickel co-catalysts. With alkali and iron co-catalysts, liquid fuels are produced. Under investigation for a number of years, notably under a contract with the Office of Coal Research, scale-up operations are proceeding.

Significant features of the Hoffman process are as follows (8), some of which have been described in the previous section.

Depending upon catalysts and conditions, either gaseous or liquid hydrocarbon products predominate. In the case of gasification using nickel catalyst, methane is the principal hydrocarbon product. Excess steam rates will yield hydrogen, principally. In this reporting, however, gasification to produce methane will be the overriding consideration.

A potential advantage to the overall reaction is that the production of methane and carbon dioxide tends to be autothermal. The large heat requirement for gasification first to only carbon monoxide and hydrogen could be avoided. The relatively small amount of heat needed to sustain the reaction would possibly be added to the reactants prior to injection into the catalyst bed proper.

In addition to the conversion of coal and lignite, other carbonaceous materials have been gasified. These include waste paper, tire rubber, polyethylene plastic, and even manure and sewage sludge.

Catalysis and Heterogeneous Reactions

DESCRIPTION OF REACTIONS

While the exact, step-wise sequence may not yet be known, observations on intermediate products obtained by operating the overall reaction in stages indicate two possibilities:

$$C + H_2O = CO + H_2$$
$$\underline{CO + H_2 = \tfrac{1}{2}CH_4 + \tfrac{1}{2}CO_2}$$
$$C + H_2O = \tfrac{1}{2}CH_4 + \tfrac{1}{2}CO_2$$

or

$$C + H_2O = CO + H_2$$
$$CO + H_2O = CO_2 + H_2$$
$$\underline{\tfrac{1}{2}CO_2 + 2H_2 = \tfrac{1}{2}CH_4 + H_2O}$$
$$C + H_2O = \tfrac{1}{2}CH_4 + \tfrac{1}{2}CO_2$$

The initiation reaction between coal and steam will occur at lower temperatures in the presence of alkali materials, chiefly potassium or sodium carbonate. The sequence is carried to completion by use of nickel catalysts. Various commercial nickel catalysts have been found to work. These catalysts in the active form are partially reduced oxides on various support materials, which are related to catalyst durability.

EXPERIMENTAL SYSTEMS

Bench-scale studies on the integrated coal-steam-catalysts systems have been carried out in a one-inch diameter, semi-continuous flow reactor as shown in Fig. 8.1. Superheated steam at reactor conditions is introduced continuously into a charge of coal and catalysts. Provision is also made for stratification or isolation of the nickel catalyst from the coal and alkali. Alternately, all components of the charge are mixed. Both distributions to a degree approximate continuous coal feed.

Studies on nickel catalyst behavior alone were carried out in a half-inch diameter flow reactor. Provision was made for introducing various gases as reactants, chiefly CO, H_2, CO_2, and H_2S in varying combinations.

In all cases, reactor pressure was maintained by back-pressure regulators, and reaction temperature may be varied by tubular heaters. Pressures from atmospheric to 800 psi have been used, and temperatures up to 1500°F. The overall reaction is largely independent of pressure, but methane production quickly falls off above 1400°F.

CATALYST BEHAVIOR

Normally, nickel methanation catalysts are effective on streams of CO and H_2 at temperatures in the neighborhood of 700°F and are easily poisoned by sulfur. The H_2/CO ratio must be kept up to avoid carbonization and breakdown of the catalyst — and may be resolved by the addition of steam.

In the integrated system, however, with all reactants mixed, operating temperatures of 1300°F have been maintained with no catalyst breakdown. Furthermore, at the higher temperature, equilibrium is away from the formation of sulfur compounds with the catalyst (9). In bringing the reactants to temperature in the presence of alkali carbonate, and during the conversion, part of the sulfur of the coal is liberated, principally as H_2S. Capture of the sulfur (H_2S) by the alkali carbonate is subject to the carbonate-sulfide-H_2S-CO_2 reaction equilibrium, whereas reaction of the H_2S with the metallic oxide catalyst is subject to the oxide-sulfide-H_2S-H_2-CO equilibrium. The former is applicable toward controlling the latter.

Hence, as a result, the effect of sulfur from the low-sulfur coals generally used has been nil for the runs made, which were of limited duration. The sulfur content of the spent catalyst has been detected as low as 0.3%, of the same order as initially present. Other runs, conversely, showed appreciable sulfur capture by the nickel, in part a matter of catalyst distribution, and suggesting regeneration — though sulfided catalyst retains a measure of activity at these conditions.

Furthermore, the presence of sulfur may allay catalyst carbonization problems (10, 11).

The low-sulfur Wyoming coals have been found to retain much of their sulfur in the ash at these reaction conditions (possibly due in part to sulfur occurring as the sulfate).

The mechanical strength of the catalyst is of some concern. It is anticipated that a fluidized or ebullating bed will be ultimately utilized, and a high degree of strength is needed.

EXPERIMENTAL RESULTS

The ultimate objective of most work on coal gasification is the production of a high-Btu gas. For a product of the order of 900 Btu/SCF, no modification of existing burners and jets would be required. The technology of CO_2-removal is already well-developed. Hence, on a CO_2-free basis, any product which is largely methane could be used as a full substitute for natural gas. With this in mind, a selected group of results are presented in Table 8.2.

Catalysis and Heterogeneous Reactions

TABLE 8.2. Gasification of coal-steam systems.

Coal	Glenrock	Reynolds	Consol	Minnkota
Rank	Sub-bit	Sub-bit	Lignite	Lignite
Catalyst	Ni-0104	Ni-0104	G-65RS	Ni-0104
Pressure (psi)	250	250	250	250
Temperature (°F)	1200°	1400°	1350°	1400°
Gas Comp.				
CO_2	48.0	46.9	44.2	44.9
CO	--	--	--	--
H_2	1.8	3.1	8.7	5.1
C_1	50.4	50.0	47.1	50.0
SCF H_2/ton	562	855	2,733	1,689
SCF Hydroc/ton	15,725	13,783	14,795	16,556
Theo. C_1 yield	16,600	16,100	15,900	18,400
Btu/SCF (CO_2-free)	977	962	899	939

The coals, as used, were air dried from 25-30% moisture to about 15-20%. Yields are reported on an "as used" basis.

A different comparison is presented in Table 8.3 on average representative yields from various organic materials.

TABLE 8.3. Yield of High-Btu Gas

(Standard Cu. Ft. per Ton)

	Actual	Theoretical
COAL	15,000	18,000
PAPER (RAW)	12,000	--
(CHAR)	20,000	--
TIRE RUBBER	24,000	25,600
POLYETH. PLASTIC	24,000	27,100
MANURE (CHAR)	14,000	--
SEWAGE SLUDGE		
(TREATED)	6,000	7,000
(CHAR)	7,000	9,000

In addition, various coal-derived chars and tars have been gasified, and also oil shale.

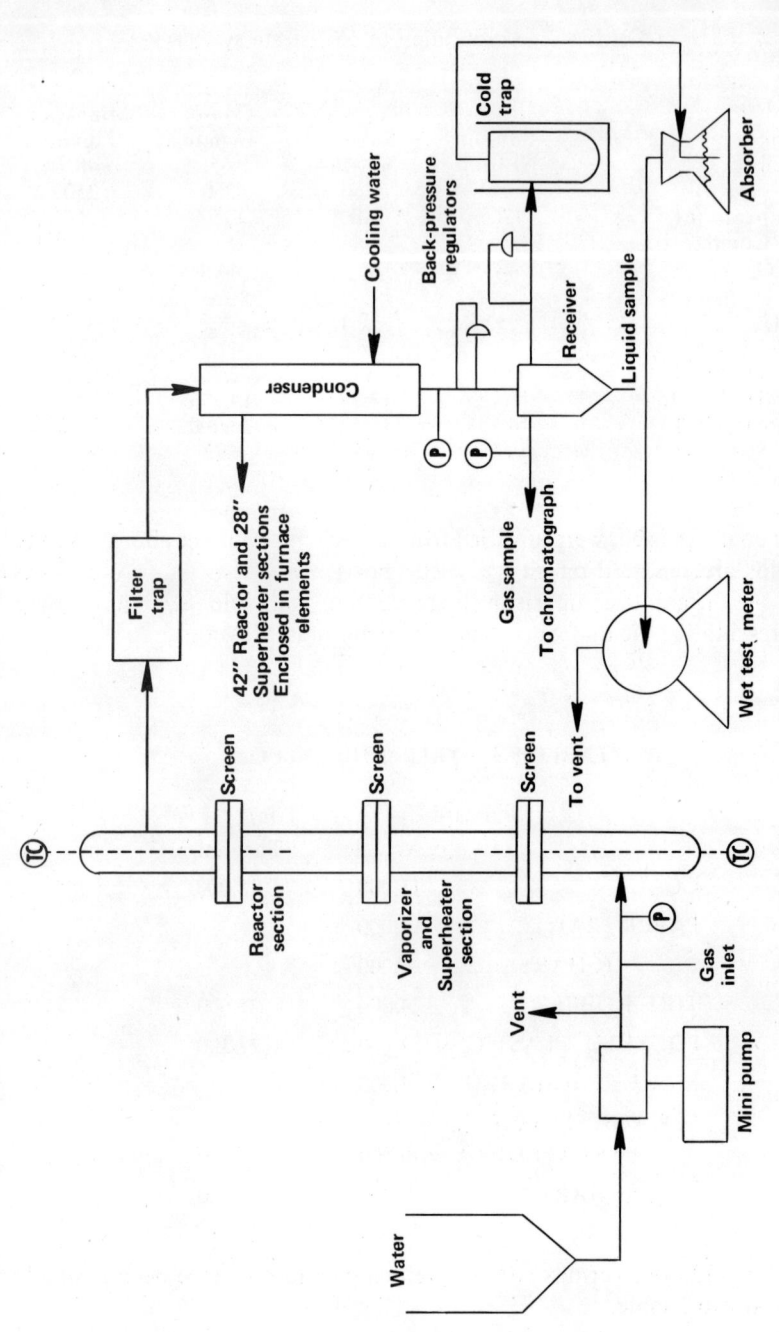

Fig. 8.1. Schematic of bench-scale setup for direct conversion.

OTHER FACTORS

The optimum pressures for processing of the cooled product gases remains to be established and will depend in part upon the final compression.

Byproduct carbon dioxide has several potential uses, aside from the limited use as dry ice. These uses include photosynthesis to accelerate plant growth under controlled conditions as in greenhouses, secondary or tertiary recovery methods for petroleum, and the production of urea for fertilizer.

Gasification with steam interfaces with the slurry pipelining of coal/water suspensions, which are 50-60% water by weight. For a coal of 60% carbon by weight, for example, an equal carbon to H_2O (steam) molar ratio corresponds to

$$\frac{0.6}{12} (18) = 0.9 \text{ lb } H_2O/\text{lb coal}$$

at 50% excess the ratio is 1.45 lb H_2O/lb coal. These feed ratios thus correspond to the proportions used in slurry pipelining.

REFERENCES

1. Hougen, O. A. and K. M. Watson, *Chemical Process Principles,* Part III, Kinetics, Wiley, New York (1947).
2. Hoffman, E. J., *The Concept of Energy: An Inquiry into Origins and Applications,* Ann Arbor Science, Ann Arbor, Michigan (1977).
3. Anderson, R. B., "The Thermodynamics of the Hydrogenation of Carbon Monoxide and Related Reactions," in *Catalysis,* Vol. IV, P. H. Emmett Ed., Reinhold, New York (1956), pp. 1-27.
4. Lewis, W. K., E. R. Gilliland and H. Hipkin, "Carbon-Steam Reaction at Low Temperatures," *Ind. Eng. Chem.,* 45, 1697-1703 (1953).
5. Ergun, S. "Kinetics of the Reactions of Carbon Dioxide and Steam with Coke," U.S. Bureau of Mines Bulletin 598 (1962).
6. Vol'kenshtein, F. F., *The Electronic Theory of Catalysis on Semiconductors,* Pergamon, New York (1963).

7. "Direct Conversion of Carbonaceous Material to Hydrocarbons," U.S. Patent 3, 505, 204, April 7, 1970. E. J. Hoffman, Inventor. Assigned to the University of Wyoming.

8. Hoffman, E. J., J. L. Cox, R. W. Hoffman, J. A. Roberts and W. G. Willson, "The Behavior of Nickel Methanation Catalysts in Coal-Steam Reactions," ACS Divisions of Petroleum Chemistry and Fuels Chemistry, American Chemical Society, 163rd National Meeting, Boston, April 9-14, 1972. *Preprints,* Volume 16, No. 2.

9. Kirkpatrick, W. J., *Advances in Catalysis,* W. G. Frankenburg et.al, Ed., Vol. III, Academic Press Inc., New York (1951), pp. 330-332.

10. Hofer, L. J. E., *Catalysis,* P. H. Emmett Ed., Vol. IV, p. 426, Reinhold Publishing Corp., New York (1956).

11. Berry, T. F., R. N. Ames and R. B. Snow, *J. Amer. Ceramic Soc., 39,* 30S (1956).

Chapter 9

THE PRODUCTION OF HYDROGEN AND CARBON MONOXIDE AS FUELS AND REACTION INTERMEDIATES

The production of either or both hydrogen and carbon monoxide is in itself of notable interest. Hydrogen and carbon monoxide and their mixtures comprise a low- to medium-Btu gaseous fuel, depending upon concentration: furthermore, mixtures in appropriate molar ratios comprise so-called synthesis gas, from which any number of compounds can be synthesized, ranging from methane and methanol to the higher hydrocarbons and oxygenated compounds, and even waxes.

Alone, hydrogen is used for direct hydrogenation of coal or for other uses. Aside from hydrogenations, hydrogen is of great significance in the fixation of nitrogen to produce ammonia and its derivatives for inorganic fertilizers. Hydrogen produced from coal would take the burden off of natural gas, the present principal source of hydrogen via reforming. While hydrogen may always be produced in two stages via gasification to produce CO and H_2 followed by the CO-shift to produce H_2 and CO_2 (which may be separated), there is also interest in one-stage direct conversions which in effect combine the two stages.

Alone, carbon monoxide may be regarded the fundamental building block of organic chemistry. Reacted with steam, hydrogen is the result, for the uses noted above. If hydrogen so produced is mixed with additional carbon monoxide in appropriate ratios, synthesis gas is the result — for conversions to organic compounds via the Fischer-Tropsch catalyzed sequence, as previously discussed in Chapter 7.

While the production, properties and usage of hydrogen is of generally accepted significance, such is not the case for carbon monoxide. Hence emphasis here is given to the latter. While readily produced as a component of producer gas (or synthesis gas), its recovery and concentration is not so simple a matter and calls for further study and development.

Though coke is generally considered the choice agent for yielding the reducing gases for metallurgical reactions, coal itself can also be used in instances (e.g., the Redox process for steel production). Another example is in the cement industry where gypsum sometimes serves as the substitute or supplement for calcium carbonate in producing the calcium oxide component of cement. Conversion of the calcium sulfate to the oxide is produced via the partial combustion products, CO (and H_2).

Reduction of the sulfate may proceed to SO_2, to sulfur, or all the way to H_2S, depending upon degree. Separation and conversion of hydrogen sulfide to elemental sulfur may be made. Gypsum has accordingly been proposed as a source for sulfur from time to time — depending upon the market for free sulfur. The prospect of overwhelming supplies as a byproduct from coal conversion more or less closes off this alternative.

An interesting and relevant aside on the production of producer gases from fixed or slowly-moving beds is the relationship to blast furnaces in making steel. Wilhelm Gumz, in *Gas Producers and Blast Furnaces*, Wiley, New York (1950), observes that "Blast furnaces may be regarded as gas producers in which other reactions such as sintering, smelting, dehydrating, deacidifying, and reducing, take place at the same time." These comments are particularly apt for the slagging gas producer, which may use additional fluxing agents, including limestone.

Thus once again we see the common generality and interrelationship for various processes using carbonaceous materials as a reactant or fuel source. Only the proportions and conditions vary, as in complete combustion versus partial combustion, in turn versus steam gasification.

9.1 HYDROGEN FROM COAL-STEAM SYSTEMS

The highly endothermic reaction of carbonaceous materials and steam to produce carbon monoxide and hydrogen,

$$C + H_2O(g) = CO + H_2$$

is catalyzed by a variety of substances, notably alkali compounds such as potassium or sodium carbonate — a phenomenon first given scientific documentation by Taylor and Neville (1). At higher temperatures, using a Na_2CO_3 catalyst, White and co-workers (2, 3, 4) found carbon monoxide to be the principal product. The subject was reviewed in Chapter 6.

Hydrogen and Carbon Monoxide

At lower tempertures circa 1200°F, Lewis et.al. (5) found that adding K_2CO_3 produced hydrogen and carbon dioxide. Thus at the less severe conditions the CO-shift also occurs:

$$CO + H_2O = CO_2 + H_2$$

At higher temperatures, say 1800°F, both carbon monoxide and hydrogen are evidently favored.

At lower temperatures, the overall reaction becomes

$$C + 2H_2O(g) \rightarrow CO_2 + 2H_2$$

This conversion has been investigated as the result of work on direct conversion (6). While conversion to hydrocarbons is obtained at lower steam rates, upon increasing the steam rate at reaction conditions the product noticeably shifts to hydrogen. This is particularly so using iron catalysts; iron catalysts also are in fact CO-shift catalysts. The same effect will occur in the presence of nickel catalysts, however, and to some extent even without catalysts (coal ash may act as a catalyst).

Additionally, complicating the situation with iron catalysts, is the fact that the steam-iron reaction also can occur:

$$Fe_xO_y + H_2O \rightleftarrows Fe_2O_3 + H_2$$

As used, iron catalysts contain partially reduced oxides, including magnetite, and may be oxidized with steam as well as reduced with hydrogen and/or carbon monoxide. They generally show an induction period when activated by a synthesis gas mixture (H_2 and CO) before being able to produce liquids via the Fischer-Tropsch reaction.

Theoretically, based on carbon content, a coal of 60 wt. % carbon could produce a hydrogen product approaching 80,000 SCF/ton by the overall reaction

$$C + 2H_2O \rightarrow CO_2 + 2H_2$$

or up to nearly 20,000 SCF/ton of methane by the overall reaction

$$C + H_2O \rightarrow \tfrac{1}{2}CO_2 + \tfrac{1}{2}CH_4$$

subject to the qualifications of equilibrium, and neglecting the hydrogen and oxygen content of the coal. Experimental results (6) on such a western coal produced an equivalent hydrogen yield of 60,000 SCF per ton. At lower hydrogen conversion, hydrocarbons were also produced. The setup was similar to that described in Reference 8 of Chapter 8 and by Figure 8.1.

While it is shown that hydrogen can be produced in one stage from coal-steam systems, the overall heat requirement remains to be considered. Though for the production of methane and carbon dioxide the overall reaction tends toward heat balance, such is not the case for the production of hydrogen and carbon dioxide:

$$C + H_2O = CO + H_2 \qquad +32.4 \text{ k cal/g-mole}$$
$$\underline{CO + H_2O = CO_2 + H_2 \qquad -8.8}$$
$$C + H_2O = CO_2 + 2H_2 \qquad +23.6$$

The overall reaction tends to remain considerably on the endothermic side.

Whether economical means could be utilized to supply the heat remains to be seen. The most obvious are external sources, and the introduction of oxygen — though the latter may have some effect on the course of the reactions. The concept of converting coal continues to be of interest, however, as per the work underway by TRW.

Other means investigated for the generation of hydrogen include electrofluidic gasification, diagrammed in Figure 9.1, and the steam-iron process, diagrammed in Figure 9.2. Both are in support of the Hygas process.

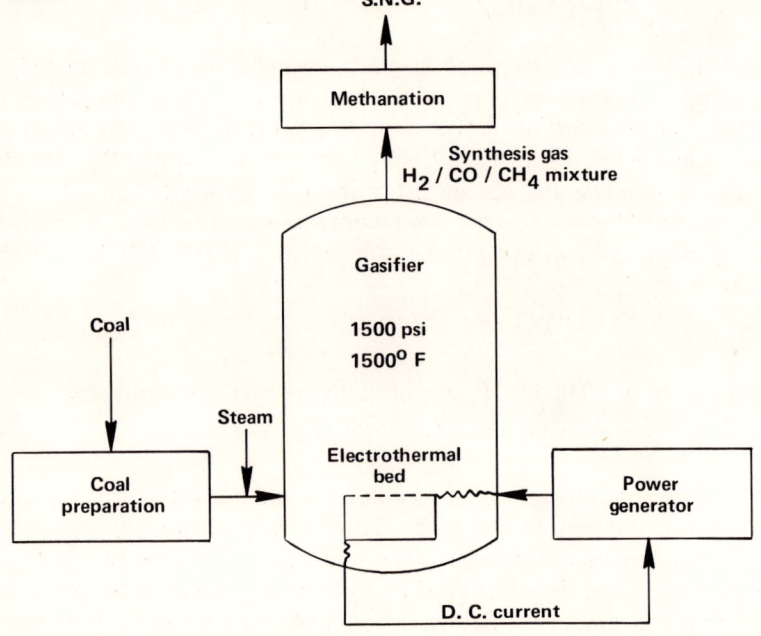

Fig. 9.1. Electrofluidic gasification.

Hydrogen and Carbon Monoxide

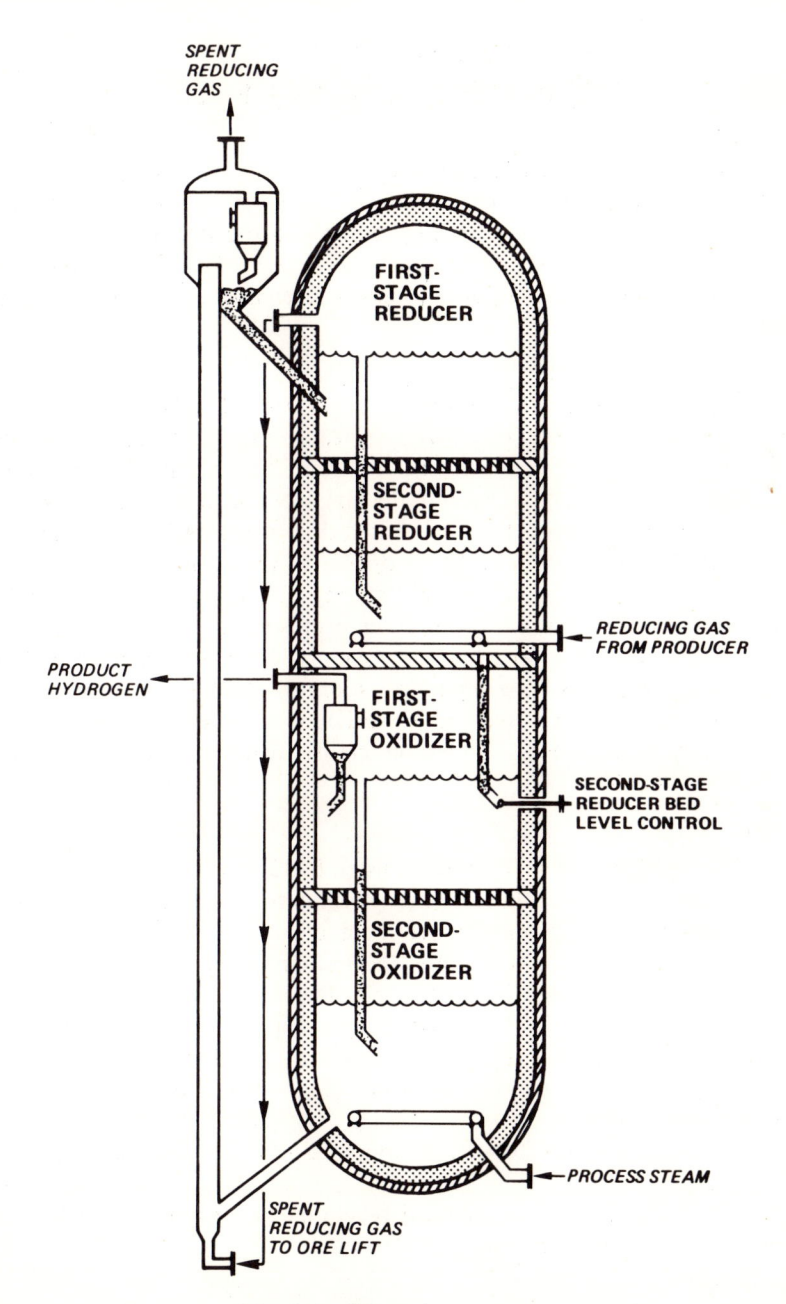

Fig. 9.2. Steam-iron process.

9.2 CARBON MONOXIDE PROPERTIES AND SOURCES

Carbon monoxide is a colorless, odorless, toxic gas, present in the atmosphere of industrial areas often to the extent of several parts per million. Even though large quantities of CO are discharged into the atmosphere by incomplete combustion of carbonaceous fuels, accumulation does not occur because it decomposes slowly according to the equation

$$2\ CO = C + CO_2$$

It has been found that in the surface waters of the western Atlantic there is a marked diurnal effect with respect to the CO concentration and that the concentration is dependent upon the CO concentration in the atmosphere (7). Despite the fact that the absolute concentration is low (5×10^{-5} ml/l of surface water) it is estimated that over a period of a year, the ocean could contribute 9×10^{12} g of CO to the atmosphere. Thus the ocean could act as a source or sink tending toward equilibrium with the carbon monoxide concentration in the atmosphere.

For the purposes here, however, the predominating concern will be with fossil fuels as raw material sources. Carbon monoxide has heretofore had some use as an intermediate in the chemicals industry. In the future it may have a considerably greater usage in a coal conversion industry.

PROPERTIES

Selected physical properties are listed by Green (8). Compressibility, heat capacity, and thermal conductivity equations as well as viscosities, flammability, and explosibility are included.

A revised and consistent set of thermodynamic properties of carbon monoxide are presented by Canjar (9). The data covers a temperature range of from 300 to 1200°F and a pressure range of from 10 to 2500 psig.

Hurst and Stuart (10) have tabulated thermodynamic values of liquid and gaseous CO for a temperature range of 70-300°K and pressures to 300 atmospheres.

Solubility data for CO in a number of compounds is given by Seidell (11).

A few representative chemical reactions of carbon monoxide with selected compounds are given as follows:

Oxidation:

$$CO + \tfrac{1}{2} O_2 = CO_2 \qquad \Delta H_0 = -123{,}642 \text{ Btu/lb-mole}$$

Hydrogen and Carbon Monoxide

Carbon monoxide may also be oxidized by oxides of several metals at elevated temperatures (8). At high CO pressures, CO reacts with certain metals, particularly nickel to form carbonyls, e.g., $Ni(CO)_4$.

Decomposition:

Carbon black acts as a catalyst in the decomposition of CO to C and CO_2 at temperatures of 1300 to 1900°F and pressures of 10 to 40 atmospheres.

Water:

CO reacts with water (steam) by the well-known shift reaction

$$CO + H_2O = CO_2 + H_2$$

CO may also react with water to produce formic acid at temperatures of 150 to 250°C and pressures of 1000 atm. (12).

Hydrogenation and Other Reactions:

A discussion of the reaction of CO with hydrogen has been presented in Chapter 7. Reaction with liquid hydrocarbons under a variety of conditions results in the formation of ketones, aliphatic acids, esters, and unsaturated compounds such as aldehydes and amides.

Carbon monoxide reacts with dimethylamine to produce dimethylformamide (DMF) (13):

$$(CH_3)_2 NH + CO = (CH_3)_2 N\overset{O}{\overset{\|}{C}}H$$

Propionic acid may be made by reacting ethyl alcohol, ethyl ether and CO under high pressure:

$$C_2H_5OH + CO = CH_3CH_2COOH$$
$$(C_2H_5)_2O + 2CO + H_2O = 2CH_3CH_2COOH$$

Considerable quantities of phosgene are produced in the Gulf Coast area by reacting CO from synthesis gas and chlorine:

$$CO + Cl_2 = COCl_2$$

Most formic acid is made by absorbing CO in caustic soda followed by neutralization.

$$CO + NaOH = NCOONa \xrightarrow{H^+} HCOOH$$

or by reaction with methanol at high pressure followed by conversion to formamide, which is hydrolyzed:

$$CO + CH_3OH = H\overset{O}{C}\text{-}OCH_3$$

$$H\overset{O}{C}OCH_3 + NH_3 = HCONH_2 + CH_3OH$$

$$HCONH_2 \xrightarrow[H_2O]{H^+} HCOOH + NH_3$$

Olefins react with CO and water (Koch reaction) in a two-step process taking place at about 300 atm. and in the presence of strong acids to form tertiary carboxylic acids. The reaction is:

$$(CH_3)_2C = CH_2 + CO + H_2O \rightarrow (CH_3)_3C\text{-}COOH \quad \text{(Pivalic Acid)}$$

HCN may be produced by the indirect reaction between CO and ammonia. That is, carbon monoxide first reacts with methanol to yield the formate. Ammonia is then reacted with methyl formate to produce formamide for dehydration to HCN:

$$CO + CH_3OH \rightarrow HCOOCH_3$$
$$NH_3 + HCOOCH_3 \rightarrow HCONH_2 + CH_3OH$$
$$\underline{HCONH_2 \rightarrow HCN + H_2O}$$
$$CO + NH_3 \rightarrow HCN + H_2O$$

Carbon monoxide is generally very reactive with copper salts, in particular copper ammonium salts. This provides methods for analysis and for selective removal.

CO of course methanates with H_2 over Ni catalyst, a means for purifying H_2 for ammonia synthesis, where the presence of CO adversely affects catalyst life.

9.3 GENERATION OF CARBON MONOXIDE

The principal method for generating carbon monoxide is as a component of producer gas via partial oxidation. The topic has been covered in Chapter 4 from a theoretical and experimental standpoint. Commercial gas producers will accordingly be discussed here, and some more novel methods presented for preparing carbon monoxide.

Hydrogen and Carbon Monoxide

GAS PRODUCERS

Gasifiers come in many and varied forms: fixed or slowly-moving beds utilizing grates, fluidized systems, and entrained systems. Air-blown systems yield a low-Btu gas; steam-oxygen blown systems yield a medium-Btu gas.

Air-blown gasifiers yielding a low-Btu gas uniformly operate at atmospheric or near-atmospheric pressures, and even under vacuum. Whereas oxygen blown gasifiers operate not only at atmospheric pressures but at higher pressures in order to increase the methane yield, an objective noted earlier.

Air-blown gasifiers are also cycled with steam, the classical method for producing both CO and H_2 in appreciable concentrations, as water gas or synthesis gas.

Several examples of air-blown producers from the past are shown in Figures 9.3 through 9.5.

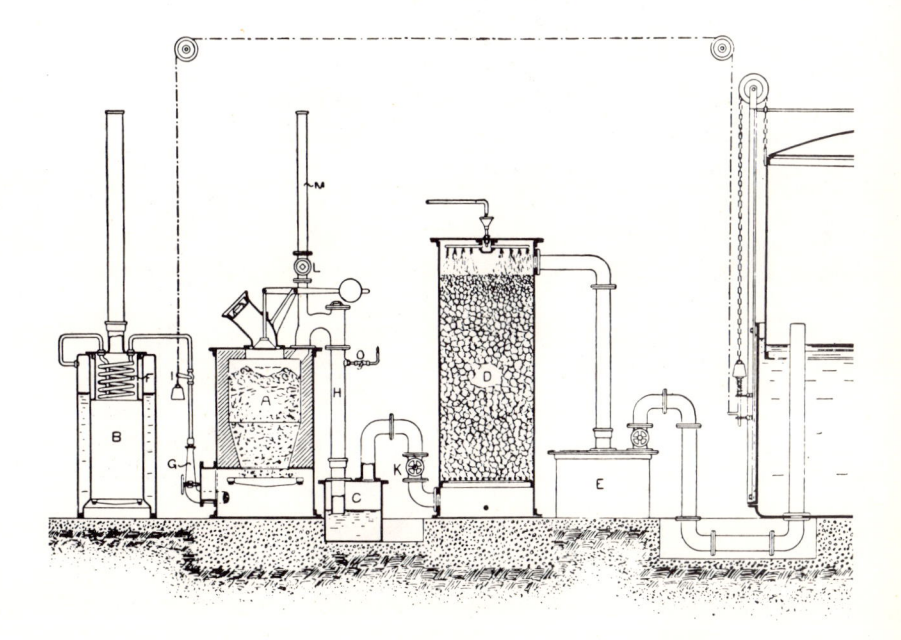

Fig. 9.3. Gas-producer [from C. W. Askling and E. Roesler, *Internal Combustion Engines and Gas-Producers*, Charles Griffin and Company, London (1912)].

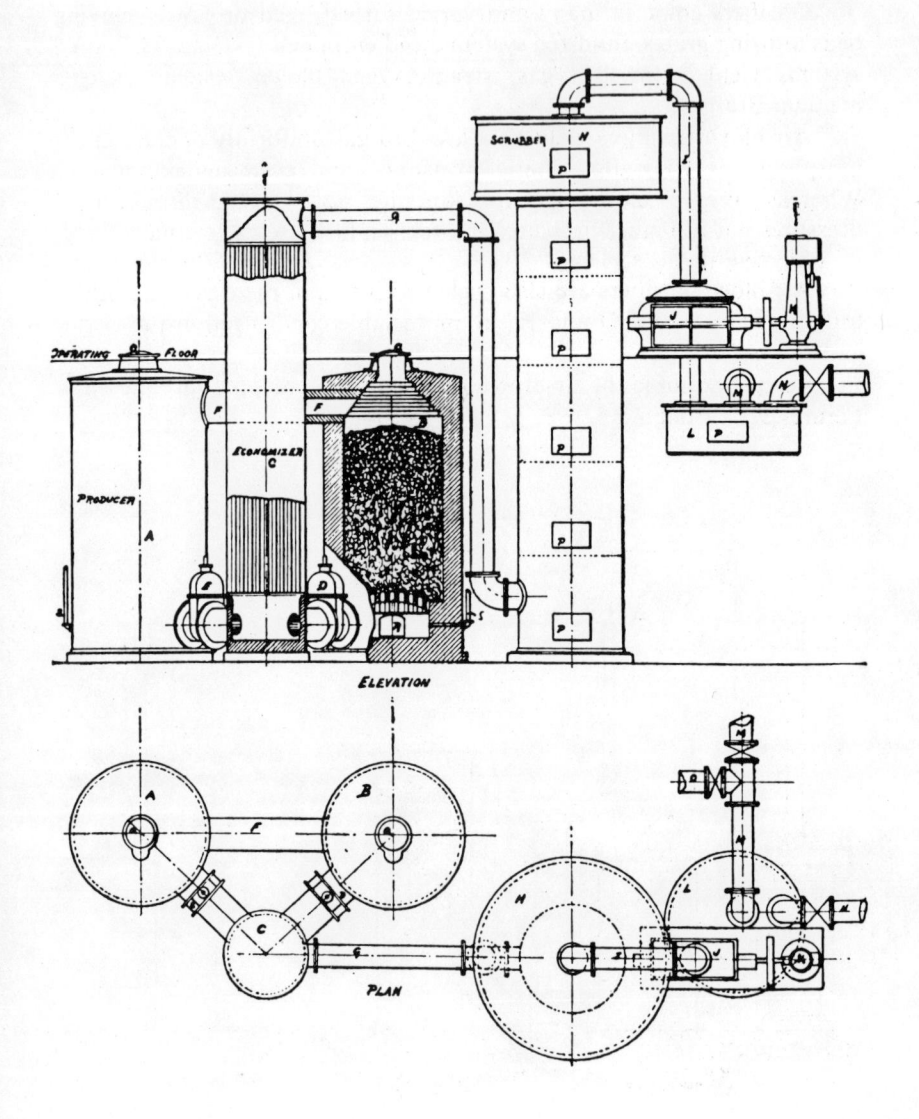

Fig. 9.4. Loomis gas-producer [from S. S. Wyer, a Treatise on Producer-Gas and *Gas Producers*, McGraw-Hill, New York (1906)].

Hydrogen and Carbon Monoxide

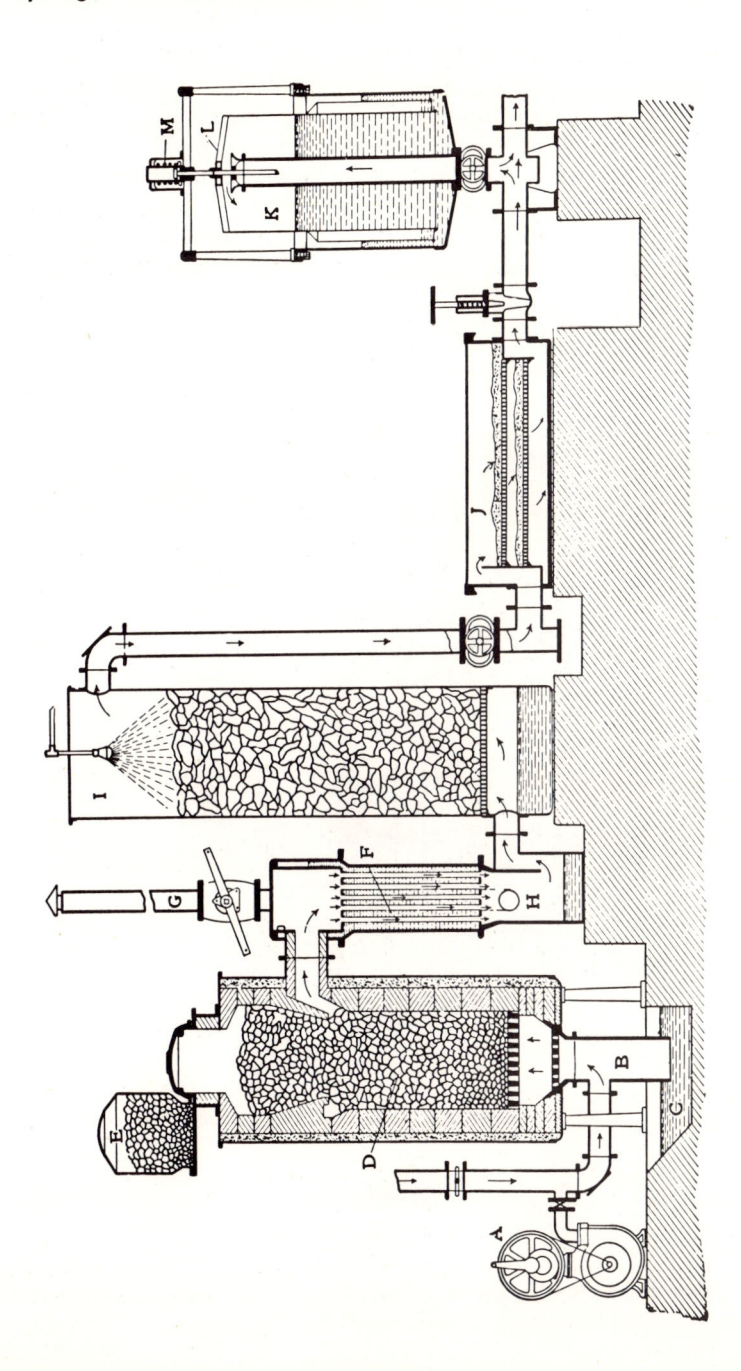

Fig. 9.5. Pintsch Suction Gas-Producer (from Wyer, *op. cit.*).

Names associated with more modern versions of gasifiers are those of Lurgi, Winkler, Koppers-Totzek, Wellman-Galusha, BASF-Flesch-Demig, etc., and the list goes on. Representative schematics are shown in Figures 9.6 through 9.14. These gasifiers are variously slowly-moving bed, fludized and entrained systems, using sized or pulverized feed.

A more unusual embodiment is the Rockgas molten salt gasifier which injects coal and air into a molten sodium carbonate. This is as much a matter of hot-gas cleanup and is described in a subsequent section (see Figure 9.15).

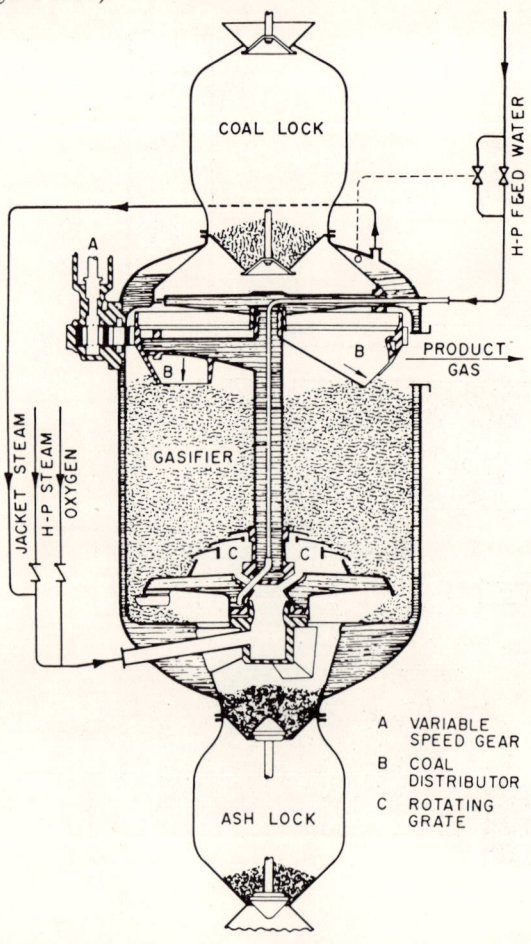

Fig. 9.6. Lurgi pressure gasifier [from J. A. Linton and G. C. Tisdall, *Coke and Gas,* 19, 402-7, 442-7 (1957). Consult *Chemistry of Coal Utilization,* Supplementary Volume, Wiley, New York (1963), p. 960.]

Hydrogen and Carbon Monoxide

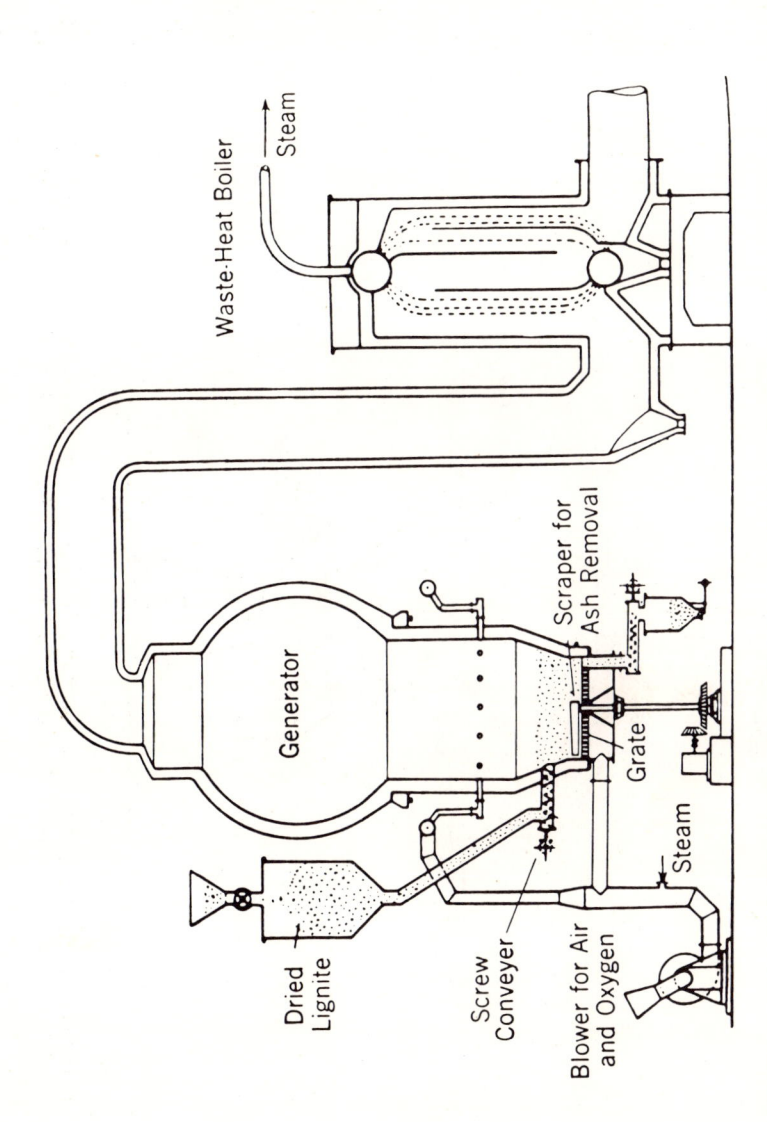

Fig. 9.7. Winkler gasifier [From C. Bosch, *Chem. Fabrik*, *1*, 1-10 (1934). Consult *Chemistry of Coal Utilization*, Vol. II, Wiley, New York (1945), p. 1666.]

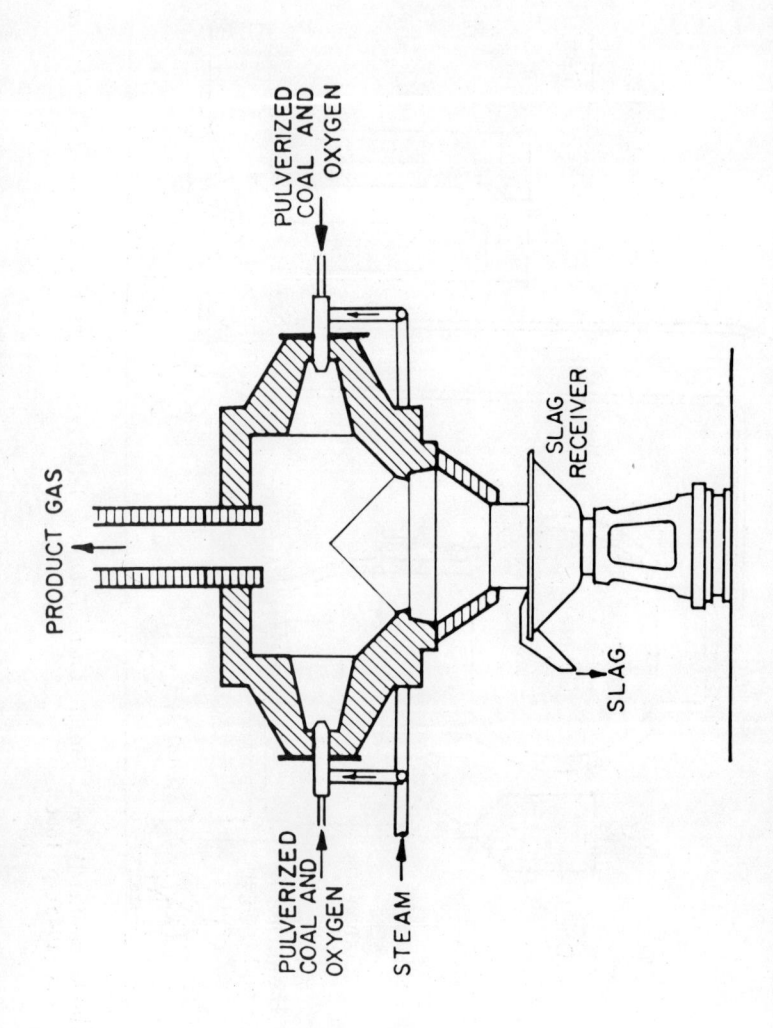

Fig. 9.8. Koppers-Totzek gasifier [From K. H. Osthaus, *Intern. Conf. Complete Gasification Mined Coal*, Liege, 1954, 255-65. Consult *Chemistry of Coal Utilization*, Supplementary Volume, Wiley, New York (1963), p. 972.]

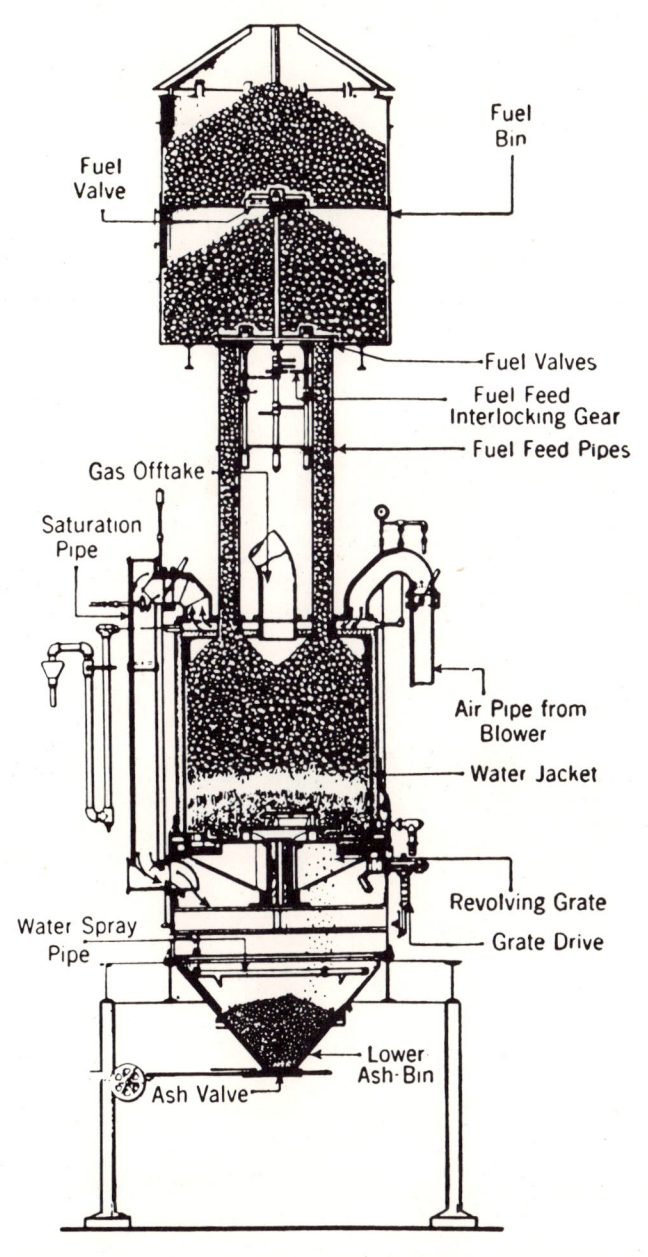

Fig. 9.9. Wellman-Galusha gasifier [From B. J. C. van der Hoeven in *Chemistry of Coal Utilization,* Vol. II, Wiley, New York (1945), p. 1659.]

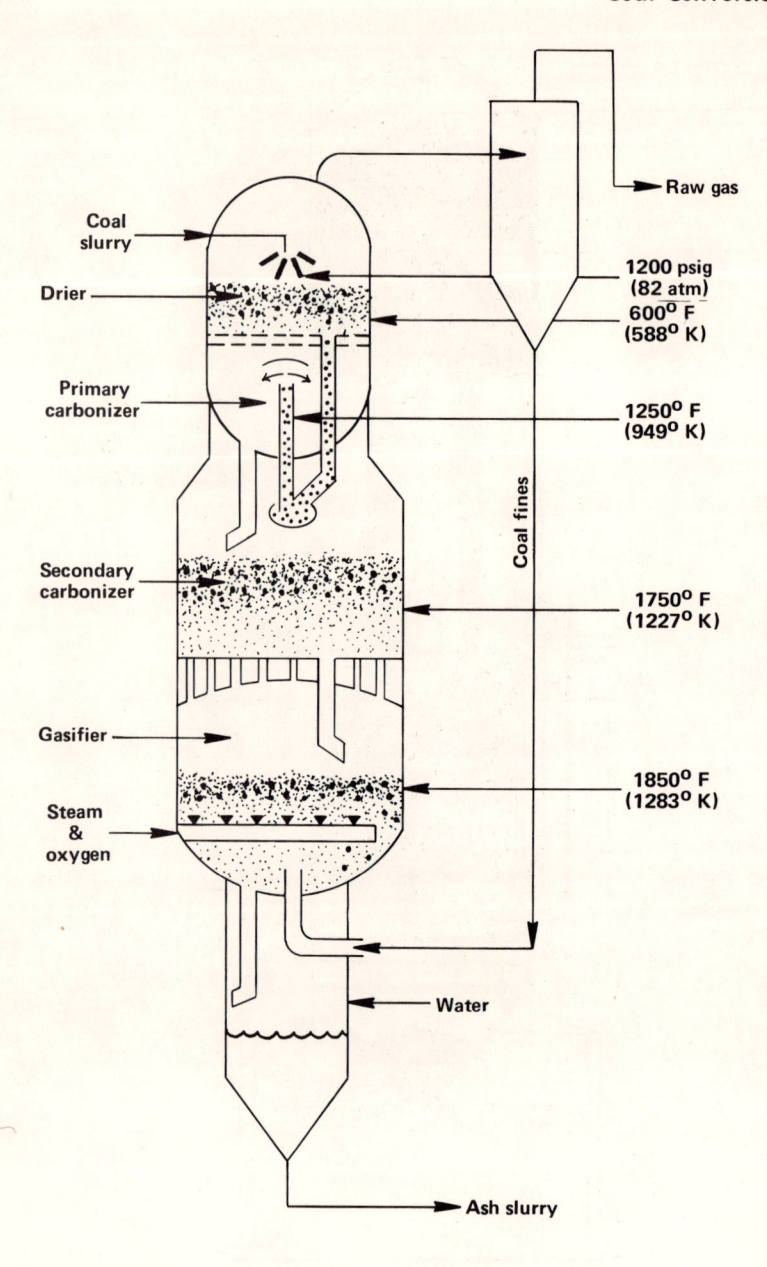

Fig. 9.10. Hygas gasifier.

Hydrogen and Carbon Monoxide

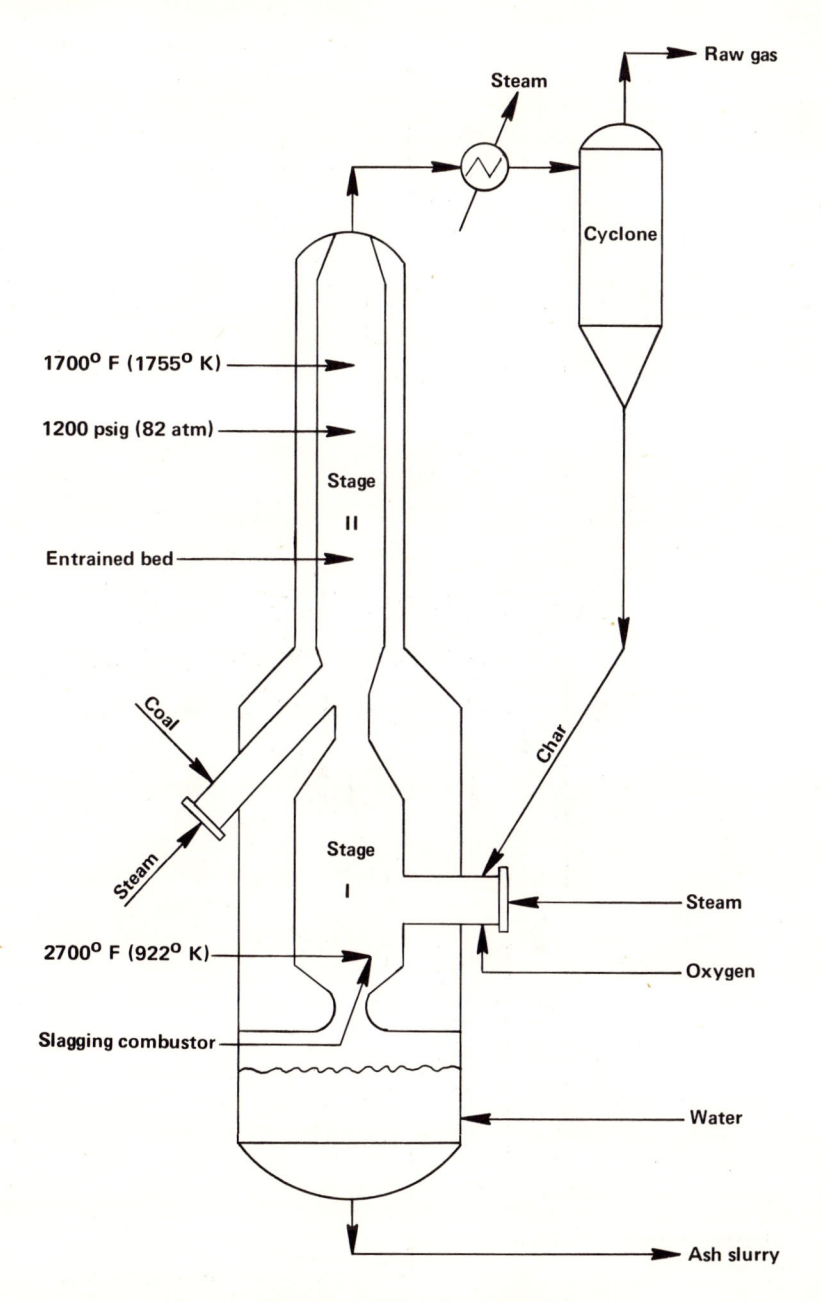

Fig. 9.11. Bi-Gas gasifier.

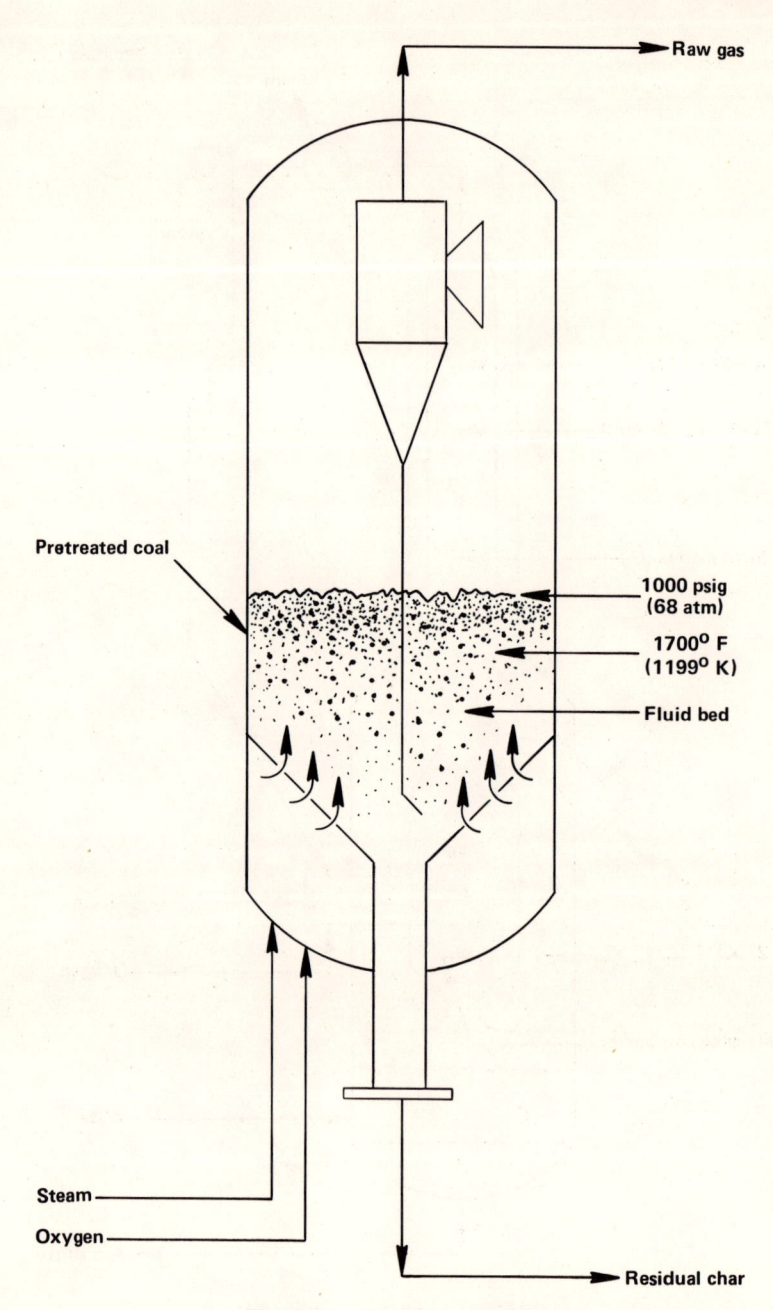

Fig. 9.12. Synthane gasifier.

Hydrogen and Carbon Monoxide

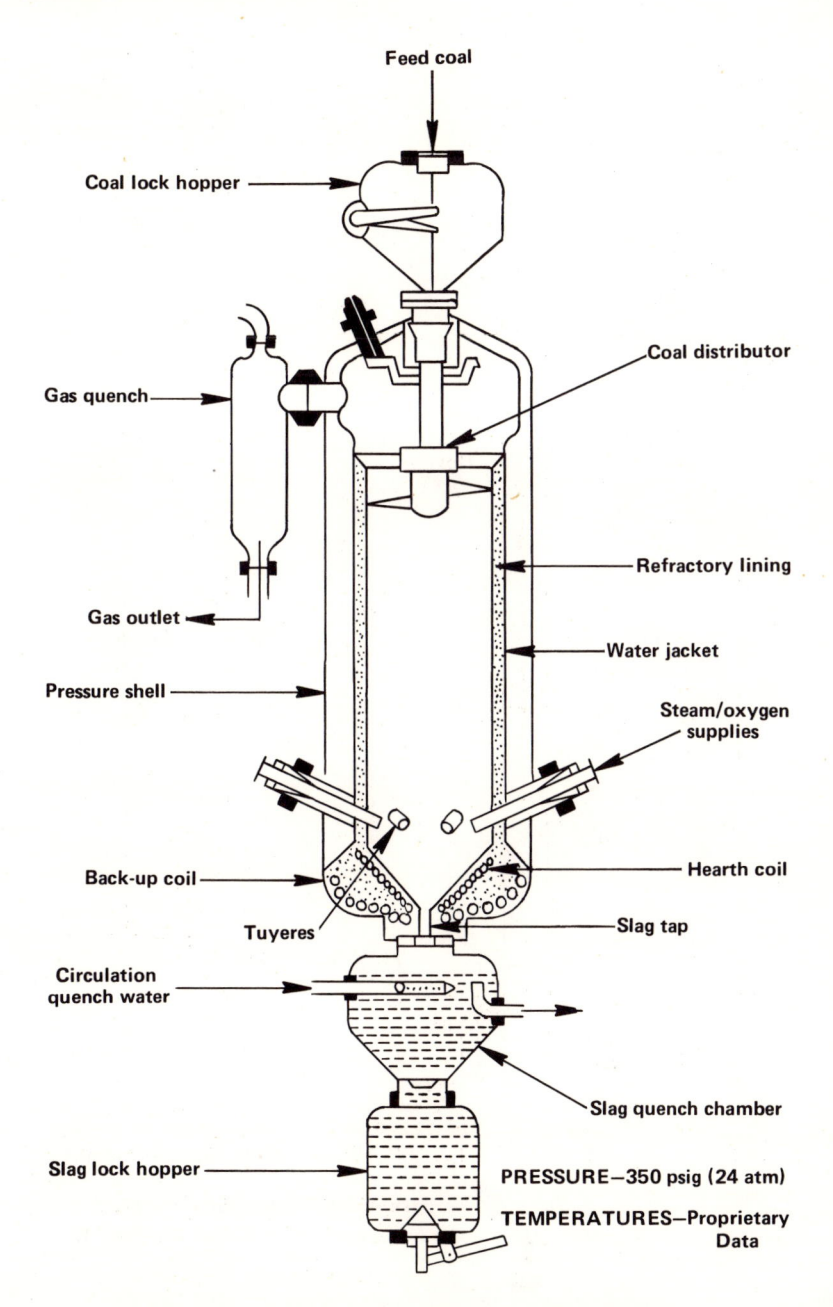

Fig. 9.13. British Gas Corporation (BGC) gasifier.

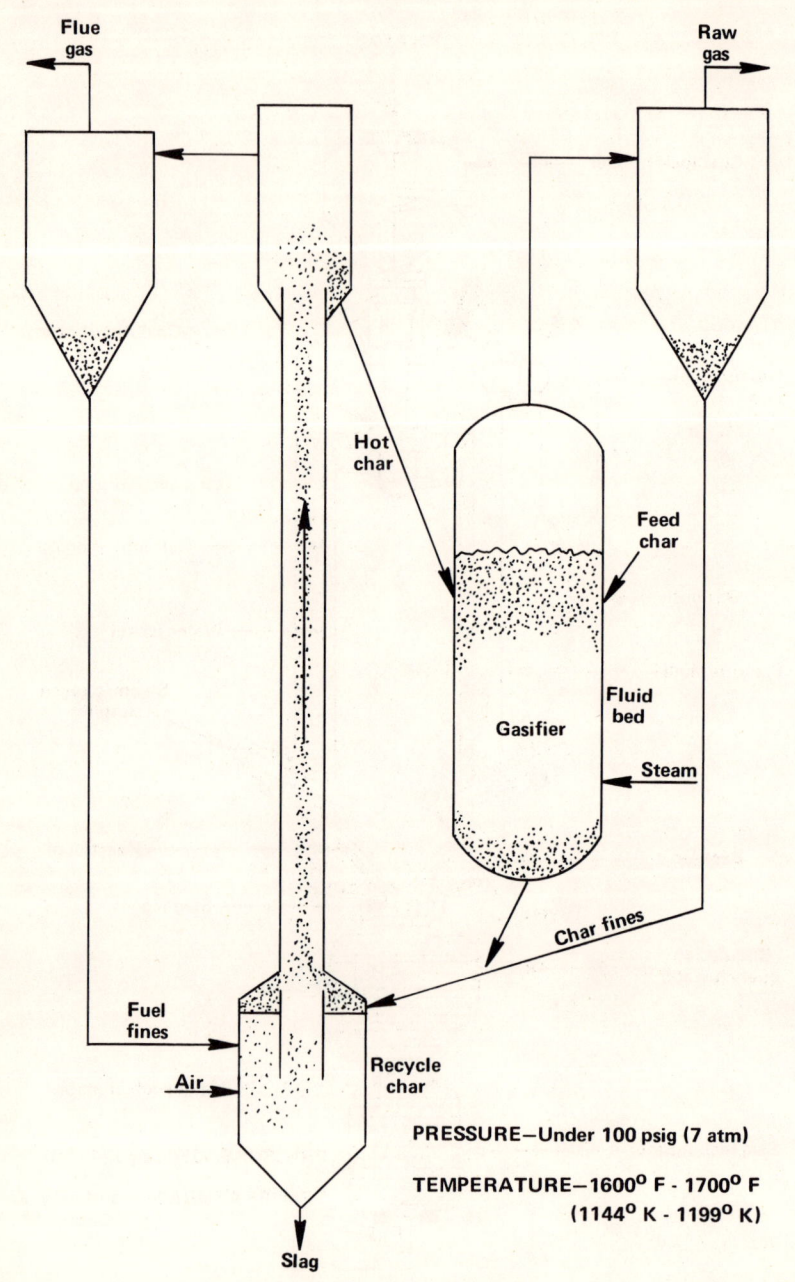

Fig. 9.14. COGAS gasifier.

Hydrogen and Carbon Monoxide

Returning to the classic embodiment, there are two main methods for accomplishing the gasification process in fixed or moving-bed gasifiers. In the countercurrent process, air is fed into the coal bed from underneath the grate supporting the coal. Gas generated in this region passes upward through the bed. Slag and ash fall through the holes in the grate. As the gas moves upward, the oxygen reacts with the coal to form chiefly CO_2. The CO_2 is subsequently reduced to CO as it passes through the burning coal. As the CO moves upward it distills volatiles in the coal above the reduction zone. Finally, it dries the coal in the topmost region of the bed. For coals of high volatile content, the product gas contains tar obtained from the distillation region. Excellent heat exchange, low exit gas temperature, low gas residence time and high solid residence time characterize the countercurrent movement.

In the inverse process, the gas and the coal move in the same direction. The air blast is admitted into the combustion zone rather than the slag region. The gases therefore pass through a combustion region rather than through a region of fresh coal. Below the oxidation region are the reduction and slagging zones. The vapors formed in the coking region pass through the oxidation zone where combustion takes place. The remaining tars are decomposed in the reduction zone. The indirect process, therefore, because of increased gas yields, is considered to be a more efficient system.

Concurrent gasification in a fluidized bed results in a marked decrease in reaction temperature. The CO/CO_2 ratio is slightly less while the nitrogen content of the product is about 2% greater. Additional preheat is required, however, to provide for drying and heating the volatiles, moisture and ash to the reaction temperature.

Whatever the system, it is desired to complete the reduction of CO_2. This is achieved by having enough carbon present for conversion and a reaction temperature high enough to approach equilibrium in a reasonable residence time. The fluidized bed reactor contains considerably less carbon and operates at fairly uniform temperatures, thereby slowing down the process.

In pulverized coal producer operations using entrained systems, an appreciable amount of unconsumed carbon remains at the end of the reaction, necessitating the washing of the gas product to free the fines. The maximum conversion attainable is about 80% without catalysts. This is with residence times of two seconds or more, whereas the residence times for complete combustion with excess air is on the order of 0.5 sec. or less, as mentioned in Chapters 3 and 4. These figures are affected by temperature and by catalyst.

The characteristic which is outstanding in the slagging producer is the removal of the ash as a liquid or melt rather than as a solid. The advantages of this type of producer include high production, utilization of low-grade fuels, adaptability to different gas demands, and the use of slag as a by-product. Because of the unusually high reaction temperatures, any addition of steam, even if superheated, results in lowering the temperature. High-pressure blasts pass through the bed undecomposed thus causing further lowering of temperatures. A successful European slagging producer is described as using high-ash fuels or refuse fuels with an air blast heated to 750-1000°F (14). Temperatures in the slagging zone were of the order of 3500°F, the exit gas was low in temperature and was low in CO_2 (about 3.5%). The capacity was high and the throughput several times the rate of normal producers. In one installation, limestone was added as a flux and the residue used for the manufacture of cement. In another, metal wastes were charged with the fuel and the product was tapped off with the slag.

Gasifiers, particularly if operating at lower temperatures, will produce condensed tars and oils along with aqueous condensate. This is affected not only by temperature but by the addition of steam. Not only does pyrolysis and reaction occur but a steam distillation effect due to the presence of moisture or steam. The tars and oils may be recycled or treated for use as a fuel. Typical liquid yields may be 10% of the coal feed. The lighter oils are composed principally of hydrocarbons whereas the heavier tars contain oxygenated materials. Furthermore, the oily and tar-like products may be only partially miscible. The more volatile fractions will appear in the product gases.

The aqueous phase contains ammonia or its derivatives, which may also constitute a byproduct of significance. Phenols, even cyanides, are present and require removal and further processing.

NOVEL GASIFICATION SCHEMES

A survey of the literature shows that there is no dearth of ideas for gasifiers. For example, a patent (15) describes the operation of an experimental overfeed updraft producer haivng gasification rates of 223 lb/ft./hr. Requirements for such a producer of full size should include a separator capable of separating about 2% of the coal feed from the gas stream. The product stream should be kept above 1000°F to insure that the tar materials do not condense in coal passages.

Another patent (16) describes a process whereby a carbonaceous fuel stream is fed intangentially into a reaction chamber where it reacts with

Hydrogen and Carbon Monoxide

an airstream fed outwardly into the column. The mixture is discharged into an enlarged combustion zone. The combustion zone is maintained at a pressure of about 3 atmospheres. The burning of the fuel-air mixture is accomplished in less than 5 seconds.

In still another (17), coal particles less than 40 mesh are subjected to a limited supply of free oxygen. The coal stream is then separated into coarse and fine fractions. The fines fraction is reacted with oxygen and steam in a fluidized system to yield CO_2 and steam. The coarse fraction is fluidized in a second column where it is reacted with the gas stream at a temperature above 1800°F and pressures greater than 6 atmospheres. The CO_2 is substantially totally reduced to CO.

A coal and water slurry in turbulent flow is passed through a heating zone to vaporize the water and form a suspension of coal (18). The suspension is reacted with free oxygen at a temperature of 2000-3000°F. The gas product is chiefly CO and H_2 mixed with free carbon. The suspended carbon may be subsequently removed and converted to CO and hydrogen.

Totzek (19) produced CO and hydrogen by using a preformed suspension consisting of 2.2 lb. of dried pulverized coal, 22 cu. ft. of oxygen and 3 cu. ft. of steam at less than 100°C injected into a reaction chamber held above the ignition temperature. Simultaneously, 30 cu. ft. of steam at 300-500°C is injected into an annular nozzle. The hot reaction mixture then enters a second reaction chamber where it is allowed to react endothermically with the superheated steam. A product containing greater than 83% combustibles (CO and H_2) was achieved.

A somewhat different method of gasification using electrodes in a fuel bed has been reported (20). In the process, a vertical retort is charged with coke from the top and heated to 2700-3500°F by passing an electric current between two electrodes while CO_2 is passed countercurrently from the bottom of the retort. The gas contact time is 1-10 seconds and the carbon content of the discharge coke is about 85%. A feed rate consisting of 144 lb. of coke/hr. and 246 lb. of CO_2/hr. resulted in the formation of about 300 lb. of CO/hr. About 5% of the product was passed around the removal conduit and served to dry and calcine the feed coke.

In a patented process (21) enrichment of producer gas is made by the addition of anhydrous $CaCO_3$ to the carbon in the producer. The equation for the reaction may be written as:

$$CaCO_3 + 3C + O_2 = CaO + 4CO \qquad -20 \text{ kcal/g-mole}$$

A thermodynamic balance, however, requires an excess of 0.75 moles of carbon. Thus 1308 kg. of CO could be produced from 1000 kg. of $CaCO_3$

and 440 kg. of carbon. The residual CaO and ash may be utilized as a fertilizer for sour soils.

Lavrov et.al. (22) have investigated the optimum conditions for enrichment of gases with CO according to the equation

$$CO_2 + H_2 = CO + H_2O$$

Experiments were carried out at different temperatures with several gas velocities under the influence of Fe-Cu catalyst. The results showed:

1. The degree of CO_2 conversion depended upon the amount of hydrogen in the original gas.
2. The Fe-Cu catalyst was very efficient and could be used effectively at a temperature of 600-800 degrees.
3. The catalyst was unstable at high temperatures.
4. The maximum conversion (30-45%) was achieved with a contact time of 0.05 to 0.25 seconds.

Enriched fuel gases may be purified by a number of methods — e.g., by washing with an organic polar solvent such as MeOH or Me_2CO at -30 degrees. In the process, water vapor, H_2S, COS, and small amounts of other gases may be separated. (See Chapter 10 — the Rectisol process, etc.)

Cost estimates for producer installation and operation are difficult to ascertain since they depend largely upon local conditions. In general however, as the size of the producer increases, the cost increases less rapidly than the output capacities. An investigation of the economics of producer operation costs has been made by Waring and Foster (23), but is now out of date.

CARBON MONOXIDE FROM BASIC OXYGEN FURNACES (BOF)

Steel converters using oxygen emit large quantities of gases including carbon monoxide mixed with iron dust. The combusted product gas is subsequently cooled, filtered to remove dust and vented to the atmosphere. By collecting the uncombusted gases, the carbon monoxide so generated can be used for other purposes.

Namy et.al. (24) determined two possible methods of recovering gases from converters without combustion. The apparatus consisted of a gas collecting system and a cooling and recovery system. The collector system was composed of a hood fitted with a water-cooled skirt that may or may not be connected directly to the converter. If connected directly to the converter, it forms a 'gas tight' system. If not connected directly,

Hydrogen and Carbon Monoxide

successful operation depends upon control of gas flow and pressure conditions to prevent in-flow of air and out-flow of gas. Washing equipment consisted of two stages of venturi scrubbing with filtration and recycling of the water. Product gases contained an average of 67% carbon monoxide.

Skelly (25), in an article describing the economic advantages of IRSID-CAFL processes, listed product gas compositions as 70% CO, 16% CO_2, 14% N_2. Heating value of the product was 230 Btu/SCF. Product value of the removed gas and dust amounted to approximately $0.43/ton. of steel.

Roederer et. al (26) describe operating data for the IRSID-CAFL installation at Dunkerque, France. Collection of gas was begun during the blow when the CO content was greater than 40% and the oxygen content less than 2%. A peak of about 70% CO was reached during the blow. Product recovery was 90%. The total volume of CO yield was 1800 SCF/ton of hot metal.

The OG process is similar to IRSID-CAFL and is described (27) as a method of recovery of oxygen top blown converter gas in the unburned state. The process was developed by the Yawata Iron and Steel Co. and Yokoyama Engineering Co. Ltd. In the process, a hood of improved design draws gas into the cooler when placed over the mouth of the converter. The cooler depends upon radiation and convection to reduce gas temperatures to 750 to 900°F. The gas then passes through a spray chamber, a venturi saturator, and the venturi to saturate and cool the gas to 122-140°F. The gas is then placed in a water-sealed gas holder.

A typical analysis of the middle blow shows 80-85% CO, 10-15% CO_2, and less than 2% oxygen. During the early and late stages of blow, when recovery is not taking place, the gas is ignited and exhausted.

The advantages of the above gas recovery system are said to be as follows:
1. Smaller volume of gas to handle than in a fully-burned system.
2. Comparatively low temperatures in the hood.
3. Dust collection relatively simple because of grain size.
4. Recovered gases are marketable and relatively free of hydrogen.

Probably the main limitation for recovery from BOF off-gases lies in the relatively small amount as compared to the needs for fuel. It becomes more a matter of in-plant usage.

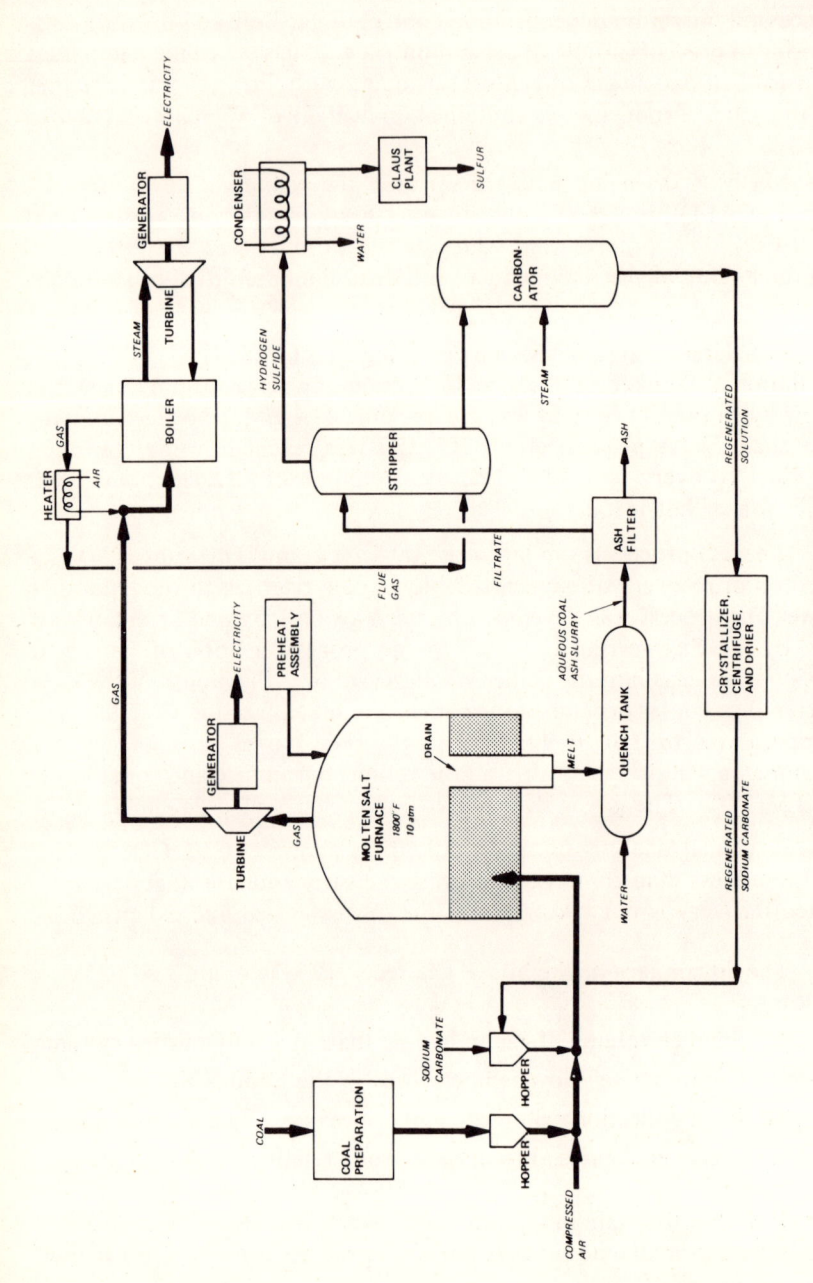

Fig. 9.15. Rockgas gasifier (ERDA).

Hydrogen and Carbon Monoxide

HOT-GAS CLEANUP

The producer off-gas is ordinarily cooled and scrubbed. Transport and use is at ambient temperatures. Scrubbing eliminates particulates and can also remove sulfur (which exists as hydrogen sulfide in the reducing conditions of the gasifier). Cooling and scrubbing are combined by using a spray tower or other direct contact methods.

Cooling of the gas exerts a thermal penalty, however, which may be as much as 25%. That is, the Btu-rating of the cold gas is less than that of the hot gas by perhaps 25%, depending upon temperature levels and composition.

This fact has led to the investigation of methods to produce a clean, high-temperature producer gas for on-site use. One such process is the Rockgas or molten salt combustion and gasification process diagrammed in Figure 9.15. Pulverized or sized coal is partially combusted with air in the presence of molten sodium carbonate which fluxes with the ash and also captures the sulfur at the temperature levels involved.

While in principle the operation appears feasible enough, there are the problems of containment of the molten alkali carbonate, and slagging reactions which occur between the aluminates and silicates of the ash, resulting in alkali losses.

CONVERSION OF CARBON DIOXIDE TO CARBON MONOXIDE

There may be instances when reduction of a stream of carbon dioxide possesses advantages over the more conventional method of first producing a producer gas — which requires the separation and recovery of carbon monoxide from an inert.

First, the generation and recovery of carbon dioxide involves an easier processing sequence. Furthermore, at times streams of relatively pure CO_2 may be available from other processes.

Second, reduction of CO_2 with coal or carbon conceivably can yield a stream essentially of carbon monoxide which can be utilized without further major purification steps.

As a disadvantage, the reduction of CO_2 is endothermic, requiring considerable quantities of heat:

$$CO_2 + C = 2CO \qquad +41{,}410 \text{ cal/g-mole}$$

This of course requires a heat source, which could be supplied by combustion to CO_2:

$$C + O_2 = CO_2 \qquad -94{,}390 \text{ cal/g-mole}$$

The net then becomes

$$CO_2 + C = 2CO$$
$$\underline{C + O_2 = CO_2}$$
$$2C + O_2 = 2CO \qquad -52{,}980 \text{ cal/g-mole}$$

which could be utilized toward power generation, etc.

OTHER METHODS FOR THE PRODUCTION OF CARBON MONOXIDE

While other raw materials or intermediates may be converted to carbon dioxide, carbonaceous sources remain the fundamental raw material for any large-scale application. A few more or less novel approaches, however, have been selected from the patent literature as follows, illustrating the diversity of methods.

Carbon Monoxide from Formic Acid: (U.S. 2,996,369)

The process consists of mixing 98% sulfuric acid with formic acid of at least 90% purity. Carbon monoxide is the result of the completed reaction carried out in the temperature range 120-130°C. The reaction time is about 10 seconds. The product gas is subsequently scrubbed to remove acid vapors, passed through an oil remover, dried and compressed.

Carbon Monoxide from Coke and Air: (U.S. 3,175,890)

Carbon monoxide may be produced by passing air through a granular incandescent carbonaceous bed at high velocity. The CO is removed from the reaction zone before it can be further oxidized. If steam is incorporated in the feed gas it will decompose in the reaction zone, thus enriching the product gas with hydrogen. The product gas is claimed to have high purity and to be of excellent yield from this highly efficient process.

Carbon Monoxide from Coke and Carbon Dioxide: (U.S. 2,921,840)

In this process, a stream of CO_2 is passed upward through a fluidized bed of fine mesh conductive carbon particles. Heat for the endothermic reaction is supplied by an electric current passed through the bed. The current is AC supplied by a transformer. The CO is recovered from above the fluidized bed.

Hydrogen and Carbon Monoxide

Carbon Monoxide from Steel Converters: (U.S. 3,177,065)

The top blowing converter is equipped with a skirt fitting over the mouth. An oxygen lance extends down into the converter. Gas from the blow is cooled by passing through a series of water-cooled ducts and then moved to a pressure regulated venturi scrubber. A motor driven blower collects the gas and discharges it into a gas holder. The high purity product gas may be used in a number of synthesis reactions.

9.4 MAXIMIZING CARBON MONOXIDE PRODUCTION

A typical fixed-bed producer may yield an off-gas at ~ 1400°F. A Lurgi gasifier (slowly-moving bed) may yield a product off-gas at ~ 1600°F. While this is a matter of reactant proportions and such things as the use or non-use of air preheating, or air-blowing vs. oxygen blowing, the product gases in any event are approximately at equilibrium. And at these temperatures, the CO/CO_2 ratio is only about 2/1. The addition of steam will, moreover, increase the concentration of hydrogen at the expense of CO, further affecting the equilibrium concentration.

Higher-temperature operations will increase the CO concentration at equilibrium. Thus if temperatures of ~ 2600°F can be reached, which is a possibility with air preheating, then the CO/CO_2 ratio can be made very high — provided additional air is not needed to drive the reaction. The use of oxygen also permits these temperature levels to be reached, e.g. the Koppers-Totzek gasifier, an entrained system.

Ordinarily, with ambient air, entrained gasifiers operate at lower temperatures, perhaps 1800-2000°F, giving a CO/CO_2 ratio of about 3 or 3.5 to 1. It is all a matter of determining an adiabatic flame temperature from the proportions of reactants used, subject to the constraints of equilibrium.

In order to maximize CO production, it is necessary that moisture be at a minimum. Otherwise, increasing amounts of hydrogen make an appearance at the expense of CO, though this is also an equilibrium affected by temperature. This would be of no consequence except that hydrogen cannot be readily recovered by selective absorption methods, whereas carbon monoxide can. The alternative, cryogenic processing, runs up the cost where such a low-boiling component is involved.

Thus low-temperature methods, which can be applied successfully to carbon monoxide, become much less attractive when applied to hydrogen recovery on a large scale. Similar remarks apply to selective diffusional methods.

Sources of water (or hydrogen) include not only water present *per se* in the coal, but the hydrogen and oxygen in the coal structure. By drying the coal, free-water is reduced. Carbonization, by driving off the water and part of the volatiles, increases the carbon content, and lowers the hydrogen content of the char. Thus chars, ideally, would be a means of increasing the CO content of the gas produced.

The maximum theoretical carbon monoxide concentration from partial combustion with air is detrminable from the equation

$$C + \tfrac{1}{2}O_2 + 2 N_2 \rightarrow CO + 2 N_2$$

The maximum concentration is thus 1/3 or 33 1/3 molar percent.

In practice, the concentration is lower due to the presence chiefly of hydrogen, carbon dioxide and water vapor. More precise values can be calculated starting with the coal ultimate analysis and the air/coal ratio used. The heats of reaction for the conversion are used simultaneous to the reaction equilibria, adjusted for the temperature of the products. The calculation becomes rather complicated, requiring tiral-and-error procedures for determining the final temperature and compositions, and is more readily handled by computer. A few selected results for specific examples are given by von Fredorsdorff and Elliott in *Chemistry of Coal Utilization,* H. H. Lowry Ed., Supplementary Volume, Wiley, New York (1963).

Finally, since the CO is to be recovered at lower (ambient) temperatures, a rapid quench is required. Otherwise, the CO can revert to CO_2 by the reverse Boudouard reaction,

$$2 CO = C + CO_2$$

if the products are slowly cooled. This reversion is explainable in terms of the reaction equilibria at lower temperatures.

Coals themselves vary greatly in moisture content. Thus younger coals contain more water, higher-ranic coals — a feature of the coalification process itself. The presence or absence of aquifers in the formation also has an effect on moisture present in the coal.

9.5 SEPARATION AND RECOVERY OF CARBON MONOXIDE

Processes for the recovery of carbon monoxide from gas streams either involve the selective absorption (or conceivably adsorption) of the CO, or low-temperature distillation methods.

While a number of liquids show affinity for carbon monoxide, the use of solutions of copper salts has become commercially significant.

Absorption of CO with Copper Ammonium Salt Solutions

The absorption of CO by copper-ammonium-salt solutions is used primarily in the purification of ammonia synthesis gas produced by the partial oxidation of hydrocarbons, the water gas reaction, or steam hydrocarbon reforming.

In general, the process includes high-pressure countercurrent absorption of CO in an aqueous solution of copper-ammonium salt (carbonate, acetate, formate, lactate, resulting in the formation of a cuprous-ammonium-carbon monoxide complex. Absorption is followed by regenerative heating thereby destroying the complex and liberating almost pure CO.

Typical difficulties which have arisen in the process are the formation of elemental copper because of certain side reactions, the loss of the active solution components, and corrosion.

There is little agreement as to the true reaction of the absorption process. However, it is generally predicted that one mole of cuprous-ammonium salt will absorb one mole of carbon monoxide for each equivalent of cuprous ion in solution.

Early experimental work with carbonates led Hainsworth and Titus (28) to believe that the reduction of cupric copper by CO took place rapidly and according to the equation

$$2 Cu(NH_3)_4CO_3 + CO + 2H_2O - (4-2n)NH_3 =$$
$$Cu_2(NH_3)_{2n}CO + 2(NH_4)_2CO_3$$

They also believed that one mole of carbon monoxide was absorbed per gram atom of copper.

Similar conclusions regarding the complex formation of CO into an unstable compound probably containing one mole of gas for each gram atom of cuprous copper was published by Larson and Teitsworth (29) after investigations of carbonate and formate salts.

Other work by Van Krevelen and Baans (30), after experimentation with very unstable chloride complexes, indicated that the reaction was

$$Cu(NH_3)_3^+ + CO + NH_3 = Cu(NH_3)_4CO^+$$

They too showed that the maximum amount of absorption of ammoniacal cuprous chloride was one mole of carbon monoxide per gram of copper.

Kohl and Riesenfield (31) indicate that there is also absorption of carbon dioxide according to the equations

$$2\ NH_4OH + CO_2 = (NH_4)_2CO_3$$

$$(NH_4)_2CO_3 + CO_2 + H_2O = 2NH_4HCO_3$$

This scheme is, however, partially reversed during the regeneration process.

Other reactions resulting in the precipitation of free copper were shown by Hainsworth and Titus (28) as

$$Cu_2(NH_3)_{2n}CO_3 + CO + 2\ H_2O - (2n-4)\ NH_3 = 2\ (NH_4)_2\ CO_3 + 2\ Cu$$

by Larson and Teitsworth (29) as

$$Cu_2\ (NH_3)_4\ CO_3 + CO + 2\ H_2O = 2(NH_4)_2CO_3 + 2\ Cu$$

and by Kohl and Riesenfeld (31) as

$$2\ CU^{++} + CO + 4OH^- + 2\ H_2O$$

resulting in a depletion of cupric ions in the solution, thereby shifting the equilibrium to the right in the reaction

$$2\ Cu^+ = Cu + Cu^{++}$$

Liquid-vapor equilibria data for the absorption process was provided by each of the above investigators. Hainsworth and Titus investigated carbonates. Larson and Teitsworth worked with carbonates and formates. Van Krevelen and Baans provided data for chlorides and formates. Van Krevelen, using data from other investigators, derived a satisfactory formula from which carbon monoxide equilibrium pressure can be calculated as a function of the temperature, the copper, ammonia, and carbon monoxide content, and the strength of the soltuion. In addition to the above, Zhavoronkov and Reshchikov (32) studied chlorides, formates, lactates, and acetates. Kvanov and Shishkov (33) made an extensive study of acetate solutions. They also derived an equation relating the amount of CO absorbed at a given partial pressure and the solution concentration. Egalon et. al (34) have constructed a monograph for determining the approximate quantity of CO absorbed in various carbonate solutions at a temperature of 25°C and one atmosphere of CO pressure.

Brown et. al. (35), investigating the thermodynamic activity of ammonia in ammoniacal solutions, constructed an automatic and continuous monitoring device for ammonia control in the TVA ammonia plant. Some of the common findings presented by the investigators were:

A. The solution of higher free ammonia content has a higher absorption capacity.
B. The absorptive capacity decreases as the temperature of the solution decreases.
C. Little difference in absorptive qualities of carbonate and formate solutions was found.
D. The times necessary to reach equilibrium in solutions of carbonate and formate are nearly equal and seem to depend primarily upon fresh surface exposure.

Heat is required for the liberation of any volatile componenents from the solution complex. Zhavoronkov (36), working with a mixed solution of formate and carbonate, listed the calculated heats of vaporization of ammonia as 16,000 Btu per lb-mole, for CO as 27,200 Btu per lb. mole, and for water as 22,700 Btu per lb-mole. For the exothermic absorption process, the heat of absorption of CO in formate, carbonate, and lactate solutions was calculated by Zhavoronkov and Reeshihikov (37), and Zhavoronkov and Chagunova (38), for a temperature range of 0-80°C. An order of magnitude value of 20,000 Btu per lb-mole is representative of the differential heat of solution. In addition to the heats of vaporization and solution, calculations demand a representative value for the heat capacity of the solution. Kohl and Riesenfeld (31) report an engineering estimate to be 0.8 Btu per lb per °F.

The composition of the absorptive solution is determined by many criteria. Some guidelines are listed below:

1. Cuprous ion (the absorptive component) should be as concentrated as possible.
2. Ammonia concentration enhances the absorptive characteristics of cuprous ion and increases the solubility of the copper.
3. An acid ion is necessary to prevent copper precipitation.
4. Cupric ion is necessary for solution but is inactive in absorption.
5. A small amount of oxygen gas in the gaseous mixture tends to prevent precipitation of copper from solution and increases the absorptive capacity of the solution.
6. Organic acids are expensive and subject to decomposition.

7. Carbonic acid is the least expensive, but it will hold less copper than the organic acids.

In practice, solutions are generally prepared by dissolving metallic copper in a solution of ammonia, water and acid. Careful attention is paid to the use of distilled water (to prevent corrosion) and the proper balance of ammonia and acid to prevent corrosion as well as to prevent the precipitation of copper compounds.

Actual compositions of the absorptive solutions are listed by each of the above investigators. Kohl and Riesenfeld (31) summarized the analysis of salt solutions in six different plants. Typical figures for solution concentrations listed are:

Cu^+	100 g/liter
Cu^{++}	15 g/liter
HCOOH	70 g/liter
CO_2	50-190 g/liter
NH_3	130 g/liter

The theoretical background of the absorption process proper is treated fully in handbooks and texts. Representative operating data from working installations are presented by Yeandle and Kline (39) [copper ammonium formate, twin absorber system operating at 1800 psi], Thomas and Watkins (40) [cuprous ammonium lactate in a packed tower pilot plant], Egalon et. al (34) [cupra-ammonium carbonate absorber together with an outline for an automatic control system for the plant], and Kohl and Riesenfeld (31).

A proposed absorber system described by Egalon et. al. (34) consists of two towers 20 inches I.D. and filled with 40 feet of iron rings. The towers are pressurized to 1420 psig. Liquor feed rate to the towers is 350 cu ft/hr at a temperature of 12°C. The feed gas rate is 280,000 SCF/hr containing 4.9% CO. Feed to the second tower was reduced to 1.5% CO and exhausts at 3×10^{-5}% CO. Rich liquor solutions exhaust at 30°C from the first tank and 20°C from the second tank. The solution make-up consists of Cu^+-100 g/liter, Cu^{++}-30 g/liter, CO_2-140 g/liter, and NH_3-170 g/liter.

Yeandle (39) describes typical conditions as 4,000 SCF/min gas feed to the first absorber operating at 1800 psig countercurrent to 75 gas/min formate solution. The liquor is cooled to 5°C. Exhaust gases pass upward through the second absorber countercurrent to a caustic or ammonia-water solution. The final exhaust contains less than 25 ppm carbon oxides. The figure of one mole of CO absorption per mole of solution for each equivalent of cuprous ion in solution is realized to only about 70%.

Regeneration is accomplished by de-pressurization and the application of heat. Unfortunately these adjustments are not conducive both to the liberation of carbon monoxide and to the recovery of ammonia from the absorbing liquor.

Because of the vaporization of ammonia during the regeneration phase, an operating limit of 180°F must be enforced. This temperature limit does not allow for the recovery of pure CO by simple evaporation.

A number of solutions are available depending on the type of operation desired. The use of the more expensive absorbents (formate, acetate, lactate) may be preferred because of the attendant low ammonia vapor pressure. If the less expensive carbonate solution is used, ammonia is recovered by a water wash or a wash with cool copper solution. The loss is made up with a liquid ammonia feed.

Regeneration plants usually consist basically of a pre-heater, an evaporator, a scrubber, and an insulated storage tank used for the removal of the last traces of CO. Kohl and Riesenfeld (31) list operating temperatures, pressures, and compositions of gas and liquid streams for the TVA ammonia recovery plant. The column is described as 61 feet high. The reflux section is six feet by 23 feet high, packed with 10-1/2 feet of Raschig rings at the top of a 7-1/2 by 38 feet high pre-heater section.

Egalon et. al. (34) describe four separate systems operating between 78-79°C. Two systems allow for the recovery of CO.

Yeandle (39) describes a recovery system used by the Spencer Chemical Company. In this system the CO is vented.

The CO obtained from regeneratin is contaminated by the presence of CO_2 and nitrogen. Relatively CO_2-free CO may be obtained by stripping in stages. This procedure, however, requires additional washes to prevent ammonia losses.

COPPER ALUMINUM CHLORIDE

The latest commercial development in carbon monoxide separation and recovery is the Tenneco COSORB process (41). Using solutions of copper aluminum chloride, the operating pressure requirement for absorption is lower than for solutions of the copper ammonium salts.

LOW CONCENTRATIONS OF CARBON MONOXIDE

Smaller amounts of carbon monoxide and hydrogen in gas streams have been adsorbed using Fischer-Tropsch catalysts (42). Two separate catalysts containing Fe-Cu-MfO-K_2O (39:8:9:0.3) and Fe-Cu-MgO-K_2O were tested at temperatures of 325 and 275°C respectively. The adsorbent proved effective in the adsorption of both CO and hydrogen. It is suggested that the CO adsorption was accompanied by carbide formation. Regeneration and selective adsorption are matters which require further resolution, if possible. Otherwise, CO is methanated by H_2.

A solvent method is described in U.S. Patent 3,317,275. In ammonia synthesis it is desired to have a hydrogen stream free of CO. CO may be freed from the gas stream by absorption in a suitable solvent. The absorption may take place at a temperature of 20°C and a pressure of 2000 psi. The CO-containing stream passes through the absorber countercurrently to the liquor. A suitable solvent is propionitrile containing 10% by weight molybdenum carbonyl complex. After absorption, the rich liquor is pumped to a desorption tower maintained at 100°C where the CO is freed from the stream. The regenerated absorbent is cooled and circulated to the absorber.

Other potential carbon monoxide solvents may be selected from an examination of Reference 11.

LOW-TEMPERATURE SEPARATION OF CARBON MONOXIDE

Substantial quantities of carbon monoxide may be obtained by subjecting a number of gases such as water gas, blast furnace gas, producer gas and coke oven gas to low-temperature separation processes. The basic requirements for the process are that the feed gas components be capable of liquefaction at low temperatures and that the gases boil at different temperatures.

Depending on the gas source, a variety of gases will be included in the feed stream. This can result in serious problems if the components solidify at temperatures higher than CO liquefies. Because of this, other non-low-temperature separation processes are often incorporated prior to the liquefaction process. For example, carbon dioxide may be removed by scrubbing with water or a caustic solution. Acetylene may be isolated almost completely by filtering gas streams with charcoal. Argon and oxygen may be separated by passing the mixture over heated sulfur. In the process of separation by low-temperature liquefaction, these measures are used primarily to reduce trace compositions to acceptable levels.

In early work, considerable attention was paid to the purification of gas streams for the production of hydrogen to be used in ammonia synthesis. Patents were issued to Claude, von Linde, and Bronn (43).

The usual source of ammonia synthesis gas was coke-oven gas. Typical impurities in the feed were of the order of 5-7% CO and 5-14% N_2. When increased purity of the synthesis gas was desired, CO was removed by scrubbing with slightly higher-boiling nitrogen. In this process, liquid nitrogen at a temperature of $-190°C$ and at a pressure of 12 to 15 atmospheres was introduced countercurrently to the stream containing CO. The rich nitrogen was run from the bottom of the tank, expanded slightly and used for pre-cooling other feed streams. Generally, the overhead contained only small amounts of CO.

In the production of synthesis gas from partial oxidation of natural gas and the hydrogen rich refinery gas, Bardin and Beery (44) describe an industrial plant using nitrogen-wash procedures for the removal of CO. The plant is capable of producing 98% hydrogen streams.

Baker (45) reports the production of 98.5% to 99.5% CO purity obtained from synthesis gas produced by steam reforming. The feed gas consists of about 30% CO, 63-73% H_2, together with quantities of methane, nitrogen, and CO_2. CO_2 is removed with MEA and a caustic wash solution. The remainder is pressurized to about 450 psig and fed to a distillation column for the removal of the methane as a bottom product. The liquid is flashed to a low pressure distillation column where CO is removed as a top product. The recovery of CO could amount to 200,000 cu. f/day.

Jacob (46), in a patent to Linde, describes a process for separating a stream of hydrogen, nitrogen and CO by cooling and rectifying the mixture using liquid nitrogen as a reflux to obtain a residual fraction rich CO. The overhead product is used in the manufacture of ammonia.

Gel'perin and Rapoport (47) present diagrams showing equilibrium compositions for mixtures of H_2, N_2, and CO at various tempertures. These data can be used for calculations of the processes in the gas scrubbing column. Data is also given for the amount of nitrogen needed for scrubbing in terms of the feed composition.

REFERENCES

1. Taylor, H. S. and H. A. Neville, "Catalysis in the Interaction of Carbon with Steam and with Carbon Dioxide," *J. Am. Chem. Soc.,* *43*, 2055-71 (1921).

2. Fleer, A. W. and A. H. White, "Catalytic Reactions of Carbon with Steam-Oxygen Mixtures, *Ind. Eng. Chem., 28,* 1301-9 (1936).

3. Fox, D. H. and A. H. White, "Effect of Sodium Carbonate Upon Gasification of Carbon and Production of Producer Gas," *Ind. Eng. Chem., 23,* 259-66 (1931).

4. Weiss, C. B. and A. H. White, "Influence of Carbonate upon the Producer Gas Reaction," *Ind. Eng. Chem., 46*, 8307 (1934).
5. Lewis, W. K., E. R. Gilliland and H. Hipkin, "Carbon-Steam Reaction at Low Temperatures," *Ind. Eng. Chem., 45*, 1697-1703 (1953).
6. Hoffman, E. J., "The Direct Production of Hydrogen from Coal-Steam Systems," American Chemical Society, Division of Petroleum and Fuel Chemistry, Los Angeles, March-April, 1971. *Preprints,* Vol. 16, No. 2, C20-C23.
7. Swinnerton, J. W., J. J. Linnenbom and R. A. Lamontagne, "Ocean, Natural Source of CO," *Science, 167*, No. 3920, 984-6 (1970).
8. Green, Ralph V., "Carbon Monoxide," in *Kirk-Othmer Encyclopedia of Chemical Technology,* Vol. 4, 2nd Ed., Interscience Publishers, New York (1964), pp. 424-445.
9. Canjar, L. N., J. Heisler, D. O'Brien and F. S. Manning, "Thermodynamic Properties of Non-Hydrocarbons. Part III Thermodynamic Properties of Carbon Monoxide," *Hydrocarbon Processing and Petroleum Refiner, 45,* (2), 157 (1966).
10. Hurst, J. and R. B. Stuart, "Thermodynamic Property Values For Gaseous and Liquid Carbon Monoxide From 70-300°K with Pressures to 300 Atmospheres," U.S. Dept. of Commerce, *Nat. Bur. Std., Tech. Note No. 202,* Nov. 30, 1963, p. 109.
11. Seidell, A., *Solubilities of Inorganic and Metal Organic Compounds,* Vo. I, pp. 453-458 (1958). 4th Ed. by William F. Linke, Amer. Chem. Socl.
12. Krase, N. V., "Carbon Monoxide as a Raw Material For Chemical Synthesis," *Am. Inst. Chem. Engr. Transactions, 32,* 493-510, (1963).
13. Hahn, A. V., *The Petrochemical Industry,* McGraw-Hill, New York (1970).
14. Waring, B. W. and J. F. Foster, *Economics of Fuel Gas From Coal,* by Battelle Memorial Inst., John J. Foster and Richard J. Lund, Ed., McGraw-Hill, New York (1950), p. 281.
15. Harlan, W. N., H. H. Kouns and B. O. Buckland, "Gas Turbine Fuel From A Pressurized Gas Producer," Gasification and Liquifaction of Coal Symposium, Annual Meeting AIME New York, Feb. 20-21 (1952), 109-121. CA 47:4581f.
16. TeNuly, Johannes A., "Process for the Production of Synthesis Gas and the Like," Patent No. 2,904,417. Appl. Sept. 3, 1957. Ser. No. 681,729, Oct. 27, 1955. Assignor to Shell Development Co., N.Y.

17. Atwell, H. V., "Process For the Production of Carbon Monoxide From Solid Fuel," Patent No. 2,879,148. Original application May 31, 1952, Serial Number 290,921. New Patent No. 2,761,772, Dated Sept. 4, 1956. Divided and this Application Dec. 6, 1955, Ser. No. 551,387. Assigned to the Texas Co., New York.

18. Carkeek, C. R. and D. M. Strasser, Patent No. 2,838,388. Continuation of Application Ser. No. 99,908, June 18, 1949. This Application Apr. 5, 1954, Ser. No. 420,848.

19. Totzek, F., "Suspension Process for the Production of CO and H from a Solid Carbonaceous Fuel, Oxygen, and Steam," Patent No. 2,905,544, Sept. 22, 1959. Application May 17, 1951, Ser. No. 226,-792. Assignor to Koppers Co., Ind.

20. Schmidt, L. D., Patent No. 3,325,253. Application May 29, 1963. Assignor to Allied Chem. Corp.

21. Burbon, E. J., "Enrichment of Producer Gas," Fr. Patent No. 1,-298,850, July 13, 1960. Application Aug. 31, 1961. CA 59:7273c.

22. Lavrov, N. V. and G. V. Grebenshchikova, "The Reconversion of CO_2 For the Enrichment of Gases with CO," *Tzvest. Akad. Nauk. Uzbk. SSR Ser. Tekh. Nauk.,* 1961. CA 56:2915h.

23. Waring, B. W. and J. F. Foster, "Economics of Fuel Gas From Coal," Battelle Memorial Inst., John J. Foster and Richard J. Lund, Ed., McGraw-Hill, New York (1950), p. 281.

24. Namy, G., J. Dumont-Fillon and P. A. Young, "Gas Recovery Without Combustion From Oxygen Converters; the IRSID-CAFL Pressure Regulation Process," Iron Steel Inst. (London), Spec. Rept. No., 8s, 98-103 (1963). Publ. 1964. CA 61:9213h.

25. Skelly, J. F., "Profits in BOF Gas Collection," *Iron Steel Engr. 43,* (3), 182-8 (1966).

26. Roederer, C., F. Mangin, S. Ogawa, I. Hamabe and G. de Cervens, "Gas Collection without Combustion," *J. Metals, 18* (7), 852-60 (1966).

27. Morita, S., M. Nishiwaki and Tagiri, "Yawata's Improved OG Equipment and its Performance," *J. Metals, 18* (7), 861-4 (1966). CA 65:10207c.

28. Hainsworth, W. E. and T. Y. Titus, "The Absorption of CO by Cuprous Ammonium Carbonate Solutions," *J. Am. Chem. Soc., 43,* 1-11, (1921).

29. Larson, A. T. and C. S. Teitsworth, "Absorption of CO by Cuprous Ammonium Carbonate and Formate Solutions," *J. Am. Chem. Soc., 44,* 2878-85 (1922).
30. Van Krevelen, D. W. and C. M. E. Baans, "Elimination of CO from Synthesis Gases by Absorption of Cuprous Salt Solutions," *J. Phys., Colloid Chem., 54,* 370-409 (1950).
31. Kohl, A. L. and F. C. Reisenfeld, *Gas Purification,* McGraw-Hill, New York (1960). Gulf Publishing Co., Houston (1974).
32. Zhavoronkov, N. M. and P. M. Reshchikov, "Absorption of CO by Solutions of Cu-NH$_4$ Salts," *J. Chem. Ind.* (USSR), 10 (8), 41-9 (1953). CA 28:1474[6].
33. Ivanov, D. G. and D. S. Shishkov, "Absorption of CO with Copper-Copper Ammonium Complex," *J. Appl. Chem.* (USSR), [237] (10), 2163-73 (Oct. 1964).
34. Egalon, R., R. Vanhille and M. Willemyns, "Purification of Gases For Ammonia Manufacture," *Ind. Eng. Chem., 47,* 887-898 (1955).
35. Brown, E. H., J. E. Cline, M. M. Felger and R. B. Howard, "Continuous Determination of Ammonia Activity in Ammoniacal Solutions," *Ind. Eng. Chem. Anal. Ed., 17,* 280-2 (1945).
36. Zhavoronkov, N. M., "Partial Pressures of Ammonia, Carbon Dioxide and Water Over Copper Ammonium Solutions," *J. Chem. Ind.* (USSR), *16* (10), 35-7 (1939). CA 34:2233[9].
37. Zhavoronkov, N. M. and P. M. Reshchikov, "Absorption of CO by Solutions of Cu-NH$_4$ Salts," *J. Chem. Ind.* (USSR), *17* (2), 25-9 (1940). CA 34:4645[3].
38. Zavoronkov, N. M. and V. T. Chagunava, "Absorption of CO by Cu-NH$_4$ Formate-Carbonate Solutions," *J. Chem. Ind.* (USSR), *17* (2), 25-9 (1940). CA 34:4645[5].
39. Yeandle, W. W. and G. F. Klein, "Purification of Ammonia Synthesis Gas," *Chem. Engr. Prog., 48,* 349-52 (1952). CA 46:8817b.
40. Thomas, W. J. and S. B. Watkins, "Absorption of CO in Aqueous Solutions of Cuprous Ammonium Lactate in a Packed Tower," *J. Appl. Chem., 15* (1), 17-28 (1965).
41. Haase, D. J. and D. G. Walker, "The COSORB Process," *Chem. Eng. Prog., 70* (5), 74-77 (May, 1974).

42. Mukherjee, P. N., N. C. Ganguli and A. Lahiri, "Adsorption of Carbon Monoxide and Hydrogen on Iron Fischer-Tropsch Catalysts," *Indian J. Technol. 3* (1), 15-18 (1965. CA 63:2416c.
43. Guillaumeron, P., "Liquefaction For Separating Hydrogen From Coke Oven Gas," *Chem. Eng., 56,* 7, 105-10 (July, 1949).
44. Bardin, J. S. and D. W. Beery, "Producing Ammonia Synthesis Gas," *Petroleum Refiner, 32,* (9), 99-102 (Feb. 1953).
45. Baker, D. F., "Low Temperature Processes Purify Industrial Gases," *Chem. Eng. Prog. 15,* (9), 399-402 (1955).
46. Jakob, F., "Gas Separation Process and Apparatus," U.S. Patent 3,251,189, May 17, 1966. Assignor to Linde A. G.
47. Gel'perin, I. I. and L. L. Rapoport, "Washing out of CO by Means of Liquid Nitrogen," *Trudy Cosudarst. Nauch-Issledovatel i Proekt Inst. Azot Prom. 5,* 249-60 (1956). CA 54: 15860a.

Chapter 10

ANCILLARY CONSIDERATIONS

The chief ancillary problems associated with the conversion of coal — including combustion — are environmentally-related. Of these, two stand out. One is the use and consumption of water in the arid West, which contains the major reserves of low-sulfur coal. The other is air pollution, notably from the combustion of pulverized coal, either in support of coal gasification and liquefaction, or for the generation of electrical power. These are apart from socio-economic problems precipitated by coal development (or the lack thereof).

While air-pollution has the potential of being controlled, there is no way to get around the use of water — at least, water entering into the conversion. In the arid West the water problem is especially touchy — and many-sided. For instance, there is controversy over rail transport vs. slurry-pipelining of coal, and again over on-site conversion or usage vs. transport to the areas of energy or fuel consumption. Along these lines, the Northern Great Plains Resources Program (an interim agency of the federal government) conducted studies which showed that less water would be used for slurry pipelining than for on-site conversion to electricity or for gasification and liquefaction. There are, however, such additional matters as direct cooling vs. using cooling water, etc.

The transport of coal to the point of use transfers the issues and burdens of air pollution from the coal-producing states to the areas of energy usage — meaning denser population. More stringent controls in these areas may mean a shift from the direct firing of coal for electrical generation over to the concept of gasifying (or liquefying) the coal first. This would presuppose that stack-gas emissions control is a less satisfactory proposition than eliminating particulate and sulfur problems during gasification or liquefaction.

There are still other factors that enter. One is that byproduct carbon dioxide from the conversion of coal or other carbonaceous materials could be used for controlled photosynthesis, a form of energy recycle. Previously mentioned, the concept is given idealization in Figure 10.1. As also mentioned, CO_2 can be used to produce urea, one of the effective means of affixing nitrogen into solid fertilizer. The other potential use mentioned is for the solution drive of heavy crudes in secondary or tertiry recovery operations.

Disposal of residues, in particular fly ash, poses problems — though there are possibilities for limited uses, e.g., incorporation into construction materials. Calcium sulfate (gypsum) sludge produced from the removal of sulfur oxides has the redeeming feature of being inert. Elemental sulfur produced from H_2S removal and conversion will constitute a marketability-storage problem. Waste liquors pose the inevitable difficulty of interfacing with the environment. The alleviation of one situation always seems to cause another. It then becomes a matter of degree.

The issues are complicatd and inter-related. No attempt will be made here to examine the spectrum. Rather, some of the underlying technological problems will be assessed as regards water and air-pollution, items of more direct bearing.

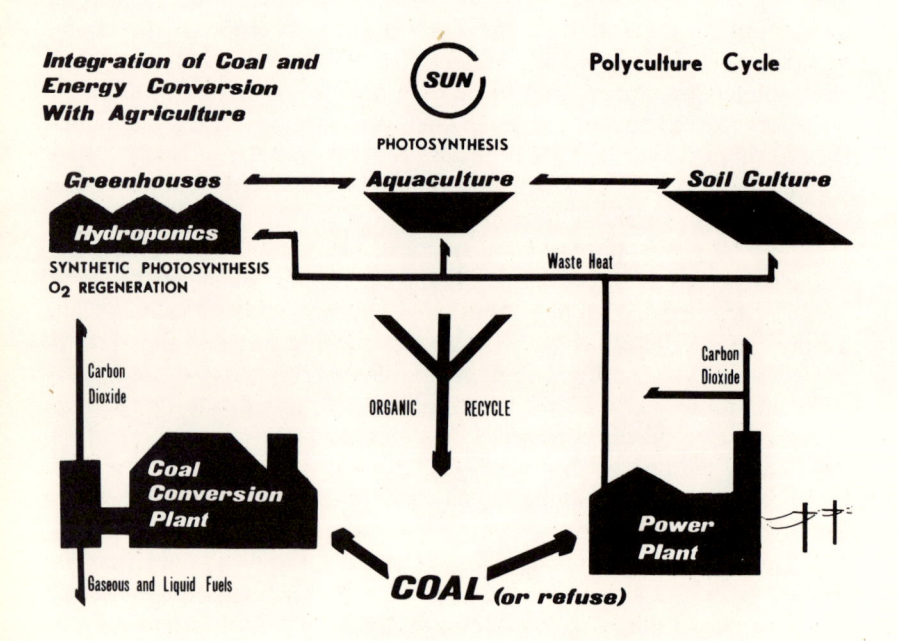

Fig. 10.1. Indirect utilization of solar energy.

10.1 WATER REQUIREMENTS FOR COAL CONVERSION

For steam-power plants the water rate is based on the steam load. While there is some makeup of feedwater for the circulating steam, the main water requirement is for the cooling and condensation of the low-pressure spent steam leaving the turbines. The energy balance relates to the enthalpic change across the condenser (i.e., the requirement for the Rankine cycle). The calculation yields the circulating cooling water rate to the condenser(s), based on an allowable temperature rise for the cooling water. This in turn is translatable into water makeup for evaporation losses in the cooling tower(s).

The above calculation is straightforward. Water requirements are not so readily obtained, however, as the complexity of processing increases. Such is the case for coal liquefaction and gasification. In last analysis, it may depend upon the thermal efficiency of processing. The matter of thermal efficiency is covered in the Appendix.

Process estimates from the several contractors to the Office of Coal Research (OCR) are here used as the basis for water requirements connected with the chemical conversion of coal to gas and/or liquids. The values are specific to the process.

This information has been assembled and entered into the accompanying table. The data is in instances obsolete, but nevertheless indicates the diversity that has appeared from time to time on the subject. Being controversial, present data is at times better left unsaid. The latest information is the aforesaid NGPRP report.

An examination of the data used in Table 10.1 shows great diversity, a matter principally of not distinguishing consumptive demand from circulating or process cooling water. As subsequent calculations will indicate, a figure of 1000-2000 gal/ton of coal may be appropriate for consumptive water whereas 30,000-60,000 may be appropriate for circulating water. The figures, however, will also depend upon the type of processing.

The major part of the circulating water is for process cooling and is not actually used up. The consumptive water is either actually converted or used as makeup.

TABLE 10.1. Water Requirements

Process	Coal Feed Rate (MT/d)	Product Rate	Water Rate (gpm)	Water Demand Water/Coal (Gal/Ton)
Co$_2$-Acceptor[1, 2]	25	250 MMSCF/D	100,000	5,800
Kellog Gasif[3]	14	250		
BCR Gasif[4]	10.6	250		
IGT Gasif[5, 6, 7]	17.8	266	120,000	9,700
Spencer[8]	10.0	7,500 TD	1,000	144
Coed[9, 10]	10.0		14,500 (c)*	2,100
				87
			610 (b)	88
				2,275
H-Coal[11, 12]	59.0	100,000 bpd	20,000	770
SASOL[13]	5.4	6,400 bpd	190,000	50,000

*(c) denotes cooling water
(r) denotes reactor water
(b) denotes boiler water

Ancillary Considerations

Unfortunately, the references consulted do not, as a rule, make a distinction between water requirements merely for cooling purposes, and water that enters into chemical reaction. The latter would be chemically used up as in the coal-steam reaction, reforming, or the CO-shift reaction.

The former, used strictly for cooling, preferably would be circulated or recycled through cooling towers and reused — the only net requirement would be for makeup. The theoretical requirements for makeup run to about two percent of the circulating water for a 20° ΔT. To this would be added other losses, such as mechanical entrainment at the cooling towers.

There is also the possibility of using atmospheric convection cooling in lieu of cooling water.

An inspection of the tabulated data indicates that gasification processes require more water than processes for producing liquids. The very high anomaly for the SASOL requirement could be a mistake in the reference article, but not necessarily. (It could refer to circulation.) Another qualification is that of efficiency.

A possible (but not necessarily applicable) reference is the cooling water requirements from average figures for a refinery or equivalent size, as presented in Nelson, "Petroleum Refinery Engineering," or about 600 gal/bbl of oil.

The figures for the H-Coal process direct conversion are found to vary considerably. Thus a 100,000 bpd plant in Illinois based on Ill. No. 6 coal requires 21,000 gpm of cooling water, while a plant in Wyoming using Wyoming subbituminous only requires 7,400 gpm. The discrepancy was not explained (air cooling?).

Other past communications regarding H-Coal water requirements, ostensibly in Wyoming, are as follows:

    Plant size: 30,000-100,000 bpd
   Natural gas: 70-200 MMSCF/D
     Water: 30,000-200,000 gpm

The foregoing would correspond to a water/coal ratio of from 2400 to 4800 in gal/ton, based on previously cited conversions. In the above, natural gas was to be used as the source of hydrogen, and part of the water for the CO-shift, as follows:

   10-15 cu ft/sec for consumption in making H_2
   50-150 cu ft/sec for cooling

These figures correspond to
> 4500-6700 gpm for consumption
>
> 22,500-67,000 gpm for cooling

The corresponding plant size was not stated, but is presumed from 30,000-100,000 bpd.

Other representative figures are as follows:

Plant Size:
> 50,000 to 100,000 bpd
>
> or 250 MMSCF/D

Water:
> 7500 to 15,000 gpm

These figures are perhaps low for gasification, but of the right order for liquefaction.

In the processing, retorting, and upgrading of oil shale to yield shale oil, the following figures have been presented (14):

> 50 gal water/bbl of shale oil

If the shale oil is also refined locally on to finished products, the total requirement becomes

> 90 gal water/bbl of shale oil

Thus nominal 10,000 bpd operations would require

> 350 gpm of water to yield oil
>
> 630 gpm of water to yield finished product

For 20 gal/ton shale, a 10,000 bpd operation would require about 20,000 T/D of shale. On this basis the respective water requirements would be

> 25 gal/ton

or
> 45 gal/ton

These figures evidently refer to consumption rather than total demand.

One source of discrepancy is that one barrel of oil does not yield a barrel of finished product. Some typical, yields for the hydrogasification of shale oil are (15):

> 5,725 SCF of $C_1 - C_3$
>
> 5.3 gal of C_4, naptha, plus recycle oil

The above is on the basis of one barrel (42 gal) of crude shale oil. On this basis, 10,000 bpd of crude shale oil would yield:

> 57.25 MMSCF/D of gas
>
> 1,300 bpd of liquid hydrocarbons

Ancillary Considerations

The yields in a given instance would depend upon the severity of hydrogenation. At less severe conditions more liquids would result.

The principal source of difference between the shale oil and coal requirements on a tonnage basis is due to the fact that the oil shale is on the order of 80 to 85 percent inert rock.

If the water requirements are examined on a product basis rather than on feed, there is less discrepancy between coal conversion and shale oil needs. The comparison is shown in the following table. In general, gasification processes require more water per equivalent of product.

TABLE 10.2. Water Requirements Based on Product Yield

Process	Product Rate	Total Demand Water/Product*
CO_2-Acceptor	250 MMSCF/D	575 gal/MSCF
Kellogg Gasif	250	
BCR Gasif	250	
IGT Gasif	266	650 gal/MSCF
Spencer	7,500 T/D	192 gal/T
Coed		
H-Coal	100,000 bpd	290 gal/bbl
SASOL	6,400 bpd	43,000 gal/bbl
Shale Oil (Recovery and Refining)		90 gal/bbl**

* One MSCF of C_1-C_2 is roughly equivalent to 0.2 bbls of gasoline on an energy basis.

** This figure is based on consumption, defined as the difference between quantity diverted and quantity returned.

Another way of looking at the matter is based on the overall thermal efficiency or net energy yield (see Appendix). Thus if a plant has an overall net thermal efficiency of such and such, and the waste heat is all ultimately rejected to cooling water with a specified allowable temperature rise, then an estimate can be made of the process cooling water circulation requirements, as follows:

$$\text{Process cooling water in gal/T} = 2000 \left[1 - \frac{\text{Eff}}{100}\right] \frac{\text{Btu-rating of coal}}{(c_p)_{H_2O} \Delta T} \cdot \frac{1}{8.33}$$

For an efficiency of 50% and $\Delta T = 20°F$, and a coal of 10,000 Btu/lb,

$$2000 \ [\ 1 - 0.50\ ] \ \frac{10000}{1.0(20)} \ \frac{1}{8.33} = \sim 60,000 \text{ gal/T}$$

If makeup is 2%, then the evaporation loss is

$$0.02 \ (60,000) = 1200 \text{ gal/T}$$

At 70% efficiency, the figures are

36,000 gal/T (circulation)

720 gal/T (evap. loss)

Beyond this is the water consumed in the conversion reactions. For the overall conversion

$$C + H_2O(g) \rightarrow \tfrac{1}{2}CH_4 + \tfrac{1}{2}CO_2$$

the theoretical water (steam) requirement for a coal of 60% carbon is

$$\frac{2000(0.6)}{12} \ \frac{18}{8.33} = 216 \text{ gal/T}$$

Liquefaction would require less, theoretically about one-half as much [the product tends to be $(CH_2)_n$-].

Presumably, from all this, water requirements might be $\sim$ 30,000 - 60,000 gal/T for circulation and $\sim$ 1000 - 2000 gal/T consumed. These numbers tie in with some of the past announced figures but not with others. But then there is much vagueness as to whether water requirements as previously given refer to circulation or to consumption.

OTHER FIGURES

Information for regional water requirements for coal conversion is available from site-specific and regional environmental impact statements. An earlier version of this type of study was a report entitled "Review and Forecast: Wyoming Mineral Industries," prepared by Cameron Engineers for the Wyoming Natural Resource Board and the State Water Planning Commission in 1969.

Perhaps the most detailed study of water requirements was prepared by the Northern Great Plains Resources Program (NGPRP). Entitled "Effects of Coal Development in the Northern Great Plains," the final report was dated April, 1975 and submitted to the U.S. Department of the Interior.

In this report, estimates on water requirements for coal gasification ranged from 8,000 to 12,000 acre-feet per year for a 250 MMSCF/D

Ancillary Considerations

plant down to 6,000 to 10,000 acre-feet per year, with a top figure of 30,000 acre-feet per year for total evaporative cooling. NGPRP adopted a mean figure of 9,500 acre-feet per year for projection purposes. This latter figure translates as follows:

$$\frac{9500(43560)7.48}{365(20000)} = 424 \text{ gal}/T$$

$$\frac{9500(43560)7.48}{365(250)1000} = 140 \text{ gal}/MSCF$$

The above assumes a plant operating 365 days per year with a feed of 20,000 T/D yielding 250 MMSCF/D.

These figures do not necessarily agree with prior figures and suggest a built-in disparity in estimating evaporation losses — that is, whether cooling towers or natural cooling is used, or both, and the proportion of loss.

For power generation, NGPRP uses the following figures:

$$19{,}000 \; \frac{\text{acre-feet}/yr}{1000 \text{ Mw}} \qquad \text{(high-water use)}$$

12,000 (medium-water use)

5,000 (dry cooling)

These figures include losses in ash handling, etc. For a 10,000 Btu/lb coal at 1/3 conversion efficiency, the medium usage figure converts to

$$\frac{12{,}000(43{,}560)(7.48)(10000)2000}{365(24)(1000)(3412.76 \times 10^3)(1/3)} = \sim 7800 \text{ gal}/T$$

This is a value considerably higher than for coal gasification — but then the thermal efficiency is less than for coal gasification, in a manner of speaking.

A theoretical number can also be backed out for 1/3 efficiency, as was done for gasification. Thus

$$2000 \left[1 - \frac{1}{3} \right] \frac{10000}{1.0(20)} \frac{1}{8.33} = 80{,}000 \text{ gal}/T$$

This is a circulation requirement. For 2% evaporation loss (20°F ΔT).

$$80{,}000(0.02) = 1600 \text{ gal}/T$$

This value is less than the aforecited result. Obviously, figures for evaporation and other losses used are higher than theoretically imposed values. It should be noted owever, that a figure of 1% evaporation loss for a 10°F ΔT is commonly used (*Chemical Engineers' Handbook*, third edition).

Finally, by comparison with slurry pipelining the coal to the locale of usage, the water requirements for a 50-50 wt percent slurry are, for instance,

$$\frac{2000}{8.33} = 240 \text{ gal/T of coal}$$

Therefore, there is the argument that, for coal produced in an arid region at least, it would require less water to slurry transport the coal out of the region to the locale of usage than to convert the coal at the mine site. There is, however, the additional argument that rail transport by unit train does not require any water, period. Then there is the matter of efficiency and cost. Etc.

10.2 REMOVAL OF SULFUR OXIDES AND PARTICULATES

The removal of sulfur oxides (SO_2, SO_3) from combustion systems undergoing essentially complete oxidation is a simple enough matter from a chemical standpoint, though not as easy from a technological or engineering standpoint on a large scale. The EPA standards are stringent with regard to particulates, less so with regard to sulfur — though some states are lowering the admissable sulfur levels. The benefits are offset by more equipment and lower efficiency, which translate into higher costs.

The upper limit for particulates is 0.1 lb of solids per million Btu's of fuel. For sulfur, the upper limit is 1.2 lb of sulfur per million Btu's. While low-sulfur western coals ($\sim$ 0.5% S) will in the main meet the standard, the higher sulfur eastern coals ($\sim$ 2-3% S) will not. Moreover, lower state standards (for example, 0.2 lb S per million Btu) cannot be met in any event without emission controls — or the use of essentially sulfur-free fuels. Lower federal standards are also in the offering.

The emissions problem is thus usually one of both particulates and sulfur when using coal. And solving one may solve the other. For example, wet scrubbers for sulfur will also tend to remove remaining particulates from flue gases.

A representative pollution-control system for a large coal-fired power plant is shown in Figure 10.2. The gases first pass through a cyclone separator where 65-70% of the fly ash particulates are removed. This is followed by electrostatic precipitation and finally wet scrubbing for removal of sulfur oxides and the final remaining particulates.

Ash resistivity peaks with temperature ($\sim$ 300-350°F) and increases as sulfur content decreases. Above 10^{10} ohm/cm, back corona may occur.

Ancillary Considerations 339

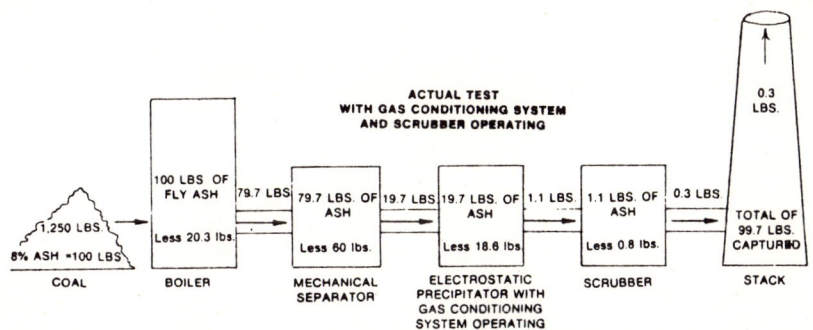

Fig. 10.2. Air-pollution control sequence (from Public Service Company of Colorado).

Hence for low-sulfur western coals the trend is away from cold precipitators toward baghouses. Alternately, a hot precipitator ($\sim$ 600-750°F) may be installed ahead of the air-preheater section of the boiler.

Wet scrubbers most generally employ limestone ($CaCO_3$) slurries or solutions of slaked lime [$Ca(OH)_2$] or some combination of the two. In either case, the final product is calcium sulfate ($CaSO_4$), which in the presence of water exists in the hydrated form known as gypsum ($CaSO_4.2H_2O$). A representative system is shown in Fig. 10.3.

Though to a great extent insoluble in water, gypsum nevertheless does not exhibit the property of precipitating from solution in an identifiable crystalline form which can be readily settled or filtered out. Rather it is formed and comes out of solution with more the amorphous consistency of toothpaste. This muck or sludge fouls up lines and equipment and at best can be handled mainly only by thickeners. And here lies the technological and engineering problem.

Chemical additives and separation procedures have been studied, and studies are still underway, with the objective of producing a sharper liquids-solids separation. Vibrators and ultrasonics have been considered for cleaning lines and equipment, as well as more mechanical means. All pay a penalty in increased complexity, cost and maintenance — and serve to reduce the overall net efficiency, as does the act of cleanup itself.

This has all led to a consideration of alternatives. Thus alkali carbonate (Na_2CO_3) solutions can be used, though the disposal of sodium sulfate or sulfite liquors is even more of a problem than separation and disposal of gypsum.

Sodium sulfite solutions can be used to remove SO_2 from flue gases by the reaction to produce the bisulfite:

$$SO_2 + H_2O + Na_2SO_3 \rightarrow 2\ NaHSO_3$$

The bisulfite is regenerated by steam stripping.

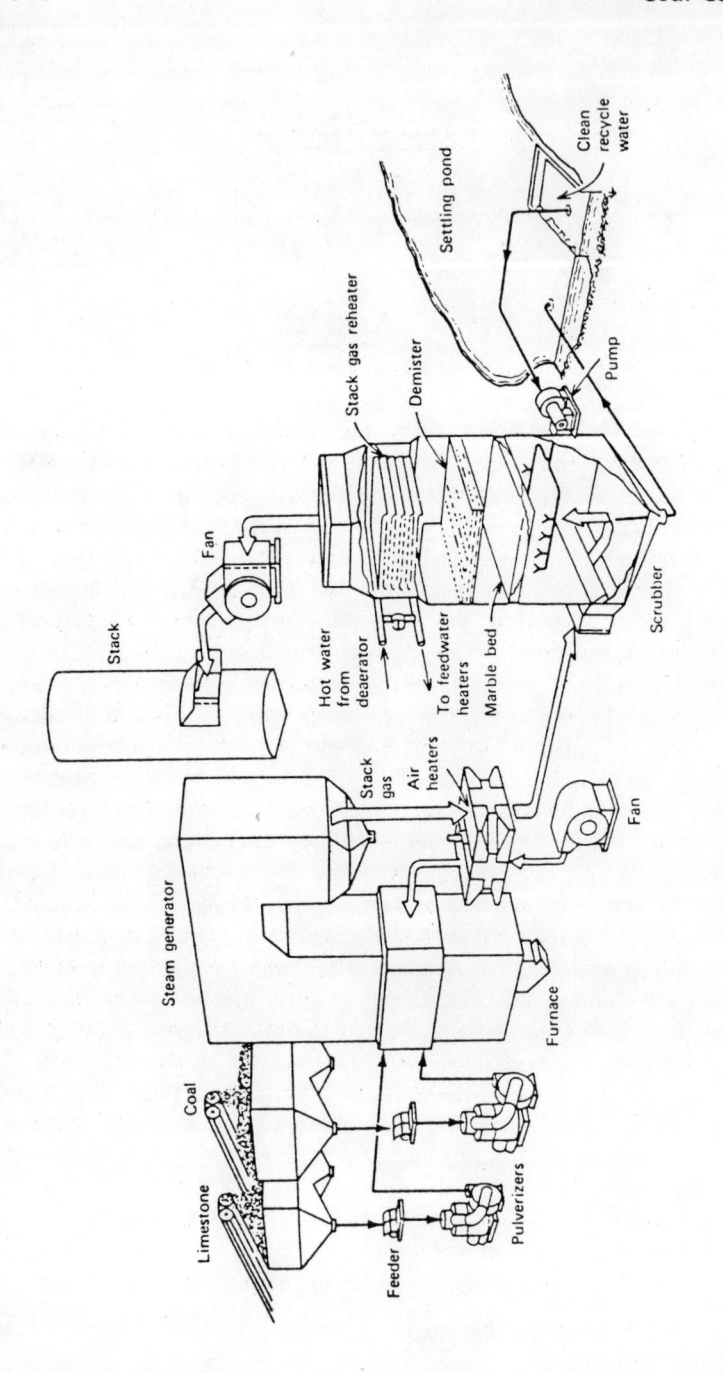

Fig. 10.3. Emission control using limestone (Combustion Engineering).

Ancillary Considerations

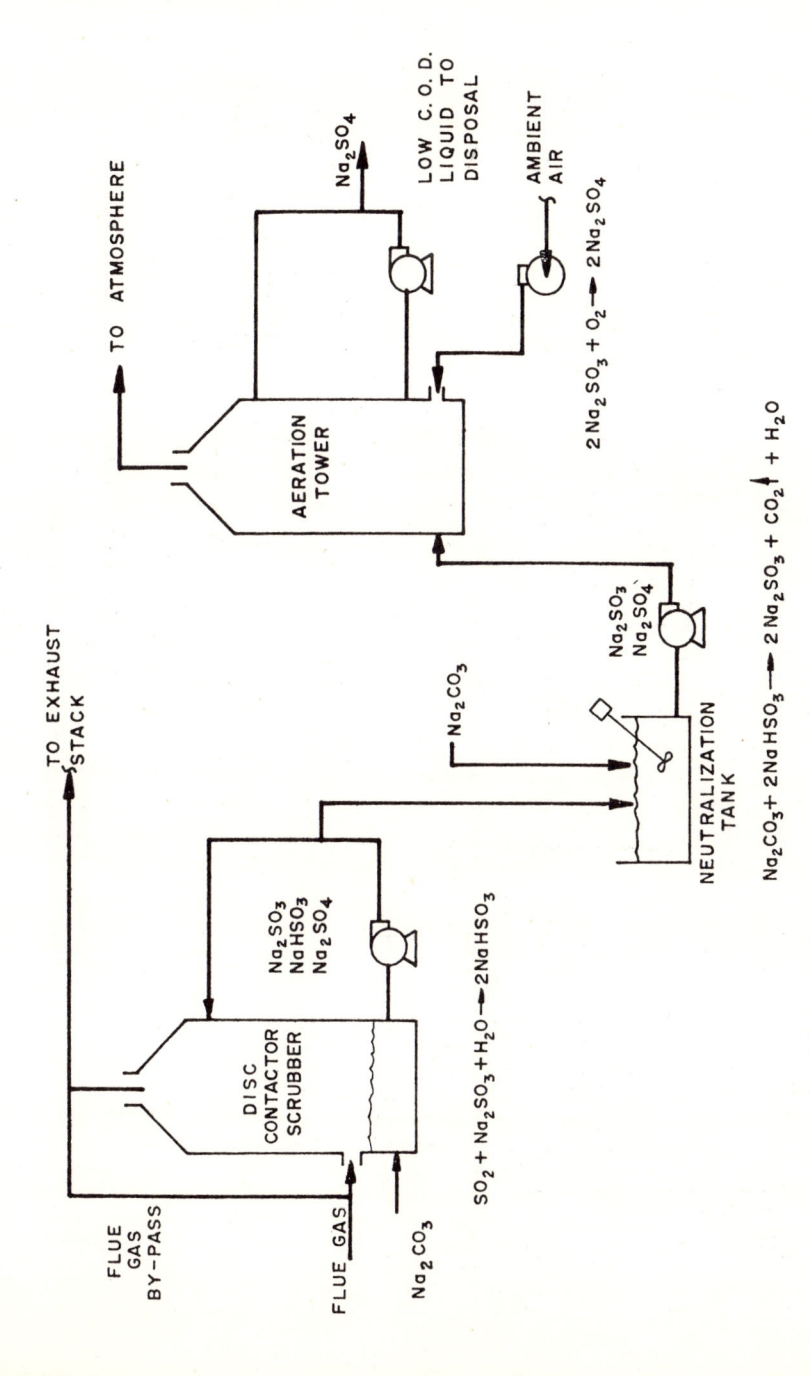

Fig. 10.4. Single-alkali process (from FMC Corp.).

Fig. 10.5. Double-alkali process (from FMC Corp.).

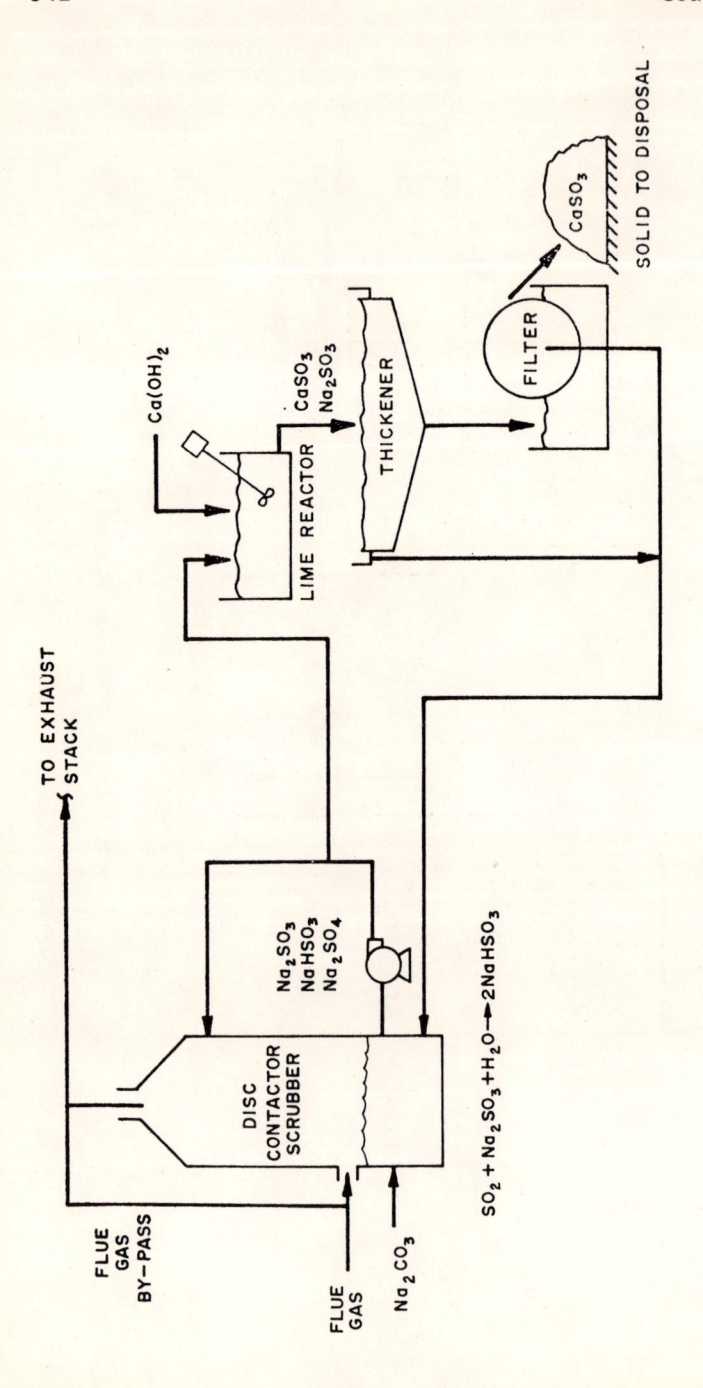

Ancillary Considerations

Two schemes for indirectly utilizing sodium carbonate or slaked lime [Ca(OH)$_2$] solutions, as developed by the FMC Corp., are shown in Figure 10.4 and 10.5. While the immediate reaction is considered to occur with the sulfite to produce the acid sulfite, the overall reaction is with sodium carbonate or Ca(OH)$_2$ as makeup, as the case may be. The double alkali process has the advantage of producing the insoluble calcium salt for disposal. UOP also has a sodium carbonate process.

Dry methods using "sorbents" or "acceptors" may turn out to be the better prospect. Thus particles of sodium carbonate will capture gaseous SO$_2$ (or SO$_3$) to form the sulfite or sulfate, depending upon the level of oxidation.

Calcium carbonate will accomplish the same objective, but is more effective if introduced calcined to the oxide, CaO. Dolomite (CaCO$_3$.MgCO$_3$) behaves similarly. Manganese dioxide (MnO$_2$) has also been used successfully, but is more expensive. Ammonia removes SO$_2$ and can be regenerated from the dissolved salts.

It is necessary to remove the alkali solids along with the fly-ash particulates. In the case of using alkali metal (sodium) compounds, it is required that the alkali be introduced *following* contact of the combustion gases with the heat transfer surfaces (e.g., boiler tubes), since the alkali will flux with the metal surfaces. Moreover, the cooler downstream temperatures favor capture of the sulfur oxides by the alkali.

Solids removal can be accomplished by using a cyclone separator followed by a baghouse. This is effective not only for fly-ash particulate removal *per se*, but for removal of sulfated alkali particles as well. Investigations in this direction have, for instance, been carried out by the Superior Oil Company and the Public Service Company of Colorado. Nahcolite ores from western Colorado were used as the alkali medium.

A dry scrubber has been developed which uses a granular media to filter particulates from gas streams. As such it would serve the same purpose as a baghouse but would be operable at much higher temperatures.

The foregoing brings up the interesting point of using cyclone burners with high-alkali (low fusion point) coals. Fluxing occurs on the burner surfaces as the ash slags and is withdrawn from a port at the burner base. The effect can be enhanced by adding additional alkali. A schematic diagram of a cyclone burner is shown in Figure 10.6. As much as 85-90% removal of particulates is claimed in instances (16, 17, 18). This compares with 50% removal with slag-tap systems (a tap at the bottom of the combustion and heat transfer chamber), and only 20% removal in dry firing — occuring as mechanical fallout.

Thus it is possible in instances to remove the coal ash during the act of combustion itself. At combustion temperatures ($\sim 3000°F$), the alkali will not capture sulfur. At lower temperatures, such as in partial combustion ($1800°F$), it can be expected that the alkali will capture the sulfur, though reducing conditions will prevail in partial combustion such that the sulfide is formed.

Energy Losses:

A difficulty in all gas scrubbing systems is that of energy loss — that is, pressure-drop, which is translated into losses in pressure-volume work. The more complex the system, the greater the losses, calling for the use of forced-draft fans in atmospheric operations.

Fluidized combustion, which operates under pressure, replaces compression of the off-gases with compression of the inlet air.

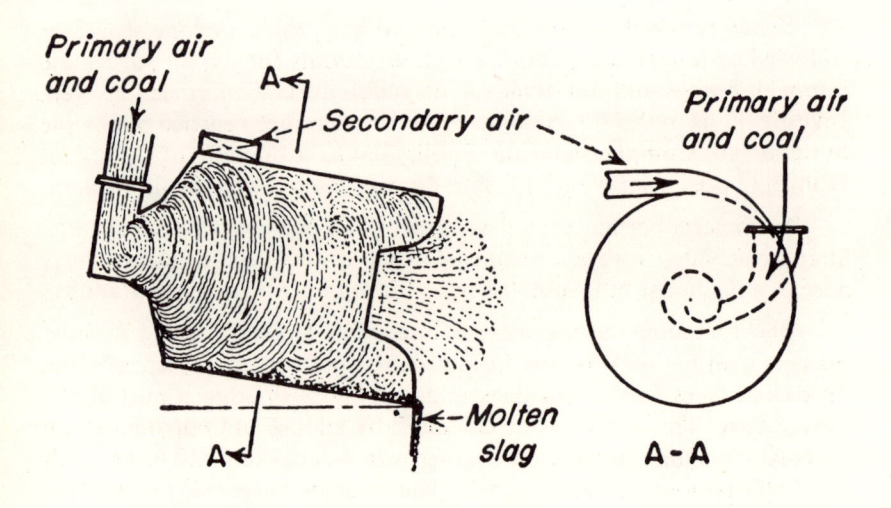

Fig. 10.6. Cyclone burner or furnace (from Babcock & Wilcox).

The off-gases can be readily scrubbed, however, since a pressure drop can be taken. More significantly, fluidized combustors operate at lower temperatures (e.g., ~ 1700°F), which permits capture of the sulfur in the combustion chamber proper.

Wet scrubbers sometimes use the venturi effect to introduce the scubbing solution into the gas. If not designed properly, losses can occur here which must be provided for upstream. In other words, it is preferable to pump the fluid and spray it into the gas stream rather than to compress the gas stream in order to pick up the fluid. Thus the design is preferably made to force the liquid into the gas by means of a pump rather than by taking compression losses.

An interesting alternative which has not received much publicity is the use of a vortex scrubber. Gas-liquid contact is made in a vortex which at the same time induces draft and provides compression. Thus two roles are combined in one: scrubbing and compression. Efficiency data is apparently not available for comparison with more conventional approaches.

10.3 ACID GAS ABSORPTION SYSTEMS

Low- and medium-Btu gas produced by partial combustion or by the steam-carbon reaction will contain hydrogen sulfide and carbon dioxide as the acid gases. Removal of only the hydrogen sulfide is necessary for most purposes if the gas is to be used as a fuel. While an oversimplification, the discussion will emphasize these gases.

If the CO-shift reaction is employed to adjust the H_2/CO ratio prior to methanation, say, then it may be advisable to remove both H_2S and CO_2.

The production of high-Btu gas requires not only the removal of H_2S but the removal of CO_2 as well since the Btu-rating is adversely affected by its presence. This is particularly true of the direct methanation route, where CO_2 may comprise up to 50% of the product gases. For low- or medium-Btu fuel gases, it is not necessary to remove CO_2.

The problem becomes more complicated if selective removal and recovery of both H_2S and CO_2 is required. This is chosen as the more general situation, therefore, with the recovery of either H_2S or CO_2 only, serving as a special case.

In the production of high-Btu gas the principal product gases for consideration are methane, carbon dioxide and hydrogen sulfide — though hydrogen and/or carbon monoxide may also be present in varying degree, as well as some more unusual sulfur compounds such as carbonyl sulfide.

The fact that ammonia or its derivatives may also be produced due to pyrolytic reactions of the nitrogen in the coal increases the complexity of the systems involved. Thus there is the probability of ammoniacal compounds as the result of reaction with the acid gases produced. For the purposes here, however, the discussion is confined to the acid gases in the form of carbon dioxide and hydrogen sulfide. For the most part, water-soluble ammoniacal compounds will appear in the condensate produced following gasifications and cooling of the gases prior to acid gas recovery.

Usually both hydrogen sulfide and carbon dioxide are removed at the same time since most solutions or solvents employed are not selective between the two. This is of no consequence if the H_2S is oxidized and recovered as sulfur (by the Claus process) since the CO_2 is then carried along only as an inert.

Thus the separations problem is viewed as that of selectively removing carbon dioxide and hydrogen sulfide. The preeminent reference in the field is Kohl and Riesenfeld, *Gas Purification* (19).

Carbon dioxide and hydrogen sulfide are both acid gases, and removal in the main involves the use of alkaline solutions or solvents in one form or another. A few organic solvents also show a degree of selectivity. The subject area is reviewed as follows.

ALKALIES

Alkaline solutions readily absorb both CO_2 and H_2S. Notable are the alkali carbonates. For instance, for sodium carbonate solutions, the reactions can be diagrammed as

$$CO_2 + Na_2CO_3 \rightleftarrows 2\ NaHCO_3$$

$$H_2S + Na_2CO_3 \rightleftarrows NaHS + NaHCO_2$$

The forward reaction indicates absorption; the reverse, regeneration or stripping.

Kohl and Riesenfeld indicate that sodium carbonate solutions are relatively inefficient for CO_2 absorption and have high thermal or steam requirements for regeneration.

Ancillary Considerations

Ordinarily, in removing H_2S — in what is called the Seaboard process — regeneration is by air-blowing, which produces an array of sulfur compounds in various stages of oxidation, which must be further captured. Alternately, steam stripping (or CO_2 stripping) will re-produce H_2S and regenerate the carbonate.

Regeneration is also accomplished by reducing the pressure, the basis of the vacuum carbonate process.

In the simultaneous absorption of CO_2 and H_2S, the H_2S absorbs faster — but the equilibrium may shift in favor of one or the other, depending upon concentration.

Certain organic and inorganic additives improve absorption, e.g., the Giammarco-Vetrocoke process which uses arsenic trioxide as an additive with improved CO_2 selectivity.

HOT POTASSIUM CARBONATE SOLUTIONS

Under pressure (~ 300 psi), hot potassium carbonate solutions are effective in removing CO_2 notably. Flashing to a lower pressure releases the CO_2 and regenerates the alkali.

It is particularly effective for high concentrations of CO_2 and H_2S (19), though backup with some other system such as one of the amines may be required in order to meet specifications.

SOLVENTS

Organic solvents exhibiting a selectivity to CO_2 and/or H_2S include methanol (the Rectisol process), propylene carbonate (the Fluor process), and the dimethyl ether of polyethylene glycol (the Selexol process). As a rule higher operating pressures are required, and there may occur the simultaneous absorption of hydrocarbons.

The steam requirements for stripping, however, are generally lower than for alkaline media. This indicates a lower heat of solution.

AMINES

Amines which have been utilized successfully include the following:

Monoethanolamine (MEA)

Diethanolamine (DEA)

Triethanolamine (TEA)

Diglycolamine (DGA)

Diisopropanolamine (DIPA)

Methyldiethanolamine (MDEA)

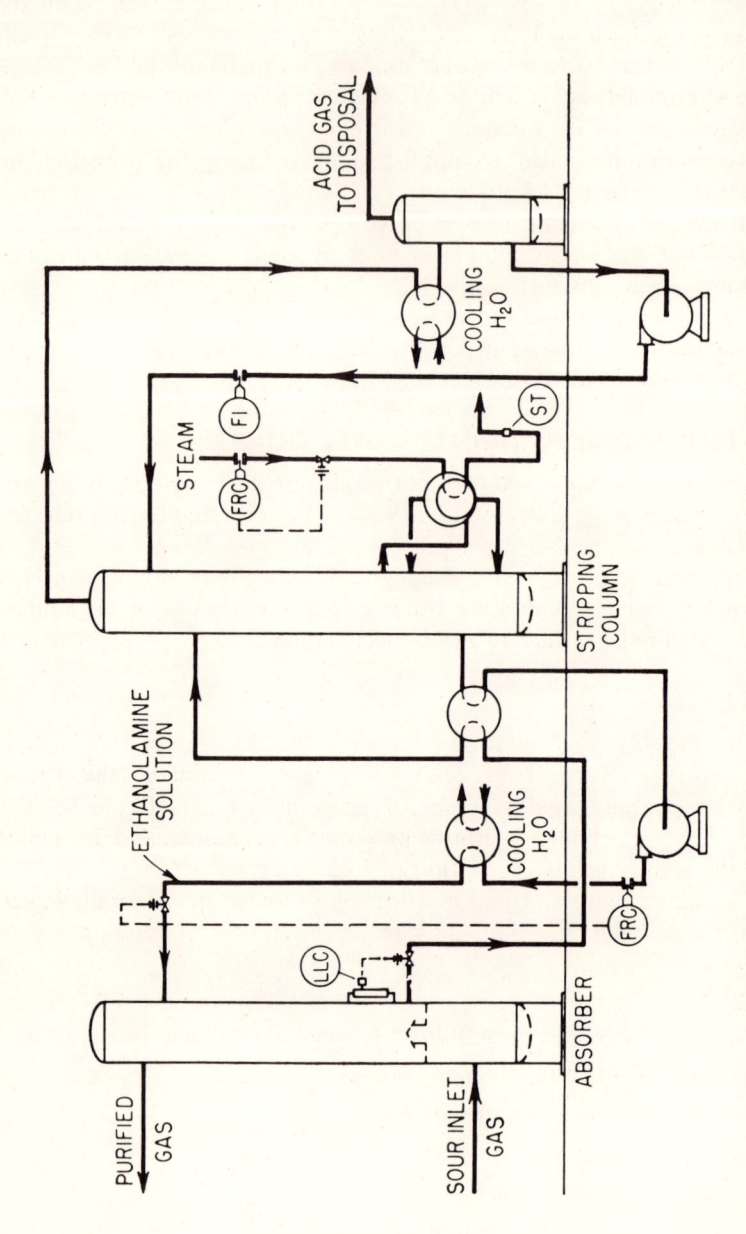

Fig. 10.7. Basic flow diagram for alkanolamine acid-gas removal system [from Kohl and Riesenfeld (19), used with permission, all rights reserved].

Ancillary Considerations

MEA, DEA, and TEA are the most commonly used amines for H_2S and CO_2 removal. DGA is also used to remove moisture and exhibits lower volatility losses. DIPA is the medium for the Adip and Sulfinol processes. MDEA exhibits a significant degree of selectivity between CO_2 and H_2S.

A typical amine absorption and regeneration cycle is shown in Figure 10.7. Direct or indirect steam stripping can be used. With use of direct steam, a partial condenser is required for the overhead leaving the stripping column. With indirect steam, a reboiler is required at the bottom of the stripping column.

The amines perform satisfactorily at low pressures as well as high pressures since a chemical complex is formed. The application of heat decomposes the complex, the function of stripping or regeneration.

Volatility losses are reduced by lowering the solution concentrations and by using a backup water wash.

Of the three common ethanolamines, there is some variation in performance. DEA provides perhaps the middle ground. It performs effectively on removing small or large amounts of CO_2 and H_2S. It has been used, for instance, to remove CO_2 from submarine atmospheres and from combustion products, as well as the more extensive use in the cleanup of sour natural gas streams (20).

MEA is more active toward CO_2 but forms stable compounds with COS and CS_2 which cannot be regenerated thermally.

While MEA and DEA are preferable to TEA from the standpoint of absorptivity, the former show low selectivity between H_2S and CO_2. TEA, on the other hand, exhibits a degree of selectivity between the two acid gases.

Vapor pressures increase in the order TEA, DEA and MEA. Thus TEA has the advantage of minimizing vapor pressure losses.

Problems in operations include foaming and corrosion, both of which are controlled in practice.

SELECTIVE ABSORPTION

If H_2S and CO_2 are each to be selectively removed, one from the other, then solvent systems with a high degree of selectivity to one or the other become advisable. Though there are some possibilities, extenuating circumstance may rule out one or another of the possibilities.

An alternative is to utilize absorption with extractive reflux, a method applicable to systems whereby the solvent phase exhibits only small or moderate selectivity. The process is described by Hoffman in *Azeotropic and Extractive Distillation* (21).

Absorption with extractive reflux introduce the feed intermediately up the column, as practiced in distillation. The solvent is separated from the stream leaving the bottom of the column, and a part of the solvent-free bottom is returned to the bottom of the column as a reflux vapor stream. The operation is diagrammed in Figure 10.8.

If a reboiler is used as in Figure 10.9, the solvent separation is incomplete and the effectiveness of the operation is reduced. The operation in this case is referred to as extractive distillation.

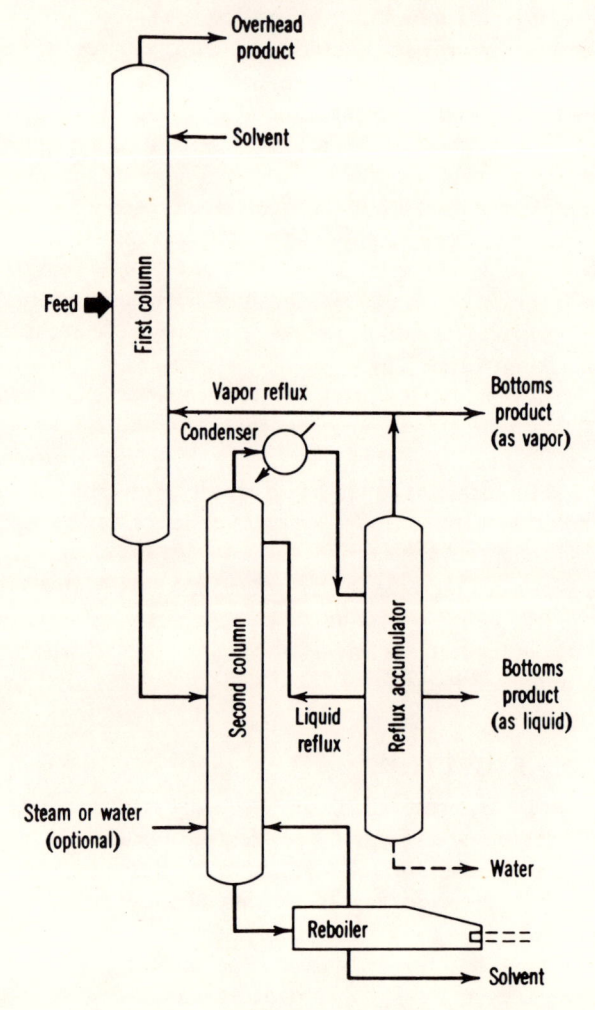

Fig. 10.8. Absorption with extraction reflux [from Hoffman (21)].

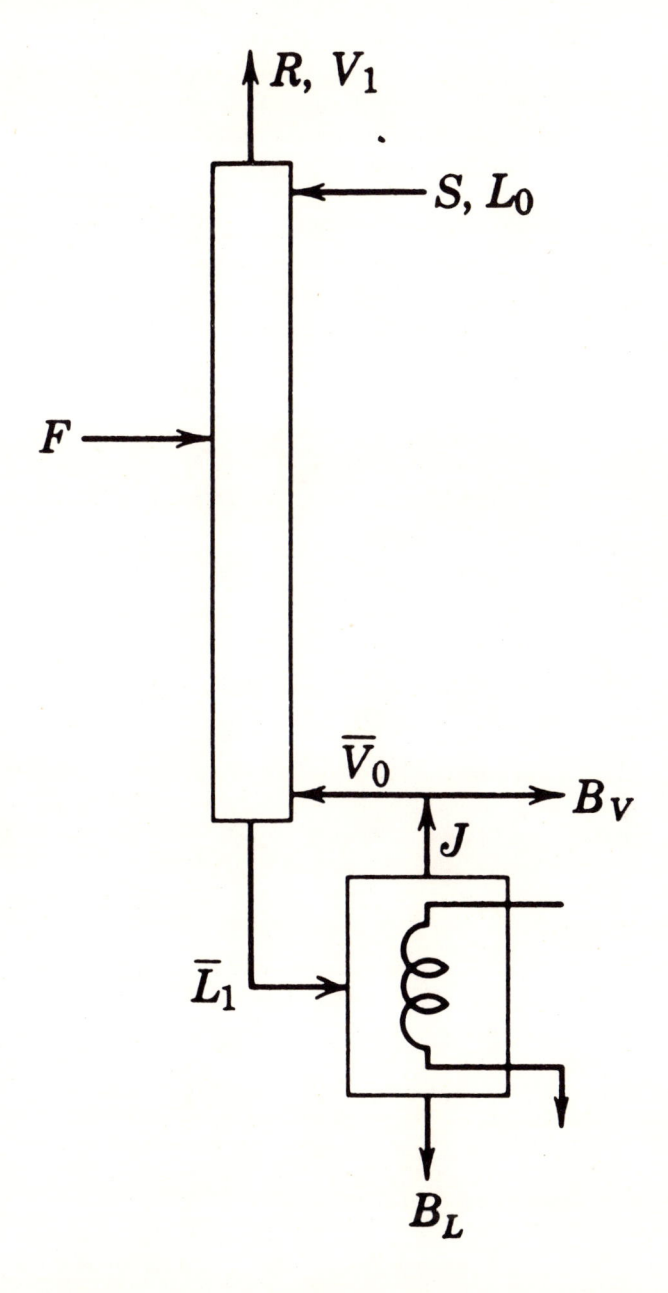

Fig. 10.9. Extractive distillation [from Hoffman (21)].

HEATS OF SOLUTION

The energy balances for absorption and stripping will relate to the heats of reaction or solution for the component systems involved. For systems with high heats of reaction, and where considerable removal takes place, heat exchangers may be necessary for control.

Representative data are as follows, principally for carbon dioxide, the acid gas in greatest conentration for the applications under consideration.

Ethanolamines:

Ref. 22:   ~ 15 kcal/g-mole of CO_2
                    27,000 Btu/lb-mole of CO_2
                    614 Btu/lb of CO_2

The above compares with 3-4 kcal per gram mole for solution in water.

Ref. 19:

	CO_2	H_2S
MEA	825 Btu/lb	820 Btu/lb
DEA	653	511
TEA	630	400
DGA	850	674
DIPA		475

Alkali Carbonates:

Evidently heats of formation can be used for determining the heats of reaction of CO_2 and H_2S with alkalis (23). The determination, however, should be based on the alkaline solution rather than the crystalline compound.

Calculations for standard conditions are as follows:

Na_2CO_3(aqueous) + CO_2(g) + H_2O(l) $\longrightarrow$ 2 $NaHCO_3$(aqueous)
    −275.13   −94.05   −68.32   2(−222.1)   kcal/g-mole

ΔH_0 = 2(2−222.1) − (− 275.13 − 94.05 − 68.32) = − 6.70 kcal/g-mole
                                                        = −12,060 Btu/lb-mole
                                                        = −31.8 Btu/SCF

Ancillary Considerations

K_2CO_3(aqueous) + CO_2(g) + H_2O(l) → 2 $KHCO_3$(aqueous)
-280.9 -94.05 -68.32 $2(-224.85)$ kcal/g-mole

$\Delta H_0 = 2(-224.85) - (-280.9 - 94.05 - 68.32) = -6.43$ kcal/g-mole
$= -11{,}574$ Btu/lb-mole
$= -30.5$ Btu/SCF

The above values are standard heats of reaction at 25°C (77°F). Adjusting the effect of temperature, assuming pressure effect negligible,

$$\Delta H_T = \Delta H_0 + \Sigma n_p (C_p)_p \Delta T - \Sigma n_R (C_p)_R \Delta T$$

where $\Delta T = T - T_0$ and

n_P = moles of a product P
n_R = moles of a reactant R
$(C_p)_P$ = molar heat capacity of a product P
$(C_p)_R$ = molar heat capacity of a reactant R

Heat capacity data from Reference 24 are:

$(C_p)_{Na_2CO_3} = 27$ cal/g-mole-°C = 27 Btu/lb-mole-°F
$= 0.25$ Btu/lb-°F
$(C_p)_{NaHCO_3} = 21$ cal/g-mole-°C = 21 Btu/lb-mole-°F
$= 0.25$ Btu/lb-°F

The heat capacity drops off rapidly as the temperature decreases but rises less sharply as temperature increases. Other values in Reference 24 are

$(C_p)_{Na_2CO_3} = 26.41$ cal/g-mole at 298°K
$(C_p)_{NaHCO_3} = 21.01$ cal/g-mole at 298°K

Kopp's rule (25) gives the following result:

$(C_p)_{K_2CO_3} = 2(6.2) + 1.8 + 3(4.0) = 26.2$ cal/g-mole-°C
$(C_p)_{KHCO_3} = 6.2 + 2.3 + 1.8 + 3(4.0) = 22.3$ cal/g-mole-°C

Calculated results for Na_2CO_3 and $NaHCO_3$ also correspond to the above.

NBS Technical Note 270-8, the latest in the series *Selected Values of Chemical Thermodynamic Properties,* will when published have information on compounds of the alkali metals. The interim report (26) lists the following:

$(C_p)_{Na_2CO_3}$ 26.41 cal/g-mole-°C

$(C_p)_{NaHCO_3}$ 20.94

$(C_p)_{K_2CO_3}$ 27.65

A value for the heat capacity of $KHCO_3$ is not given, but by inference,

$(C_p)_{KHCO_3} \sim 22.18$

For water,

$(C_p)_{H_2O(l)} = 18$ Btu/lb-mole-°F

For carbon dioxide,

$(C_p)_{CO_2(g)} = 11$ Btu/lb-mole-°F

Assuming a temperature of 250°F, which could apply to stripping, or absorption with hot potassium carbonate solution,

$$\Delta H_{250°F} = \Delta H_{77°F} + 2(20.94)(250-77) -$$
$$(26.41+18+11)(250-77)$$
$$= -12,060 + 7,245 - 9,586 = -15,401 \text{ Btu/lb-mole } CO_2$$
$$= -40.6 \text{ Btu/SCF of } CO_2$$

The heat of reaction becomes more exothermic with temperature.

For CO_2 with potassium carbonate solutions,

$$\Delta H_{250°F} = \Delta H_{77°F} + 2(22.18)(250-77) - (27.65+18+11)(250-77)$$
$$= -11,574 + 7,674 - 9,454 = -13,354 \text{ Btu/lb-mole } CO_2$$
$$= -35.2 \text{ Btu/SCF of } CO_2$$

The effect of temperature is not great. Kohl and Riesenfeld (19) cite a value of 32 Btu/SCF for the heat of solution of CO_2 with hot carbonate solution.

The heat of solution of CO_2 in water is 5000-7000 Btu/lb-mole of CO_2 versus about 27,000 Btu/lb-mole of CO_2 for absorption by ethanolamine solutions. It would be expected that the reaction with alkali carbonate solutions would be still higher.

Ancillary Considerations

THERMAL INPUT

While heat requirements for stripping are related to heats of reaction, there are other factors which influence the net heat input. These include operating conditions, which vary.

Some rough guidelines are provided by Kohl and Riesenfeld (19). For ethanolamines, a figure of about 0.25 lb of stripping steam per SCF of acid gas removed can be backed out of the data presented. Catacarb additives improve the performance.

For hot carbonate removal of acid gases, the figure is about 0.2 lb of stripping steam per SCF of acid gas removed. The requirement will vary with concentration and other conditions.

ADSORPTIVE METHODS

For the removal of low or trace concentrations, solid adsorbents or "sorbents" are effective. The phenomenon may involve physical adsorption at the particle surfaces and pores, as with molecular sieves, or chemical reaction at the surfaces and pores, as in the case of iron oxide particles. An embodiment of a commercial process using iron oxide is shown in Figure 10.10.

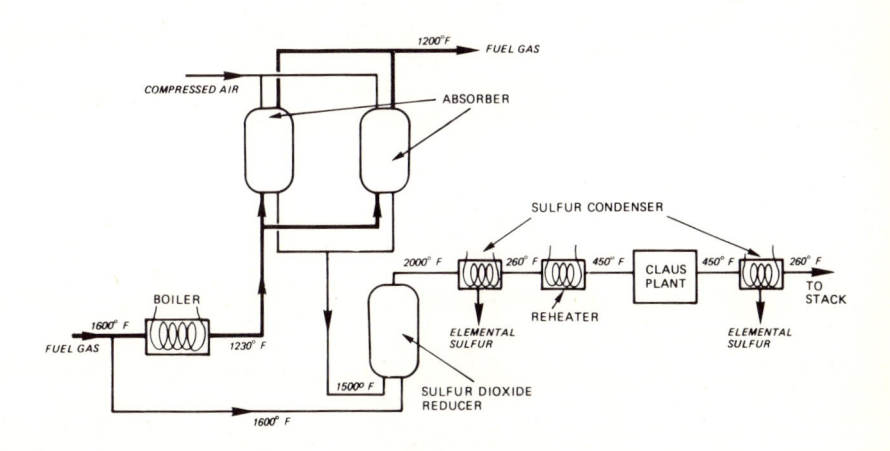

Fig. 10.10. Iron-oxide desulfurization.

NOVEL METHODS FOR CO_2 REMOVAL AND RECOVERY

In addition to the established practices of using ethanolamines and hot potassium carbonate solutions or such other systems as propylene carbonate as mentioned in Kohl and Riesenfeld (19), a few more novel methods and findings merit mention.

Alkali metal solutions as well as ammonium carbonate, phosphate, borate, and phenolate containing As_2O_3 have been used as absorbers. The influent gas stream is washed at 25 atm. at a temperature of less than 1000°C. The rich liquor is regenerated isothermally with steam at a pressure of one atmosphere (27).

Kuo (28) has shown that in the absorption of CO_2 with LiOH, the removal efficiency is not affected by the initial gas concentration. Under conditions of axial flow, the rate constant is independent of velocity. When the rate is governed by a surface reaction, contact time is a significant parameter. When the rate is diffusion-controlled, the bed volume is a significant parameter.

A selective solvent containing an ether of a carbonitrile has proved to have more capacity than propylene carbonate, triacetin, or methyl alcohol. The ether carbonitrile, $CH_3OCH_2CH_2CN$, removed 90% of the CO_2 from a natural gas stream containing about 19% CO_2. Rich absorbent was flashed at 90°F and 200 psia (29).

Removal of CO_2, H_2S and COS from liquids or gases has been accomplished by means of alkanolalkylamines or morpholine derivitives. Gases may be purified by countercurrent flow through the absorption column. The absorbent is regenerated by heating and lowering the pressure and stripping. Data indicate that an influent stream containing 48% H_2, 47% CO, 4.0% CO_2, 0.8% H_2S, and 0.03% COS, was reduced to less than 0.1% CO_2, 0.0005% H_2S, and 0.005% COS (30).

Activated carbon has been used in experimental studies of the absorption of CO_2 from air streams. The contaminated gas is passed countercurrently through a column packed with duraluminum rings. Carbon particles ranged in size from 315 microns to 1500 microns. Without duraluminum packing the mass transfer coefficient was 1.3-1.5 times greater (31).

Adsorption isotherms for CO_2 have also been measured at pressures of 0-200 atm. under conditions of both static and dynamic conditions for Soviet-type zeolites NaA, NaX, CaX, CaA, and MgA and American types 4A and XW (32).

Sodium and calcium forms of A-type zeolite have been employed in fluidized beds (33). Sodium A zeolite proved to be mechanically stable. Calcium A undergoes considerable attrition. Simultaneous adsorption of

Ancillary Considerations 357

CO_2 and H_2O resulted in a CO_2 concentration decrease of 30-35% in one stage. Desorption was achieved at 300-320°C by blowing air containing 1.5-2.4 g/m³ of moisture content.

Conviser (34) reports that the molecular sieves best suited for CO_2-removal in natural gas purification are 4A, 5A and 13X. Results showed a water dew point of − 100°F and a CO_2 concentration of less than 50 ppm.

Investigations of artificial zeolites 5A, CaA, 13X and NaX as CO_2 adsorbers indicated that NaX absorbed slightly better than the others (35). At elevated presures the adsorption of CO_2 first increases, then decreases. Under dynamic conditions the adsorption rate for water significantly exceeded the adsorption rates of other gases. This provides the element of selectivity.

Adsorption isotherms for the Linde 4A and 5A artificial zeolites showed a high activity for CO_2, attaining about 11% for 4A and 15% for 5A (36). Other sections of the investigations showed that:

1. Artificial zeolites are much better than natural ones.
2. The adsorption capacity is temperature dependent.
3. The activity decreases with increasing temperature.
4. The activity decreased as the number of cycles increased.
5. In regeneration, most of the CO_2 is eliminated at temperatures less than 200°C.
6. With intensive heating, regeneration takes less than one hour.

Carbon dioxide in gas streams produced by the burning of carbonaceous fuel may be adsorbed by artificial zeolites by first cooling the stream to 100°F. CO_2 is subsequently desorbed by heating (37).

The subject of adsoprtion, above, is akin to chromatography.

Korsh et.al. (38) reviews thirty references having to do with current trends of elimination of CO_2 and water from fuel gases.

Jedrezejczyk (39) presents a review of methods of purification of chemical gas from dust, CO_2 and sulfur compounds.

Life-support systems in manned space vehicles demand the removal of CO_2 from the cabin atmosphere. Two schemes are described as follows.

Major et.al. (40) conducted studies of the absorption of CO_2 using activated carbon in Linde molecular sieves. They report that carbon is insensitive to water vapor in the air-CO_2 mixtures. It is further stated that the use of activated carbon for the removal of CO_2 in space craft cabins merits attention.

Brown et.al. (41) describe a solid state electro-chemical device which removes CO_2 from the cabin atmosphere and regenerates oxygen. The apparatus performing chemisorption and electrolysis has a capacity of 9 pounds per day with the regeneration rate of eight pounds per day. It operates in a 100% oxygen atmosphere of 5 psia with CO_2 partial pressure of 5 mm.

Colombo and Mills (42) report the recovery of CO_2 by the use of metallic oxides. Investigations covered the use of Li, Na, K, Rb, Cs and Mg. The absorbent capacity of Mg is reported to be 1.092 lb of CO_2/lb oxide. The equations for the reaction are as follows.

$$MgO + H_2O = Mg(OH)_2$$
$$CO_2 + H_2O = H_2CO_3$$
$$Mg(OH)_2 + H_2CO_3 = Mg\,CO_3.2H_2O$$

The energy of regeneration was shown to be 1.153×10^3 Btu/lb CO_2. In summation, the authors state that:

1. The affinity of MgO is adequate.
2. The presence of water is necessary for efficient operation.
3. The absorptive quality of MgO is superior to molecular sieves.
4. High rates of absorption are possible with further development of the catalyst.

For the most part, the foregoing may be regarded as in the experimental stage or for small-scale use only — at least as far as development has gone.

NO_X ABSORPTION

High-temperature combustion processes produce nitrogen oxides. This is a problem in particular for MHD power generation, much less so for fluidized combustion, which operates at less than normal combustion temperatures.

While nitrogen oxides are expected to be soluble in more basic solutions, it has also been observed that amine solutions will pick up NO_x. This extends to solutions of ammonia and also to 2-aminoethanol.

10.4 CONVERSION OF HYDROGEN SULFIDE

Hydrogen sulfide and other sulfur compounds recovered offer the possibility for conversion by the Claus process or other processes. While sulfur is marketable up to a limit, large-scale coal conversion operations

Ancillary Considerations

would saturate the market. Conversion to sulfur in this context becomes but a convenient way to store a (waste) byproduct.

The Claus conversion, which utilizes alumina catalysts, yields sulfur directly by the reaction

$$H_2S + \frac{1}{2} O_2 \rightarrow S(g) + H_2O(g)$$

Conversion may alternately be effected in two stages:

$$H_2S + \frac{3}{2} O_2 \rightarrow SO_2 + H_2O(g)$$

$$\underline{SO_2 + 2 H_2S \rightarrow 3 S(g) + 3 H_2O(g)}$$

$$3 H_2S + \frac{3}{2} O_2 \rightarrow 3 S(g) + 3 H_2O(g)$$

The sulfur vapor is condensed and eventually solidified. The inertness of elemental sulfur to the ambient atmosphere permits unconfined storage.

Byproduct sulfur compounds are produced simultaneously in the Claus process. These compounds are preferably oxidized to SO_2 and absorbed as the sulfate for disposal. Alternately the SO_2 may be reacted with H_2S to yield further sulfur.

Combined processes selectively remove H_2S and oxidize it to sulfur. An example is the Stretford process, which employs a solution of sodium carbonate with sodium vanadate plus additives. The vanadate oxidizes the absorbed H_2S to elemental sulfur, which is collected. Regeneration of the vanadate is accomplished with air. The process is more applicable to dilute concentrations of H_2S.

10.5 DRYING AND COMPRESSION

High-Btu gas is required to meet certain moisture or dew-point specifications, particularly for pipeline transmission. Drying is usually accomplished by using diethylene glycol absorption or by using solid sorbents such as silica gel or molecular sieves. Both types of system are regenerable.

Compression depends upon use. Pipeline transmission over long distances involves nominal pressures of ~ 1000 psi. Intra-city transmission and distribution may involve pressures of 100-150 psi maximum. Whereas in-plant use may require only 15 psig or less. The subject is covered further in the Appendix.

10.6 OTHER COMPOUNDS

The system of gases and liquids which result from coal conversion may be very complex, exhibiting a bewildering array of compounds free and in combination, both stable and unstable. The removal or separation of some of these materials is discussed in Reference 43.

Free ammonia is removed by water scrubbing followed by distillation. The NH_3 recovered may be burned, converted to anhydrous ammonia product, or further converted to ammonium salts (for fertilizer). (NH_3 may of course become a major product by producing hydrogen from coal, which is then used to fix nitrogen.)

Hydrogen cyanide, if present, is picked up during acid gas recovery. It may be selectively absorbed by an acidic wash of controlled pH and ultimately converted to sodium or potassium cyanide (for use in extractive metallurgy), or otherwise converted to NH_3, etc. — or simply oxidized.

Crude phenols are removed from water-phase residues by extraction with light oil fractions, followed by neutralization or recovery of the phenols. The phenolic fraction may be recycled as feed or mixed with liquid fuels.

REFERENCES

1. Clancey, J. T. and J. A. Phinney, "Status of CO_2 Acceptor Process Development," Presented at the American Gas Association Production Conference, Buffalo, May 25, 1965.
2. "Pipeline Gas from Lignite Gasification, A Feasibility Study," Consolidation Coal Company, Report No. 1 to the Office of Coal Research, Project No. 180, January 13, 1965.
3. Skaperdas, G. T. and W. H. Heffner, "Preliminary Evaluation of the Kellogg Coal Gasification Process," Presented at the American Gas Association Production Conference, Buffalo, May 25, 1965.
4. Donath, E. E. and R. A. Glenn, "Pipeline Gas from Coal by Two-Stage Entrained Gasification," Presented at the American Gas Association Production Conference, Buffalo, May 1965.
5. Hueble, J. and F. C. Schora, "Development of the IGT Coal Hydrogasification Process," 58th Annual Meeting, American Institute of Chemical Engineers, Philadelphia, 1965.

6. Tsaros, C. L., S. J. Knabel and L. A. Sheridan, "Process Design and Cost Estimate for Production of 266 Million SCF/Day of Pipeline Gas by the Hydrogasification of Bituminous Coal — Hydrogen by the Steam-Iron Process," Institute of Gas Technology, Report to the Office of Coal Research, OCR Contract No. 14-01-0001-381, August, 1966.

7. Schora, F. C. and J. Heubler, "Current Status of Coal Hydrogasification Process Development Work," Presented at the American Gas Association Production Conference, Buffalo, May 25, 1965.

8. Kloepper, D. L., T. F. Rogers, C. W. Wright and W. C. Bull, "Solvent Processing of Coal to Produce a De-Ashed Product," Spencer Chemical Division, Gulf Oil Corporation, Report to the Office of Coal Research, OCR Contract No. 14-001-275, February, 1965.

9. Eddinger, R. T., J. F. Jones, M. R. Schmid, L. D. Friedman and L. Seglin, "Char Oil Energy Development," FMC Corporation, Report to the Office of Coal Research (PCR 469), OCR Contract No. 14-01-0001-235, December 1965.

10. Jones, J. G., M. R. Schmid, M. E. Sacks, Y. Chen, C. A. Gray and R. T. Eddinger, "Char Oil Energy Development," FMC Corporation, Report to Office of Coal Research (PCR-534), Contract No. 14-01-0001-235, January, 1967.

11. Hellwig, K. D., E. S. Johansen, C. A. Johnson and S. C. Shuman, "Make Liquid Fuels from Coal," *Hydrocarbon Processing, 45* (5), 165-169 (May, 1960).

12. Hellwig, K. C., M. C. Chervanek, E. D. Johanson, H. H. Stotler and R. H. Wolk, "Convert Coal to Liquid Fuels with H-Coal," 62nd National Meeting, American Institute of Chemical Engineers, Salt Lake City, May 1967.

13. Johnson, W. B., "Coal Beats Oil Here," *Pet. Refiner, 35* (12), 222-228, December, 1965. Also appears as "Gasoline in Coal" in the *Kelloggram,* The M. W. Kellog Company, New York, 1957.

14. Ely, N., "The Oil Shale Industry's Water Problems," *Fourth Oil Shale Symposium,* Denver, April 6, 1967.

15. Carpenter, H. C. and P. L. Cottingham, "Pipeline Gas by Hydrogasification of Shale Oil," 58th National Meeting, American Institute of Chemical Engineers, Dallas, February, 1966.

16. Sherman, R. A. and B. H. Landry, "Combustion Processes," in *Chemistry of Coal Utilization,* Supplementary Volume, H. H. Lowry Ed., Wiley, New York (1963).

17. Ely, F. G. and D. H. Barnhart, "Coal Ash — Its Effects on Boiler Availability," in *Chemistry of Coal Utilization,* Supplementary Volume, H. H. Lowry Ed., Wiley, New York (1963).

18. *Steam,* the Babcock and Wilcox Company, New York (1955, 1960, 1963).

19. Kohl, A. L. and F. C. Riesenfeld, *Gas Purification,* Gulf Publishing Company, Houston (1974).

20. Updegraff, N. C. and R. M. Reed, "25 Years of Progress in Gas Purification," *Petroleum Engineer, 26* (10), C-57-63 (Sept., 1954).

21. Hoffman, E. J., *Azeotropic and Extractive Distillation,* Interscience, New York (1964). Krieger, Huntington, New York (1977).

22. Lyudkovskaya, M. A. and A. G. Leibush, "Solubility of Carbon Dioxide in Solutions of Ethanolamines Under Pressure," *Zhur. Priklad. Khim.* (J. Applied Chem.), *22,* 558-67 (1949). CA 45:2291d.

23. Dryden, I. G. C., "Equilibrium Between Gaseous Carbon Dioxide and Solutions of Alkali Carbonates, Bicarbonates, and Hydrosulfides, I. Potassium Salts." *J. Soc. Chem. Ind., 66,* 59-62 (1947). CA 41:5003b.

24. Mellor, J. W., *A Treatise on Inorganic and Theoretical Chemistry,* Longmans, Green & Co., London (1922-1937).

25. *Handbook of Chemistry,* N. A. Lange Ed., Handbook Publishers, Sandusky, Ohio (1934-   ).

26. Wagman, D. D., W. H. Evans, V. B. Parker and R. H. Schumm, *Chemical Thermodynamic Properties of Sodium, Potassium and Rubidium: An Interim Tabulation of Selected Values,* National Bureau of Standards Information Report 76-1034, April, 1976.

27. Societa Generala per L'industria Mineraria e chimica and vetrocoke societa per Azioni, "Removal of CO_2 from Gases-Montecatini," Fr. Patent No. 1,412,789, Oct. 1, 1965. Ital Appl., Oct. 29, 1963. CA 65:5071e.

28. Kuo, K. L., "Parameters Governing Performance & Design of Reactors For CO_2 Absorption by Lithium Hydroxide Particles," *Soc. Auto. Engr. Pre-print No. 912F* (1964), pp. 22. CA 62:4931e.

Ancillary Considerations

29. Woertz, B. B., "Process for CO_2 Removal From Natural Gas," Patent No. 3,266,220, Aug. 16, 1966. Appl., Dec. 27, 1962. Assignor to Union Oil Co. of Calif. CA 65:16759g.
30. Sharma, M. J., "Purification of Liquid and Gases," Belg. Patent No. 660,875, Sept. 10, 1965. Brit. Appl., Mar. 12, 1964. Assignor to Shell Internationale Research. CA 64:6141b.
31. Kavetskii, E. D., L. A. Ahopyan and A. N. Planovskii, "Mass Transfer During Adsorption of CO_2 and Activated Carbon in a Column," *Khim i neff. Mashiivosti, 4,* 23-5 (1965).
32. Katina, N. F., A. J. Moroz and E. V. Vagin, "Purification of Air by Adsorption of Synthetic Zeolites. I. Adsorption of CO_2 From Air At High Pressures," *Tr. vses. nauchn. Issled. Inst. Kislorodn. Machinosti,* No. 10, 132-9 (1965). CA 69:11755f.
33. Kozlova, T. I., V. N. Lepilin and P. G. Pomankov, "Absorption Kinetics in a Counter Current Column of a Fluidized Adsorbent," *Zh. Prikl. Khim., 39,* (8), 1719-23 (1966). CA 65:1487a.
34. Conviser, S. A., "Removal of CO_2 from Natural Gas with Molecular Sieves," *Proc. Gas. Conditioning Cong.* (1964), 1F-12F. CA 63:17750h.
35. Torocheshnikov, N. S., A. I. Sidorov and N. V. Kel'tsev, "Fine Durification of Gases from CO_2 with Simultaneous Removal of Water Vapor as the First Phase of the Process of Separation and Processing of Gases," *Materialy Vses Soveshch. po Tseolitan 2nd Leningrad.,* 240-9 (1964). Publ. 1965. CA 64:18456e.
36. Bark, S. E., N. V. Kel'Tsev, I. P. Ogloblina, N. M. Seryieva, M. I. Skvortsova and N. S. Torocheshividov, "Application of Synthetic Zeolites As Molecular Sieves for the Preparation of Protective Atmospheres," *Sintetich Tseolity Issled. Primenenie, Akad, Nauk. SSSR. Otd Khim Nauk.,* 276-83 (1962). CA 58:8721h.
37. Duffy, J. G., "Method For Recovery of CO_2," Patent No. 3,100,-685. Filed Oct. 30, 1959, Ser. No. 849,805. Assignor to Arnold Equipment Co.
38. Korsh, M. P., I. F. Bogdanov and N. V. Lavrov, "Current Trends in the Elimination of H_2S and CO_2 From Fuel Gases," *Tr. Inst. Gorguch. Iskop. Akad. Nauk. SSSR,* No. 16, 367-87 (1961). CA 57:16959d.
39. Jedrzeiczyk, B., "Purification of Chemical Gas From Dust, CO_2, and Sulfur Compounds," *Gaz. Woda, Tech. Saint., 35,* (8), 289-92 (1961). CA 59:13615g.

40. Major, C. J., B. J. Sollami and K. Kammermeyer, "Carbon Dioxide Removal from Air by Adsorbent," *Ind. Eng. Chem. Process Design, 4,* 327-33 (July 1965).
41. Brown, D. L., W. Glass and J. L. Greatozex, "Performance of an Electro-Chemical Devide for Simultaneous CO_2 Removal and Oxygen Regeneration," *Chem. Eng. Prog. Symposium Ser. 62,* No. 63, 50-54 (1966).
42. Colombo, G. V. and E. S. Mills, "Regenerative Separation of Carbon Dioxide via Metalic Oxides," *Chem. Eng. Prog. Symposium Ser. 62,* No. 63, 89-94 (1966).
43. Muder, R. E., "Light Oil and Other Products of Coal Carbonization," *Chemistry of Coal Utilization,* Supplementary Volume, Wiley, New York (1963).

Appendix

THE DETERMINATION OF NET ENERGY EFFICIENCIES FOR COAL CONVERSION PROCESSES

Net energy analysis is becoming of increasing importance as a means for evaluating and rating processes and methods for converting and otherwise utilizing energy resources. Its significance was in fact noted by a clause in the act creating the Energy Research and Development Administration (ERDA) whereby new processes were to be subjected to such a net energy analysis — though the term was not further defined.

Economic cost information, in spite of its implied accuracy and exactness, is sometimes necessarily vague. The detail obscures the basics — another case of the forest and the trees. There are also other obliquities: for instance, mere size affects the economic analysis. That is, the product value goes up linearly while plant investments and operating costs go up by some fractional exponent. Thus if a plant is assumed large enough, a profit can ultimately be shown, as indicated in the figure below:

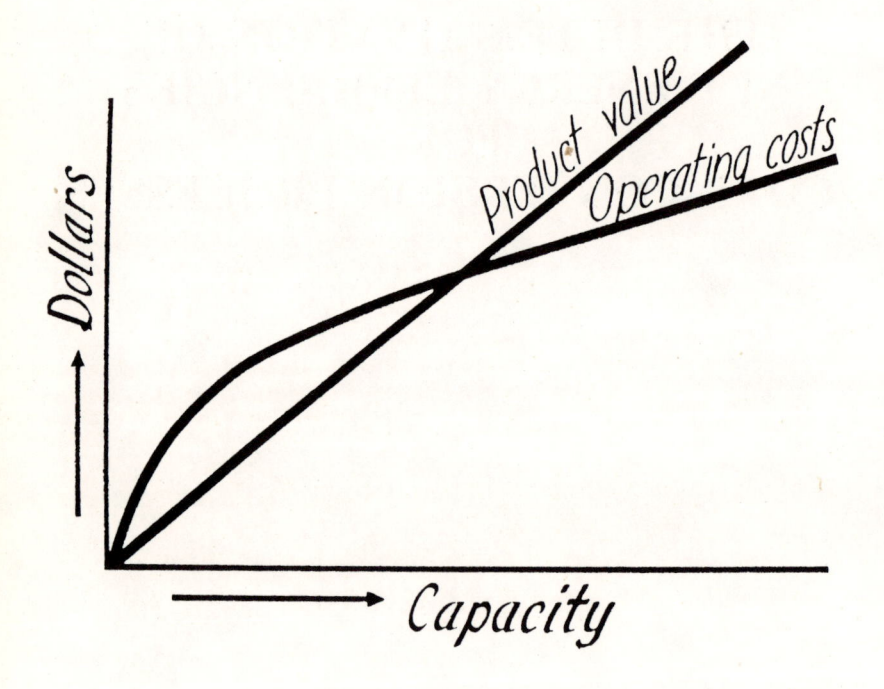

Obscured are such matters as risk, the difficulties of obtaining capital, and the complexities of large operations. For these reasons, net energy analysis can be viewed as a more abstract and fundamental criterion in making judgments.

Appendix

For the applications of coal conversion, the determination of overall net efficiency or net energy yield is more readily made for the production of gaseous compounds — e.g., methane — than for liquids, since the analysis and stoichiometry are more straightforward.

The methodology will consequently be demonstrated mainly for the production of methane — that is, high-Btu or pipeline quality gas, usually called synthetic natural gas (SNG).

Direct methanation, being the simplest route for the production of methane from carbon (coal) and steam, is used as the reference base.

Before proceeding to the detail of the calculations, the terminology is further reviewed as follows.

NET ENERGY ANALYSIS

The following description was written by the author in connection with a meeting and workshop on the subject of net energy ("Minority Report," Net Energy Analysis Workshop, Stanford University, August, 1975, sponsored by the National Science Foundation):

> Energy is that function of the variables necessary and sufficient to describe the behavior of a system such that, under change, the total energy of the system is conserved (remains constant) though the various contributing forms or sub-elements may vary and be transformed or converted one to another.
>
> Through the years, for physical and chemical systems, these various sub-forms have been established and identified in terms of the appropriate contributing variables, and denoted by such designators as kinetic and potential energy, thermal energy, pressure-volume energy or work, electrical energy, etc. These relationships provide the means for evaluating and establishing equivalency between forms.
>
> In application, the absolute values of these entities are not necessary — only the quantitative change from one form to another is of interest, and becomes the basis for computing the degree of conversion or utilization, and the efficiency or net yield for a conversion.
>
> The study of transformation or utilization sequences resulting in a more useful or convenient energy form, the determination of all factors and inputs which affect each sequence, and the relative evaluation of the designated energy form so realized constitute the purposes of net energy analysis.

The consideration may be made on a steady-state basis, on a transient or instantaneous basis, or on an integrated basis over an interval of time. While strictly speaking the entirety of the system would be all-encompassing, in practice there must be a limit to the system boundaries. A value judgement can be made not to consider factors approaching "zero opportunity cost", whereby an energy input, say, does not require an equivalency — that is, has a "zero" equivalency in terms of either the energy source or the resulting energy form of interest. For example, in certain applications, solar energy can be regarded as "free".

The question of consistency in units naturally arises when considering transformations within the system, or inputs and outputs of the system. This must be handled by assuming an equivalency (e.g., the mechanical equivalent of heat, or electrical equivalent of heat) and an efficiency of conversion. This assumption of efficiency becomes arbitrary, but may be based on prior experience and will be related to the particular system and its elements. The use of consistent units — by common practice, a thermal unit of energy — will, by providing a common denominator, permit more ready comparison of the performance of one system against another.

Inasmuch as value judgements are called for in the efficiency of converting from one form of energy to another and in proceeding from one step to another in a conversion and utilization sequence, it becomes highly desirable to analyze the chemistry, physics, and engineering involved in the sequence or process. In other words, the "why" and "how" should be established. In this way, the known performance of systems can be correlated and used to predict the performance of future systems. At the same time credibility is induced, since results are in a sense "explained" and verified. As in the practice of medicine, it is not enough that a drug or treatment is effective; the reasons for the efficacy must also be known.

Lastly, the results of any and all analyses should be presented in a simple and direct manner. One obvious means is merely to give the percentage ratio of the desired "product" of the transformation in terms of the specified source. The units as presented do not even have to be in the same units. Thus in the production of electrical power, we can merely ratio Btu(e) to the Btu(t) represented by the fuel.

At the same time, it is desirable to acknowledge other inputs or outputs, which may be referenced to either the source or "product." Thus in the nuclear power cycle, the energy inputs can be lumped or

Appendix

summed in terms of, and in the units of, the electrical power generated. The net effect is one of energy recycle. As another example, the oxygen requirement for coal gasification can be expressed equivalently in terms of either the coal required or the gas produced, on a Btu basis or on a materials basis, and adapted to either the feed or product.

The final objective becomes that presentation of results, stripped of jargon, whereby the uninitiated as well as the initiated can immediately grasp the overall efficacy of any conversion and utilization sequence or process, and the relative merit of one compared to another.

Or in other words, "It takes energy to get energy".

For more on the subject of energy and applications of net energy analysis (particularly for the nuclear cycle), consult the author's *Concept of Energy*, previously cited.

Efficiency determinations are not always on the same basis and may not even be defined in the same manner. Comparisons thus often become meaningless.

For instance, all energy inputs may not be included or unrealistic energy recycle may be assumed. Overly optimistic conversions are sometimes used, or else extravagant claims are made for byproduct utilization.

Moreover, in defining the efficiency or energy yield, the calculation may be based on at least two alternatives. As the one instance, the energy inputs in thermal units are subtracted from the thermal value of the product and the difference is then divided by the thermal rating of the feed. In the other instance, the thermal value of the energy inputs is added to the thermal rating of the feed, and the sum is divided into the thermal value of the product. Or some combination of these limiting cases can be used.

As a further parameter, the thermal value or rating of the feed and product can alternately be expressed on a gross or net heating value basis if hydrogen and water are involved.

In any event, the discrepancies can be considerable just in the manner that the data is calculated and reported. The first-mentioned definition is emphasized herein.

For the above reasons and others, care is required in exercising comparisons between various efficiency determinations, especially if independently made. The deployment of a common base is essential.

Obviously, the more simple and elemental the calculations, the less diffuse the answer and the greater confidence in the results. Therefore, the attempt is made in the following sections to base the determinations primarily on a carbon balance and standard heats of reaction which are adjusted for temperature, with common conversions and efficiencies of waste-heat utilization as the parameters.

The idealized sequence in a conversion involves heating (or cooling) the reactants to reaction temperature, maintaining the reaction at temperature, and the cooling (or heating) of the products. Each leg will result in efficiency losses due to the addition or removal of heat. It is this procedure that is followed in the subsequent calculations, based on the production principally of high-Btu gas (methane) by several possible routes. As an adjunct, the production of low- or medium-Btu gas is also examined.

I. COMPARISON OF REACTION SEQUENCES FOR THE PRODUCTION OF HIGH-BTU GAS (METHANE)

Irregardless of the route taken, the overall conversion of carbon and steam to produce methane is theoretically as follows:

$$C + H_2O(g) \rightarrow \tfrac{1}{2} CH_4 + \tfrac{1}{2} CO_2$$

If, say, 50% excess steam is involved, the conversion reads

$$C + \tfrac{3}{2} H_2O(g) \rightarrow \tfrac{1}{2} CH_4 + \tfrac{1}{2} CO_2 + \tfrac{1}{2} H_2O(g)$$

This commonality is illustrated by comparing several different routes to producing methane:

    Direct methanation (the Hoffman process)

    Indirect methanation

    Direct hydrogenation (hydrogasification)

It is noted that in steam gasification, irregardless of the kind of gasifier, about 50% excess steam is required to drive the conversion. This is ascertainable from an inspection of operating data for the varius processes (e.g., W. M. Bodle and K. C. Vyas, "Clean Fuels from Coal — Introduction to Modern Processes," *Clean Fuels from Coal,* Symposium II, Institute of Gas Technology, Chicago, June, 1975). The excess steam must be condensed following gasification, along with any oils or tars formed.

The comparison is as follows.

DIRECT METHANATION

The conversion is as already written above, with 50% excess steam. Excess steam, in principle, is recyclable.

For carbon in the form of β-graphite, the conversion reaction is slightly endothermic at standard conditions. For coals, depending upon composition and structure, the conversion may be neutral or even slightly exothermic. This subject is examined in Section II.

The mechanism is not known. It may be viewed as forming intermediate activated species such as [H] and [CO] or by the formation of activated complexes with the catalyst. Even CO_2 may be an intermediate.

No oils or tars are produced.

INDIRECT METHANATION

Three steps are involved: gasification, CO-shift and methanation. The sequence is as follows.

Gasification:

Assuming 50% excess steam,

$$C + \frac{3}{2}H_2O(g) \rightarrow CO + H_2 + \frac{1}{2}H_2O(g)$$

The reaction is highly endothermic and is supported by adding oxygen in the case of the Lurgi gasifier, but may also be supplied indirectly by external combustion. At pressure, some methane is also formed. These adjustments are not entered in this idealized or theoretical consideration, but are carried in subsequent calculations, and affect the product distribution and yield.

The excess steam is condensed along with oils and tars that are distilled over. The oils and tars may be recycled or used as a fuel.

CO-Shift:

Further steam is added to adjust the H_2/CO ratio:

$$\frac{1}{2}CO + \frac{1}{2}H_2O(g) \rightarrow \frac{1}{2}H_2 + \frac{1}{2}CO_2$$

This reaction is somewhat exothermic at standard conditions.

The gases are cooled and the CO_2, along with H_2S that is produced, is removed prior to methanation.

The cumulative sequence to this point reads

$$C + \frac{3}{2}H_2O(g) \rightarrow \frac{1}{2}CO + \frac{3}{2}H_2 + \frac{1}{2}CO_2$$

Methanation:

Methanation is a highly exothermic reaction, requiring heat removal and temperature control:

$$\frac{1}{2}CO + \frac{3}{2}H_2 \rightarrow \frac{1}{2}CH_4 + \frac{1}{2}H_2O(g)$$

The steam produced may be recycled.

Overall Conversion:

The overall conversion ideally reads:

$$C + \frac{3}{2}H_2O(g) \rightarrow \frac{1}{2}CH_4 + \frac{1}{2}H_2O(g) + \frac{1}{2}CO_2 \uparrow$$

Appendix

In principle, the overall conversion is the same as for direct methanation, except that H_2O and CO_2 are produced and removed in different parts of the sequence.

The heating and cooling of reactants between successive steps, however, results in cumulative efficiency losses. Other differences also occur in the actual processing.

DIRECT HYDROGENATION

Three steps are again involved. The steps are gasification and CO-shift as before, followed by hydrogenation (hydrogasification). A difference from indirect hydrogenation is that all the carbon monoxide is converted to hydrogen by the CO-shift reaction.

The steps are as follows.

Gasification:

Again assuming 50% excess steam,

$$C + \frac{3}{2} H_2O(g) \rightarrow CO + H_2 + \frac{1}{2} H_2O(g)$$

Gasification is supported internally with oxygen or externally with air, as noted before. Gasification may be at pressure or at atmospheric pressures. The products are cooled to remove excess steam plus oils and tars.

CO-Shift:

Conversion of all the CO is made:

$$CO + H_2O(g) \rightarrow H_2 + CO_2$$

The products are cooled and CO_2 (and H_2S) removed prior to hydrogasification.

The cumulative conversion to this point reads

$$C + \frac{5}{2} H_2O(g) \rightarrow 2 H_2 + \frac{1}{2} H_2O(g) + CO_2$$

Hydrogenation:

The exothermic hydrogasification reaction proceeds at higher pressures and temperatures:

$$C + 2 H_2 \rightarrow CH_4$$

Less severe conditions produce liquids and bitumens. Unreacted char is always a consideration.

Overall Conversion:

For the overall conversion,
$$2\,C + \tfrac{5}{2} H_2O(g) \rightarrow \tfrac{1}{2} CH_4 + \tfrac{1}{2} H_2O(g) + CO_2$$
or, on the basis of one mole of carbon,
$$C + \tfrac{3}{2} H_2O(g) \rightarrow \tfrac{1}{2} CH_4 + \tfrac{1}{2} H_2O(g) + \tfrac{1}{2} CO_2$$

The overall stoichiometric conversion thus appears ideally the same. Discrepancies will appear, however, as more detail is included.

Appendix

II. STOICHIOMETRY BASED ON COAL ANALYSIS

The previously determined idealized conversions were based on carbon only. A somewhat different result is obtained if the hydrogen and oxygen content of a coal is also considered. This circumstance is examined for the overall conversion to methane, in which hydrogen can appear in the product as well as carbon dioxide. Conversion to liquids is also inspected.

A representative coal composition — appropriate for a western coal — is chosen as the basis for the calculation.

COAL ANALYSIS

The assumed analysis is as follows, on an as-received, air-dried basis:

C	60.0% by wt.
H	5.0
O	23.5
S	0.5
N	1.0
Ash	10.0
	100.0%

The moisture is included in the above ultimate analysis. The original moisture content, which was probably ~ 25% by weight, is regarded as reduced to ~ 15%.

On a molar basis the carbon, hydrogen and oxygen contents convert as follows:

C	0.5 lb-mole/lb of coal
H	0.025
O	0.0074

The heat of combustion is determinable from the respective weight fractions using Dulong's formula:

$$- \Delta H = 14{,}544\, C + 62{,}028\, (H - \tfrac{8}{0}) + 4.050\, S$$

While accurate to perhaps 2-3% for high-rank coals, lower-rank western coals may show calculated values ~ 5% lower than experiment.

(Dulong's formula, incidentally, more or less regards all oxygen as already combined to form water.)

Substituting, the calculated value for the heat of combustion is 10,004.2 Btu/lb exclusive of the sulfur, which adds another 20.3 Btu/lb. If this value is maybe 5% low as indicated, a rounded true value is approximately

$$10,500 \text{ Btu/lb}$$

This is a gross rating, meaning the net heating value is obtained by subtracting the latent heat of the total H_2O content obtained after combustion — which includes free moisture and combustion product water.

Another relation which may be used to calculate the gross heat of combustion is the Grummel-Davies formula. A more recently proposed formula is that of Subramanium. The latter is said to provide a closer approximation, expecially for the younger western coals, which contain more oxygen. [The subject has been reviewed and comparisons made by T. K. Subramanium, *Coal Age, 82,* No. 12 (Dec., 1977).]

ANALYSIS OF THE CONVERSION

The conversion of coal and steam may directly or ultimately yield methane, heavier hydrocarbons and/or oxygenated compounds, depending upon catalysts, conditions and routes. We are here interested in the overall conversions only.

Assumptions are as follows. All oxygen in the coal is regarded as already combined with water (as per Dulong's formula, noted in the previous subsection). The difference between total hydrogen and that combined with the oxygen is considered the net hydrogen available for reaction. The combined water, which here contains all the oxygen, is also available for reaction. Additional water is added to the system, and the combined water plus the added water constitutes the total water available for reaction.

Per mole of carbon, the reactants are designated as

$$C + x\ H_2O(g) + y\ [H]$$

where

x = total moles of water (steam) available for reaction

$[y]$ = atoms of hydrogen available for reaction

For the purposes here, the water (steam) is stoichiometric. No excess is carried.

Two cases are considered:

Appendix

Conversion to Methane:

$$C + x\, H_2O + y\, [H] \rightarrow$$
$$(1 - \tfrac{x}{2}) CH_4 + \tfrac{x}{2} CO_2 + (\tfrac{y}{2} + 2x - 2) H_2$$

where the number of moles of hydrogen appearing in the product is designated as

$$n_{H_2} = \tfrac{y}{2} + 2x - 2$$

Conversion to Octane:

$$C + x\, H_2O + y\, H \rightarrow$$
$$\tfrac{1}{8}(1 - \tfrac{x}{2}) C_8H_{18} + \tfrac{x}{2} CO_2 + (\tfrac{y}{2} + \tfrac{25}{16}x - \tfrac{9}{8}) H_2$$

where

$$n_{H_2} = \tfrac{y}{2} + \tfrac{25}{16}x - \tfrac{9}{8}$$

In the conversion, "x" includes free and formed water from the coal, plus any additional water added, as previously noted. For convenience, no distinction is made between water and steam.

The equivalence for the reference coal is:

	lb / lb of coal	lb-mole / lb of coal	lb-mole / lb-mole of C	lb-atom / lb-mole of C
C	0.600	0.050	1.0	1.0
H	0.050	0.025	0.5	1.0
O	0.235	0.00734	0.14688	0.29376

The maximum water (steam) available from the coal for reaction is

$$0.2938 \text{ lb-mole/lb-mole of C}$$

The hydrogen available is

$$y = 2(0.5 - 0.2938) = 0.4124 \text{ lb-atom/lb-mole of C}$$

Calculations are tabulated in Table A for the production of methane for different assumed moles of hydrogen in the product gas. Btu-ratings are based on 1013 Btu/SCF for methane and 325 Btu/SCF for hydrogen. Thermal recovery is based on 10,5000 Btu/lb for the coal.

Note that as the assumed hydrogen content of the product gas increases the thermal recovery increases. In other words, the overall conversion becomes more endothermic — requiring more heat input to sustain the conversion.

Coal Conversion

Note also that the proportions of CH_4 are greater than for CO_2, results found experimentally for direct conversion (see Chapter 8).

The thermal recovery calculations make no attempt to enter the effects of latent heat change for the water (steam), or to adjust for other effects, and are thus necessarily relative.

TABLE A
Coal to Methane

	$n_{H_2} = 0$	$n_{H_2} = 0.05$	$n_{H_2} = 0.30$
y	0.4124	0.4124	0.4124
x	0.8969	0.9219	1.0469
Add. H_2O	0.6031	0.6281	0.7531
Moles			
CH_4	0.5515	0.5390	0.5235
CO_2	0.4485	0.5610	0.4765
H_2	--	0.0500	0.3000
	1.0000	1.0500	1.3000
SCF/Ton of Coal			
CH_4	20,900	20,400	19,800
CO_2	17,000	17,500	18,100
H	--	1.900	11,400
Btu/SCF (CO_2-free)	1013	954	762
Thermal Recovery (Gross)	100.8%	101.3%	113.2%

Reference Conversion:

$$C + H_2O(g) \rightarrow \tfrac{1}{2} CH_4 + \tfrac{1}{2} CO_2$$
$$-94.0518 \quad -10.5195 \quad \tfrac{1}{2}(-212.798) \quad \text{kcal/g-mole}$$

$\Delta H_0 = 1.8277$ kcal/g-mole or 3,290 Btu/lb-mole

Calculations are tabulated in Table B for the conversion to octane for two parameters of hydrogen product. The gross heat of combustion for octane liquid is 20,591 Btu/lb, for octane vapor, 20,747 Btu/lb.

Appendix

Again, as the hydrogen content increases, the value of the thermal recovery also increases, denoting increased heat input to sustain the conversion. The thermal recovery is based on both octane and hydrogen.

At $n_{H_2} = 0$, however, the thermal recovery value is less than 100%, indicating the conversion is slightly on the exothermic side at standard conditions.

TABLE B
Coal to n-Octane

	$n_{H_2} = 0$	$n_{H_2} = 0.30$
y	0.4124	0.4124
x	0.588	0.780
Add. H$_2$O		
Moles		
C$_8$H$_{18}$	0.08825	0.07625
CO$_2$	0.294	0.390
H$_2$	--	0.300
	0.382	0.766
Lb/Ton of Coal		
C$_8$H$_{18}$	1006	869
Bbl/Ton of Coal		
C$_8$H$_{18}$	4.09	3.53
SCF/Ton of Coal		
CO$_2$	11,100	14,800
H$_2$	--	11,400
Thermal Recovery (Gross)	98.6%	102.6%

Reference Conversion:

$$C + \tfrac{18}{25} H_2O(g) \rightarrow \tfrac{2}{25} C_8H_{18} + \tfrac{9}{25} CO_2$$

$$-94.0518 \quad \tfrac{18}{25}(-10.5195) \quad \tfrac{2}{25}(-1317.45) \quad \text{kcal/g-mole}$$

$$\Delta H_o = 3.770 \text{ kcal/g-mole or } 6,786 \text{ Btu/lb-mole}$$

The thermal recovery for liquid benzene is

$$\frac{1147(17986)}{2000(10500)} = 98.2\%$$

The conversion is slightly on the exothermic side at standard conditions.

CONVERSION TO BENZENE

In comparison, the conversion to benzene, an aromatic, is

$$C + x\,H_2O + y\,[H] \rightarrow \frac{1}{6}(1 - \frac{x}{2})C_6H_6 + \frac{x}{2}CO_2 + (\frac{y}{2} + \frac{5}{4}x - \frac{1}{2})H_2$$

For $n_{H_2} = 0$, the theoretical yield of benzene is

$$\frac{1}{6}(1 - \frac{0.235}{2}) = 0.147 \text{ mole/mole of C}$$

or 11.47 lb/lb-mole of C

This translates to

$$2000\,(0.05)\,(11.47) = 1147 \text{ lb/ton of coal}$$

or

$$\frac{1147}{0.879(8.33)} = 156.6 \text{ gal/ton}$$

or 3173 bbl/ton

CONVERSION TO OXYGENATED COMPOUNDS

Using methanol as the example for the overall conversion, in lb-moles.

$$C + \frac{4}{3}H_2O(g) \rightarrow \frac{2}{3}CH_3OH + \frac{1}{3}CO_2 \qquad 27{,}900 \text{ Btu}$$

The conversion is considerably endothermic at standard conditions.

On the other hand, in lb-moles,

$$C + H_2O(g) \rightarrow \frac{1}{2}CH_4 + \frac{1}{2}CO_2 \qquad 3{,}300 \text{ Btu}$$

Combining,

$$\frac{2}{3}CH_3OH \rightarrow \frac{1}{3}H_2O(g) + \frac{1}{2}CH_4 + \frac{1}{6}CO_2 \qquad -24{,}600 \text{ Btu}$$

Appendix

The conversion of methanol to methane is thus considerably exothermic, as is also noted for the Mobil process of converting methanol to hydrocarbons in the gasoline boiling range.

PYROLYSIS

If instead all the hydrogen were converted to methane by air-free destructive pyrolysis, the methane yield is

$$\frac{0.5}{2} = 0.25 \text{ mole/mole of C}$$

This is a maximum value and assumes no moisture occurs in the pyrolysis products. On a tonnage basis, the methane yield is

$$0.25(379)(0.05)(2000) = 9{,}475 \text{ SCF/ton of coal}$$

$$\text{or} \sim 400 \text{ lb/ton of coal}$$

The unconverted carbon remaining is the difference

$$(1 - 0.25)\, 12\, (0.05)\, 2000 = 900 \text{ lb/ton of coal}$$

To this would be added the ash: $0.10(2000) = 200$ lb/ton of coal. The total char would thus be 1100 lb/ton of coal.

The oxygen in the coal is not accommodated by the above calculation.

Therefore, as a lower limit, assuming the oxygen is all converted to water in pyrolysis, the net hydrogen available is, as determined previously,

$$0.4124 \text{ lb-atom/lb-mole of C}$$

or

$$0.2062 \text{ lb-mole/lb-mole of C}$$

The methane yield decreases to

$$\frac{0.2062}{2}(379)(0.05)(2000) = 3911 \text{ SCF/ton of coal}$$

There is also the strong probability, however, that part of the oxygen combines with carbon to form CO_2 and CO, and also appears in the oils and tars formed.

The actual behavior is both rate and equilibrium dependent, though pyrolysis techniques have been in evidence which tend to maximize CH_4 production (e.g., N. Wayne Green, "Synthetic Fuels from Coal — the Garrett Process," *Clean Fuels from Coal,* Symposium II, Institute of Gas

Technology, Chicago, June, 1975). The objective of the COED multistage pyrolysis process, on the other hand, is to maximize liquids, as is the case also in a variation of the Garrett or Occidental process. These cited processes both involve fast pyrolysis of pulverized fuel with significant departures from equilibrium. Moreover, very active chars are also produced, rivaling the combustibility of the coal volatiles. [This latter effect also occurs in the TOSCOAL process, a moving-bed process which involves pyrolysis by indirect methods (ceramic spheres).]

Other chemical reactions take place in so-called "fast fluidized beds". An example is fluidized catalytic cracking (FCC) in petroleum refining. The preponderant conversion ensues in the riser tube where the catalyst is entrained in the hot, vaporized gas oil feed. The reactor vessel *per se* is more of a disengagement section [e.g., Strother, et. al., "Riser Cracker Gives Advantages," *Hydrocarbon Processing* (May, 1972)]. Combustion or more particularly partial combustion of pulverized coal may also proceed in fast fluidized beds (see A. Squires, "Gasification of Coal in High-Velocity Fluidized Beds," *Clean Fuels from Coal,* Symposium II, Institute of Gas Technology, June, 1975). We are speaking of residence times of a few seconds down to fractional seconds.

Appendix

III. CARBON BALANCES AND HEATS OF REACTION

Use of the carbon balance provides a convenient basis for comparing the stoichiometric sequences of coal conversion processes. The matter was dealt with idealistically in Section I. The subject is now examined from the standpoint of the several simultaneous reactions and equilibria which may occur during gasification. Application is made to the production of methane.

Any reacting system involving variously carbon (coal), water (steam), oxygen (or air), and hydrogen, carbon monoxide, carbon dioxide, and methane will necessarily involve the following reactions, equilibria and heats of reaction, expressed on a lb-mole basis:

$$\Delta H_0$$

$C + H_2O(g) = CO + H_2$	56,488 Btu
$CO + H_2O(g) = H_2 + CO_2$	$-17,709$
$CO + 3 H_2 = CH_4 + H_2O(g)$	$-88,687$
$C + O_2 = CO_2$	$-169,293$

Other dependent reactions may be derived from these four independent reactions, as mentioned in Chapters 3 and 4. Means for calculating the standard heats of reaction from heats of combustion and heats of formation are presented in the supplement to this section.

If the other reactions are based on the complete conversion of one lb-mole of carbon by the first reaction, then in rounded numbers,

$C + H_2O(g) = CO + H_2$	56,500 Btu
$x [CO + H_2O(g) = H_2 + CO_2]$	$x [-17,700]$
$y [CO + 3 H_2 = CH_4 + H_2O(g)]$	$y [-88,700]$
$z [C + O_2 = CO_2]$	$z [-169,300]$

where x, y, z denote the fractional molar conversions for the respective reactions.

Thus, for the overall conversion,

$$(1+z) C + (1+x-y) H_2O + z O_2 \rightarrow$$
$$(1-x-y) CO + (1+x-3y) H_2 + y CH_4 + (x+z) CO_2$$

where the change in enthalpy at standard conditions is given by

$$\Delta H_0 = 56{,}500 + x\,(-17{,}700) + y\,(-88{,}700) + z\,(-169{,}300)\ \text{Btu}$$

Alternately,

$$a\,C + b\,H_2O + c\,O_2 \rightarrow d\,CO + H_2 + f\,CH_4 + g\,CO_2 + h\,H_2O$$

where

$a = 1+z$
$b-h = 1+x-y$
$c = z$
$d = 1-x-y$
$e = 1+x-3y$
$f = y$
$g = x+z$

$z = a - 1 = c$
$y = f$
$x = g-(a-1) = g-c$
$ = f+(b-h)-1$
$ = 1-d-f$
$ = e-1+3f$

The quantity h is the excess moles of H_2O making an appearance in the product. If h>0., it denotes that excess water is used as a reactant.

If part of the CO_2 is produced externally in thermal support of the steam-carbon reaction, let this amount be represented by g″ such that

$$g = g' + g''$$

where g′ is the amount generated internally. Most usually, if external combustion only is involved, g″ = z.

Since complete conversion of C with H_2O is assumed by the first reaction, the total moles of product is the sum

$$n_T = 2 + x - 3y + z + h$$
$$= d + e + f + g + h$$

where h is the excess moles of water, as indicated.

The application of this set of equations, in toto or in part, is presented for the following cases:

    Direct methanation (the Hoffman process)
    Indirect methanation
        Oxygen-blown pressure gasification
        External combustion
    Direct Hydrogenation (hydrogasification)

Appendix

The features will correspond to those developed and described in Section I, with accommodations made for heat balances and equilibrium.

DIRECT METHANATION

The reaction, as previously presented, is

$$C + H_2O(g) \rightarrow \tfrac{1}{2} CH_4 + \tfrac{1}{2} CO_2 \qquad 3{,}300 \text{ Btu}$$

Low-pressure conversion is assumed. In order to support the reaction at standard conditions by external combustion,

$$z\,[C + O_2 \rightarrow CO_2] \qquad z\,[-169{,}300]$$

Accordingly,

$$z = \frac{3300}{169300} = 0.0195 \text{ or } \sim 0.020 \text{ lb-mole}$$

Therefore, for the overall reaction, assuming 50% excess steam,

$$1.020 + 1.5\, H_2O(g) + 0.020\, O_2 \rightarrow$$
$$0.5\, CH_4 + 0\%.5\, CO_2 + 0.5\, H_2O(g) + 0.02\, CO_2$$

The net enthalpy change is zero overall: $\Delta H_0 = O$.
The molar yield is 0.490 mole CH_4/mole of C.

INDIRECT METHANATION (Oxygen-Blown Pressure Gasification)

Indirect methanation involves gasification, CO-shift and methanation. Oxygen-blown pressure gasification (O-BPG), using steam and oxygen under pressure, favors the appearance of methane. The Lurgi gasifier is an embodiment. The sequence is as follows:

Gasification:

Assuming the same set of four equations to apply, the H_2/CO ratio is given by

$$\frac{H_2}{CO} = \frac{1+x-3y}{1-x-y} = \alpha$$

Solving for y,

$$y = \frac{1+\alpha}{3-\alpha} x + \frac{1-\alpha}{3-\alpha}$$

A common mode of operation yields an H_2/CO ratio of about 2/1 — though this will depend upon steam rate, temperature and other factors.

Substituting $\alpha = 2$,
$$y = 3x-1 \text{ or } 3x-y = 1$$

On a water-free basis the mole fraction of methane is

$$\frac{y}{n_T} = \frac{y}{2+x-3y+z} = \beta$$

$$y = \frac{2+x+z}{1/\beta + 3}$$

At 350 psi, say, operating data indicate a methane concentration typically of ~ 10% in the raw gas on a moisture-free basis. Hence $\beta \sim 0.1$ and

$$y = \frac{2+x+z}{13}$$

Still higher pressures increase the methane concentration further in the off-gases, as in the Synthane and Bi-Gas gasifiers, which operate at ~ 1000 psi.

For the gasifier to be in heat balance at standard conditions requires that

$$0 = 56{,}500 + x\,(-17{,}700) + y\,(-88{,}700) + z\,(-169{,}300)$$

or, rearranging,

$$1 = 0.313\,x + 1.570\,y + 2.976\,z$$

Solving the three relations simultaneously yields

x = 0.400	or	x ~ 0.400
y = 0.199		y ~ 0.200
z = 0.188		z ~ 0.200

For the raw gasifier product gases, on a moisture-free basis,

		Moles	Percent
CO	= 1−x−y	0.401	20.14
H_2	= 1+x−3y	0.803	40.33
CH_4	= y	0.199	9.99
CO_2	= x+z	0.588	27.53
		1.991	99.99

Appendix

In rounded numbers,

	Moles	Percent
CO	0.400	20
H_2	0.800	40
CH_4	0.200	10
CO_2	0.600	30
	2.000	

At higher temperatures the CO/CO_2 ratio increases, and the other proportions are affected. The use of coal rather than carbon affects the analysis of course. The results, however, are characteristic of Lurgi pressure gasifier performance in the range, say, of 1600-2000°F.

The oxygen/feed ratio is

$$\frac{0.188}{1.188} \sim 0.16 \text{ or } 1/6 \text{ moles } O_2/\text{mole of C}$$

The oxygen/steam ratio is

$$\frac{0.188}{1.201} \sim 0.16 \text{ or } 1/6 \text{ moles } O_2/\text{mole of steam}$$

The oxygen/product gas ratio is

$$\frac{0.188}{1.991} \sim 0.10 \text{ or } 1/10 \text{ moles } O_2/\text{mole of product gas}$$

The oxygen/steam ratio calculated is similar to that in practice. Since 50% excess steam is characteristically used, however, the actual oxygen rate should be ~ 50% over the actual value. (Consult the operating data provided by Bodle and Vyas cited in Section I.)

The above ties in with the fact that the oxygen/product ratio of 0.10 is also less than the ratio of 0.15 sometimes cited. [Consult von Fredersdorff and Elliott, "Coal Gasification," in *Chemistry of Coal Utilization,* Supplementary Volume, H. H. Lowry Ed., Wiley, New York (1963).]

Thus the efficiency of use of oxygen can be regarded as about 2/3, based on the theoretical or calculated value.

Using 50% excess steam, the theoretical gasifier reaction appears as

$$1.2\ C + 1.8\ H_2O(g) + 0.2\ O_2 \rightarrow$$
$$0.4\ CO + 0.8\ H_2 + 0.6\ CO_2 + 0.2\ CH_4 + 0.6\ H_2O(g)$$

The standard heat of reaction is zero by definition.

The gases are quenched and cooled to remove excess water, oil and tar preparatory to the CO-shift conversion. Oils and tars affect the CO-shift catalyst. Significant quantities of CO_2 may also suppress the CO-shift, and CO_2 is sometimes removed at this point.

CO-Shift:

For the CO-shift to yield an H_2/CO ratio of 3/1,

$$0.4\ CO + 0.8\ H_2 + 0.1\ H_2O(g) \rightarrow 0.3\ CO + 0.9\ H_2 + 0.1\ CO_2$$

The heat of reaction is

$$\Delta H_o = 0.1\ [-17{,}700] = -1{,}800\ Btu$$

The reaction is favored by a catalyst (usually iron), and the equilibrium moves to the left at higher temperatures, say $> 1000°F$.

Following reaction the product gases are cooled to remove H_2S and CO_2 prior to methanation. Nickel methanation catalysts are particularly sensitive to sulfur poisoning at the methanation conditions used.

Methanation:

The methanation step is as follows:

$$0.3\ CO + 0.9\ H_2 \rightarrow 0.3\ CH_4 + 0.3\ H_2O$$

$$\Delta H_o = 0.3\ [-88{,}700] = -26{,}600\ Btu$$

Temperatures are typically 650-750°F, requiring removal of the exothermic heat of reaction.

Overall:

On adding the reactions for the three steps, the overall conversion appears as

$$1.2\ C + 1.9\ H_2O(g) + 0.2\ O_2 \rightarrow$$

$$0.5\ CH_4 + 0.7\ CO_2 + 0.9\ H_2O(g)$$

$$\Delta H_o = -28{,}400\ Btu$$

The molar yield is 0.42 mole CH_4/mole of C.

INDIRECT METHANATION (External Combustion)

The feature distinguished here is that the heat in support of gasification is supplied by external combustion, with direct or indirect heat transfer.

The steps are again gasification, CO-shift and methanation.

Appendix

Gasification:

It is assumed that gasification takes place at near-atmospheric pressures such that no methane is formed. Furthermore, higher temperatures (~ 1800-2000°F) produce a CO/CO_2 ratio of ~ 2.5-3.5/1.0. Any liquids produced would be recycled.

Using the first two equations as previously enumerated, where $y = 0$,

$$\frac{CO}{CO_2} = \frac{1-x}{x} = 2.5 \text{ to } 3.5$$

Using the lower value, $x \sim 0.3$.

For the gasifier to be in heat balance requires that

$$0 = 56{,}500 + x(-17{,}700) + z(-169{,}300)$$

For $x = 0.3$, $z = 0.3$. The fourth reaction occurs outside the system.

Excess steam (~ 50%) is used, representing about 0.6 moles. Hence the reactions are

$$C + 1.9\ H_2O(g) \rightarrow 0.7\ CO + 1.3\ H_2 +$$
$$0.3\ CO_2 + 0.6\ H_2O(g)$$
$$0.3\ C + 0.3\ O_2 \rightarrow 0.3\ CO_2$$

where the latter reaction takes place outside the gasifier proper.

CO-Shift:

An H_2/CO ratio of 3/1 is required:

$$0.7\ CO + 1.3\ H_2 + 0.2\ H_2O(g) \rightarrow$$
$$0.5\ CO + 1.5\ H_2 + 0.2\ CO_2$$
$$\Delta H_o = 0.2\ [-17{,}700] = -3{,}540 \text{ Btu}$$

Gas cleanup is employed prior to methanation.

Methanation:

$$0.5\ CO + 1.5\ H_2 \rightarrow 0.5\ CH_4 + 0.5\ H_2O(g)$$
$$\Delta H_o = 0.5\ [-88{,}700] = -44{,}350 \text{ Btu}$$

The same comments apply as before.

Overall:

The overall sequence is represented by

$$C + 2.1\ H_2O(g) \rightarrow 0.5\ CH_4 + 0.5\ CO_2 + 1.1\ H_2O(g)$$

$$\Delta H_0 = 3{,}310\ \text{Btu}$$

and

$$0.3\ C + 0.3\ O_2 \rightarrow 0.3\ CO_2$$

$$\Delta H_0 = -51{,}200\ \text{Btu}$$

Combining,

$$1.3\ C + 2.1\ H_2O(g) + 0.3\ O_2 \rightarrow$$

$$0.5\ CH_4 + 0.5\ CO_2 + 1.1\ H_2O(g) + 0.3\ CO_2$$

$$\Delta H_0 = -47{,}890\ \text{Btu}$$

The greater exothermicity over the preceding case is due to the fact that no methane was formed in the gasifier.

The molar yield is 0.38 mole CH_4/mole of C.

DIRECT HYDROGENATION

Direct hydrogenation (hydrogasification) involves gasification, CO-shift and hydrogasification.

Gasification:

Atmospheric gasification with external combustion is assumed, though pressure gasification with oxygen could be used as well. The gasification reactions are as determined for the preceding case:

$$C + 1.9\ H_2O(g) \rightarrow 0.7\ CO + 1.3\ H_2 + 0.3\ CO_2 + 0.6\ H_2O(g)$$

$$\Delta H_0 = 51{,}200\ \text{Btu}$$

$$0.3\ C + 0.3\ O_2 \rightarrow 0.3\ CO_2$$

$$\Delta H_0 = -51{,}200\ \text{Btu}$$

CO-Shift:

All the CO is converted:

$$0.7\ CO + 1.3\ H_2 + 0.7\ H_2O(g) \rightarrow 2.0\ H_2 + 0.7\ CO_2$$

$$\Delta H_0 = 0.7\ [-17{,}700] = -12{,}390\ \text{Btu}$$

Appendix

Hydrogasification:

Under sufficient pressure and temperature, 1000-2000 psi or higher and ~ 1200°F or higher,

$$C + 2 H_2 \rightarrow CH_4$$
$$\Delta H_0 = -32,200 \text{ Btu}$$

Overall:

$$2 C + 2.6 H_2O(g) \rightarrow CH_4 + CO_2 + 0.6 H_2O(g)$$
$$\Delta H_0 = 6,610 \text{ Btu}$$

and

$$0.3 C + 0.3 O_2 \rightarrow 0.3 CO_2$$
$$\Delta H_0 = -51,200 \text{ Btu}$$

Combining,

$$2.3 C + 2.6 H_2O(g) + 0.3 O_2 \rightarrow CH_4 + CO_2 + 0.6 H_2O(g) + 0.3 CO_2$$
$$\Delta H_0 = -44,590 \text{ Btu}$$

The theoretical molar yield is 0.435 mole CH_4/mole of C.

SUPPLEMENT TO SECTION III

Heats of reaction may be determined from the heats of combustion or the heats of formation of the reactants and products. The heats of combustion for the substances involved are as follows:

Heats of Combustion:

$C + O_2 \rightarrow CO_2$	94,051.8	Cal/g-mole
	169,293	Btu/lb-mole
$CO + \frac{1}{2} O_2 \rightarrow CO_2$	67,636.1	Cal/g-mole
	121,745	Btu/lb-mole
$H_2 + \frac{1}{2} O_2 \rightarrow H_2O(g)$	57,797.9	Cal/g-mole
	104,036	Btu/lb-mole
$CH_4 + 2 O_2 \rightarrow CO_2 + 2 H_2O(g)$	191,759	Cal/g-mole
	345,166	Btu/lb-mole

The above are used to calculate the several heats of reaction.

Direct Methanation:

$C + O_2(g) \rightarrow CO_2$	-169,293	Btu/lb-mole
$-\frac{1}{2}[CH_4 + 2O_2 \rightarrow CO_2 + 2H_2O(g)]$	$-\frac{1}{2}[-345,166]$	Btu/lb-mole
$C + H_2O(g) \rightarrow \frac{1}{2}CH_4 + \frac{1}{2}CO_2$	3,290	Btu/lb-mole

Water-Gas Reaction:

$C + O_2 \rightarrow CO_2$	-169,293	Btu/lb-mole
$-[CO + 1/2 O_2 \rightarrow CO_2]$	$-[-121,745]$	Btu/lb-mole
$-[H_2 + 1/2 O_2 \rightarrow H_2O(g)]$	$-[-104,036]$	Btu/lb-mole
$C + H_2O(g) \rightarrow CO + H_2$	56,488	Btu/lb-mole

CO-Shift:

$CO + 1/2 O_2 \rightarrow CO_2$	-121,745	Btu/lb-mole
$-[H_2 + 1/2 O_2 \rightarrow H_2O(g)]$	$-[-104,036]$	Btu/lb-mole
$CO + H_2O(g) \rightarrow H_2 + CO_2$	-17,709	Btu/lb-mole

Methanation:

$CO + 1/2 O_2 \rightarrow CO_2$	-121,745	Btu/lb-mole
$3[H_2 + 1/2 O_2 \rightarrow H_2O(g)]$	$3[-104,036]$	Btu/lb-mole
$-[CH_4 + 2O_2 \rightarrow CO_2 + 2H_2O(g)]$	$-[-345,166]$	Btu/lb-mole
$CO + 3H_2 \rightarrow CH_4 + H_2O(g)$	-88,687	Btu/lb-mole

Methanation (CO$_2$ Product):

$CO + 1/2 O_2 \rightarrow CO_2$	-121,745	Btu/lb-mole
$H_2 + 1/2 O_2 \rightarrow H_2O(g)$	-104,036	Btu/lb-mole
$-1/2[CH_4 + 2O_2 \rightarrow CO_2 + 2H_2O(g)]$	$-1/2[-345,166]$	Btu/lb-mole
$CO + H_2 = 1/2 CH_4 + 1/2 CO_2$	-53,198	Btu/lb-mole

Appendix

Direct Hydrogenation:

$C + O_2 \rightarrow CO_2$	-169,293	Btu/lb-mole
$2[H_2 + 1/2 O_2 \rightarrow H_2O(g)]$	$2[-104,036]$	Btu/lb-mole
$-1[CH_4 + 2 O_2 \rightarrow CO_2 + 2 H_2O(g)]$	$-1[-345,166]$	Btu/lb-mole

$C + 2 H_2 \rightarrow CH_4$	-32,199	Btu/lb-mole

Heats of Formation

In principle, the same results are obtained from the heats of formation. For example:

$C + H_2O(g) \rightarrow CO + H_2$

-57,797.9	-26,416		Cal/g-mole
-104,036	-47,549		Btu/lb-mole
	Net	56,487	Btu/lb-mole

$CH_4 + 2 O_2 \rightarrow CO_2 + 2 H_2O(g)$

-17,889	- 94,052	2(-57,797.9)	Cal/g-mole
-32,200	-169,294	2(-104,036)	Btu/lb-mole
	Net	345,166	Btu/lb-mole

IV. THERMODYNAMIC EFFICIENCIES FOR METHANE PRODUCTION

Determinations of net overall efficiency or net energy yield for coal conversion processes are most readily made on the basis of carbon, a procedure that does not take into account the hydrogen and oxygen contained in the coal. The discrepancies, however, will be minor and are justified by the greater simplicity of the stoichiometry.

In this section the several routes to gasification set forth in the preceding sections are investigated with regard to net overall efficiency. Emphasis is on the waste-heat losses which occur in heating the reacting system, sustaining the reaction, and cooling the reactants.

The thermodynamic approach is utilized, based on the thermodynamic properties of the reacting system — notably the heats of reaction and the changes that occur with temperature. This provides the abstraction for the conversion sequence.

A more exacting analysis would require a detailed process design in each case, with final judgement reserved for the data from actual operations.

CONVERSION

The conversion sequence has been written previously as

$$a\ C + b\ H_2O(g) + c\ O_2 \rightarrow$$
$$d\ CO + e\ H_2 + f\ CH_4 + g'\ CO_2 + h\ H_2O(g) + g''\ CO_2$$

where, in Btu's,

$$\Delta H_0 = 56{,}500 + [g-(a-1)]\ [-17{,}700] + f\ [-88{,}700] + (a-1)\ [-169{,}300]$$

and

$$g = g' + g''$$

where

g' = moles of CO_2 produced internally

g'' = moles of CO_2 produced externally

For the several processing routes considered, the tabulation in Table C follows from the previous section, Section III.

With regard to the overall conversion, interstage heating and cooling is required, as delineated in previous secitons. The temperate profile, ideally, would appear as follows for a multi-stage process:

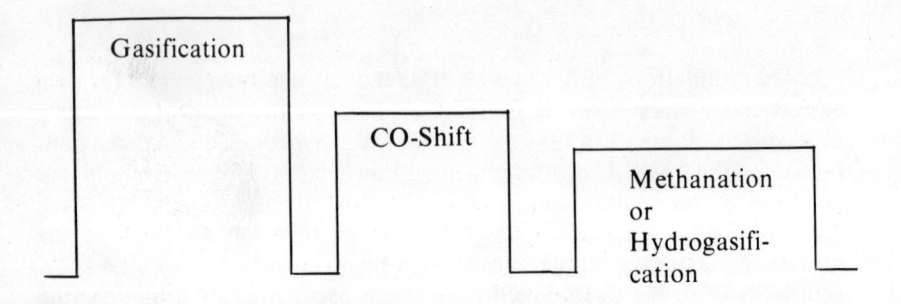

```
       Gasification
                      CO-Shift
                                   Methanation
                                   or
                                   Hydrogasifi-
                                   cation
```

Thus it is necessary to cool after gasification in order to remove tars and oils plus excess water prior to the CO-shift. In turn, the CO-shift products are cooled in order to remove H_2S (and CO_2) prior to methanation or hydrogasification. The final product is cooled to remove condensables preparatory to drying and compression.

If interstage cooling were not required, then the temperature would merely drop from one level to another (with a savings in waste-heat utilization).

With direct methanation, only one stage is involved, avoiding the cumulative efficiency losses of waste-heat utilization for interstage cooling and heating.

Non-ideally speaking, the temperature profile would look more like the following:

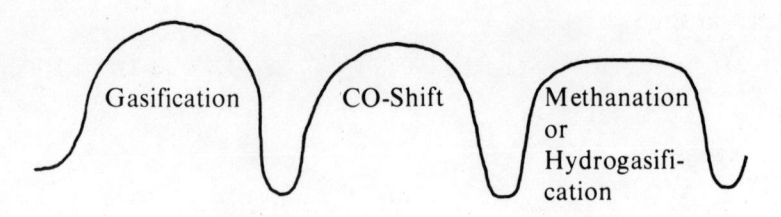

```
    Gasification    CO-Shift    Methanation
                                or
                                Hydrogasifi-
                                cation
```

Heating and cooling losses would be totally dependent upon the plant and could be analyzed only from the specifics of actual plant performance. Hence idealized behavior is instead assumed for the purposes at hand.

Appendix

TABLE C

	c	H$_2$O (in)	O$_2$	CO	H$_2$	CH$_4$	CO$_2$*	H$_2$O (out)		Overall Standard Heat of Reaction (Btu)
	a	b	c	d	e	f	g'	h	g''	
DIRECT METHANATION	1.020	1.5	0.020	0	0	0.5	0.5	0.5	0.020	0
INDIRECT METHANATION										
Lurgi	1.2	1.9	0.2	0	0	0.5	0.7	0.9	0.3	-28,400
External Comb	1.3	2.1	0.3	0	0	0.5	0.5	1.1	0.3	-47,890
DIRECT HDROGENATION	2.3	2.6	0.3	0	0	1	1.00	0.6	0.3	-44,590

EFFICIENCY

The overall efficiency or net energy yield is defined here based on gross heats of combustion for the feed carbon and the combustible products. In round numbers,

$$\text{Eff.} = \frac{f(379)(1000) - \Sigma \Delta H_{in}}{n_C (168{,}000)}$$

where

f = lb-moles of methane product

ΔH_{in} = an energy input

n_C = the total moles of carbon required for the conversion. It is the carbon gasified with steam plus the carbon combusted internally or externally in support of gasification. It also includes all conversion losses, whether from gasification, CO-shift, or methanation, or from combustion in support of gasification.

Though the heats of combustion are defined here as the gross values, net values could as well be used — and would more closely simulate actual performance since the latent heat of water is not ordinarily recouped in practice. (See Section V.)

It may be noted also that the thermal energy inputs could alternately be expressed in terms of a carbon requirement or equivalent, putting the efficiency determination on a carbon basis rather than a thermal basis. This, however, would require the introduction of still other efficiency values for the conversion of carbon to a thermal input.

The following categories are included herein for energy inputs:

Oxygen
Production of pure oxygen for internal combustion or compression of air for external combustion.

Steam
Latent heat effects.

Reaction
Adjustments to standard heat of reaction

Size reduction

Appendix

Acid gas removal
  Utilization of waste heat from the condensation of steam.

Compression of product

Additional plant energy requirements
  A contingency item.

No credit is given for byproducts: chars, oils, or tars. Rather, common conversion efficiencies are used as a parameter, which infers recycle. Catalyst regeneration is not considered. Neither are lock-hopper pressure-volume compression losses for pressure operations. Etc.

FEED CARBON

The total moles of carbon reacted as appearing in the overall conversion equation and in Table C is designated as "a". Of this, "c" moles react with oxygen or air in support of gasification with steam. This leaves "a-c" moles of carbon that react with the steam.

EFFICIENCY OF CONVERSION

The overall efficiency of conversion for a sequence of successive steps or stages is the product of the efficiencies for each step: e. g.,

$$E_{conv} = (E_{gasif})(E_{CO\text{-}shift})(E_{meth})$$

Thus if the successive efficiencies as enumerated on the right were 90%, 95%, and 95%, respectively, the overall conversion efficiency would be ~ 81%. Etc.

In view of the above, 80% is probably a reasonable lower limit for the overall conversion efficiency for the production of methane per se. Consequently, in subsequent calculations values of E_{conv} of 80%, 90% and 100% will be used as parameters in comparing the processes under consideration.

EFFICIENCY OF OXYGEN OR AIR UTILIZATION

The efficiency of utilization of oxygen in the Lurgi gasifier (internal combustion) has been shown in a previous section to be about 2/3 or 66-2/3 % as compared to the theoretical value. This is probably due to a number of reasons: heat losses, reaction with ash-components, etc.

Similarly, for using air for external combustion an efficiency of 2/3 will be adopted.

TOTAL ACTUAL CARBON REQUIREMENT

If n_C represents the total actual number of lb-moles of carbon required, as previously designated, then

$$n_C = \frac{a-c}{E_{conv}} + \frac{c}{E_{comb}}$$

where $E_{comb} \sim 2/3$ or 66-2/3% in all cases.

OXYGEN

The use of oxygen or air in support of the steam gasification reaction requires an energy input to produce the oxygen or compress the air. This input is designated as

$$\frac{c}{E_{comb}} \Delta H_{O_2}$$

where ΔH_{O_2} is in Btu/lb-mole of oxygen used, whether pure or in air.

For processes that require pure or essentially pure oxygen (the Lurgi gasifier, in this study), a figure of 500 kwh/ton of oxygen is assigned for oxygen production. This is a U.S. Bureau of Mines figure and agrees with private estimates. At a thermo-electric efficiency of 30%, this translates to $\sim$ 5,700,000 Btu(t)/ton of O_2 or

$$\Delta H_{O_2} = 91,000 \text{ Btu(t)/lb-mole of } O_2$$

For processes that require air for combustion in support of gasification, there is a power requirement for compression of the air. If the outlet pressure of the compressor-blower is $\sim$ 1.5 times the inlet pressure, the formulas for isentropic compression give the following:

$$\frac{T_2}{T_1} = \left(\frac{1.5}{1}\right)^{\frac{k-1}{k}} = 1.5^{0.28} = 1.12$$

where k = 1.4 for air at standard conditions. If $T_1 \sim 460+60 = 520°R$, then

$$T_2 = 520 (1.12) = 582°R \text{ or } 122°F$$

and

$$-W_{theo} = 5 (7.2) (582-520) = 2232 \text{ Btu(mech)/lb-mole of } O_2$$

Appendix

That is, since there are ~ 5 moles of air per mole of oxygen in the air, a factor of 5 is introduced above. Note that the theoretical work is mechanical work expressed in thermal units.

For an electro-mechanical efficiency of 80% for compression (*Chemical Engineers' Handbook*, 3rd ed.) and a thermo-electric efficiency of 30% for power generation, the above translates to

$$-W_{actual} = \frac{2232}{0.80 (0.30)} = 9300 \text{ Btu(thermal)/lb-mole of } O_2$$

The above gives the actual energy input as heat which would ultimately produce 2232 Btu(mech) or 2232 (778) = 1.74×10^6 ft-lb as mechanical or shaft work.

Hence for air compressor-blower requirements,

$$\Delta H_{O_2} = 9300 \text{ Btu(t)/lb-mole of } O_2$$

LATENT HEAT

Steam is part of the reacting system and, as such, changes from standard conditions are carried in the adjustments for the heats of reaction, to be derived in the next subsection.

If the H_2O is furnished and retrieved as the liquid phase, however, adjustment must also be made for latent heat changes.

If "b" is the total lb-moles of steam used and "h" is the steam occurring as product and/or excess — according to the overall balance as written previously — then

$$\left[\frac{b}{E_b} - h\, E_h \right] 18{,}000 = \text{net thermal input in Btu}$$

where E_b is the boiler efficiency for vaporizing the water, and E_h is an efficiency for condensing the saturated steam.

It is noted that boiler efficiencies are generally rated on the basis of the gross heating value of the fuel, as observed in Chapter 3 and presented in Figures 3.1-3.4 and Table 3.1. A typical value is ~ 80%. This is a lower efficiency value than would be obtained based on the net heating value of the fuel. The latter is a truer assessment since the combustion water is not condensed in practice.

Hence, apropos of the above, it is assumed that

$$E_b \sim 90\%$$

If the latent heat of condensation of the steam cannot be recovered, then $E_h = 0$. This value is assumed for this portion of the calculation.

However, a credit will be given for part of this waste heat in the determination for acid gas removal (stripping).

REACTION HEAT REQUIREMENTS

In a chemical conversion the difference in the enthalpy function of the reacting mixture between any two states is independent of the path. This is inherent in the derivation of the enthalpy function [e.g., refer to E. J. Hoffman, *The Concept of Energy,* Ann Arbor Science Publishers (1977)].

If energy transfer occurs with the surroundings during the reaction, then the system behavior becomes dependent upon the path of change. For the purposes of this study, a rectangular path of heating, reaction, and cooling has been chosen as the simplest means of ideally representing the system behavior. This was described in the introductory portions of this section.

For a chemical conversion in steps 1-2-3-4, as was indicated, the temperature profile is represented as follows:

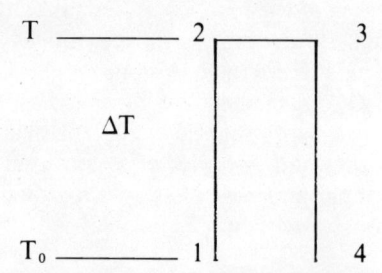

$$\begin{array}{c} T \underline{\qquad\qquad} 2 \quad\quad 3 \\ \\ \Delta T \\ \\ T_0 \underline{\qquad\qquad} 1 \quad\quad 4 \end{array}$$

For an isolated system, assuming constant heat capacities at constant pressure,

$$H_3 - H_2 = H_4 - H_1 + \Sigma n_P(C_p)_P \Delta T - \Sigma n_R(C_p)_R \Delta T$$

where "P" refers to a product and "R" to a reactant. If $H_4 - H_1 = \Delta H_0$, the standard heat of reaction, then

$$H_3 - H_2 = \Delta H_0 + \Sigma n_P(C_p)_P \Delta T - \Sigma n_R(C_p)_R \Delta T$$

This is the familiar formula for the effect of temperature on the heat of reaction at constant pressure.

If $H_3 - H_2$ is positive (endothermic at temperature T), the heat to be added to sustain the reaction at temperature T is

$$\frac{H_3 - H_2}{E_{\Delta H}} = \frac{H_0}{E_{\Delta H}} + \frac{1}{E_{\Delta H}} \left\{ \Sigma n_P(C_p)_P \Delta T - \Sigma n_R(C_p) \Delta T \right\}$$

Appendix

where $E_{\Delta H}$ or E_H is the fractional efficiency of heat input.

If on the other hand $H_3 - H_2$ is negative (exothermic at temperature T), the heat to be added is

$$E_{\Delta H}(H_3 - H_2) = E_{\Delta H}(\Delta H_0) + E_{\Delta H}\left\{\Sigma n_P(C_p)_P \Delta T - \Sigma n_R(C_p)_R \Delta T\right\}$$

where in this case $E_{\Delta H}$ or E_H is the efficiency of heat retrieval. Furthermore, in this case, the heat added will have a negative value.

If $H_3 - H_2 = 0$, the reaction is thermally neutral or autothermal at temperature T. For an adiabatic reactor, no heat transfer is involved.

Turning to the heating and cooling of reactants and products, if heat is added in leg (1-2), the amount is

$$\frac{1}{E_R} \Sigma n_R(C_p)_R \Delta T$$

where E_R is the efficiency of heat input to the reactants.

If heat is removed in leg (3-4), the amount as heat added is

$$- E_P \Sigma n_P(C_p)_P \Delta T$$

where E_P is the efficiency of heat retrieval. The minus sign denotes the quantity is negative in the context of heat added.

If $\Delta H = H_3 - H_2$ is positive at temperature T, the net heat added during the cycle is

$$\frac{1}{E_{\Delta H}}\left\{\Delta H_0 + \Sigma n_P(C_p)_P \Delta T - \Sigma n_R(C_p)_R \Delta T\right\} +$$

$$\frac{1}{E_R} \Sigma n_R(C_p)_R \Delta T - E_P \Sigma n_P(C_p)_P \Delta T$$

$$= \frac{\Delta H_0}{E_{\Delta H}} + \left[\frac{1}{E_{\Delta H}} - E_P\right] \Sigma n_P(C_p)_P \Delta T -$$

$$\left[\frac{1}{E_{\Delta H}} - \frac{1}{E_R}\right] \Sigma n_R(C_p)_R \Delta T$$

If $E_R = E_P = E_H$ the above reduces to

$$\frac{\Delta H_0}{E_H} + \left[\frac{1}{E_H} - E_H\right] \Sigma n_P(C_p)_P \Delta T$$

If $\Delta H = H_3 - H_2$ is negative at temperature T, the net heat added is

$$E_{\Delta H} \left\{ \Delta H_0 + \Sigma n_P(C_p)_P \Delta T - \Sigma n_R(C_p)_R \Delta T \right\}$$

$$+ \frac{1}{E_R} \Sigma n_R(C_p)_R \Delta T - E_P \Sigma n_P(C_p)_P \Delta T$$

$$= E_{\Delta H} \Delta H_0 + \left[E_{\Delta H} - E_P\right] \Sigma n_P(C_p)_P \Delta T -$$

$$\left[E_{\Delta H} - \frac{1}{E_R}\right] \Sigma n_R(C_p)_R \Delta T$$

If the efficiencies are equal the above reduces to

$$E_H \Delta H_0 + \left[\frac{1}{E_H} - E_H\right] \Sigma n_R(C_p)_R \Delta T$$

Most usually the heat added will be negative in sign.

In general, however, the sign and value of the heat added depends upon the relative magnitude of ΔH_0 as compared to the temperature adjustment using the heat capacities.

If the reaction at T proves autothermal or thermally neutral, and the reactor is adiabatic at temperature T, then the net amount of heat added is

$$\frac{1}{E_R} \Sigma n_R(C_p)_R \Delta T - E_P \Sigma n_P(C_p)_P \Delta T$$

or, at equal efficiencies,

$$\frac{1}{E_H} \Sigma n_R(C_p)_R \Delta T - E_H \Sigma n_P(C_p)_P \Delta T$$

For the purposes of simplifying the initial calculations, the gasifier reaction has been regarded as autothermal or in heat balance at the reference or standard temperature T_0. A more precise analysis would require the reaction to be autothermal at temperture T. That is $H_2 - H_2 = 0$. The difference in most instances is slight, since the effect of the heat capacities of the reactants and products is in the main to cancel one another out.

Appendix

For an overall sequence of conversions, the total net energy input is given by the following inclusive formula:

$$\text{TOTAL NET} = \sum\left\{\frac{\Delta H_0}{E_H} + \left[\frac{1}{E_H} - E_H\right]\Sigma n_P(C_p)_P \Delta T\right\}$$
end.

$$+ \sum\left\{E_H \Delta H_0 + \left[\frac{1}{E_H} - E_H\right]\Sigma n_R(C_p)_R \Delta T\right\}$$
exo.

$$+ \sum\left\{\frac{1}{E_H}\Sigma n_R(C_p)_R \Delta T - E_H \Sigma n_P(C_p)_P \Delta T\right\}$$
neutral

The preceding relationships will be used in determining the net heat added for the appropriate chemical conversions. To assist in the calculations, a compilation of mean or average heat capacities has been assembled for the several reactants and products (and inerts) variously involved. This is presented in Table D. Pressure effects are assumed negligible for the purposes of calculations.

TABLE D

Average Heat Capacities

	SCF-°F BTU	lb-mole-°F BTU	lb-°F BTU
C	0.019	3.0	0.25
Air	0.019	7.2	0.25
N_2	0.19	7.2	0.26
O_2	0.19	7.2	0.23
CO	0.19	7.2	0.26
CO_2	0.029	11.0	0.25
$H_2O(g)$	0.024	9.1	0.50
H_2	0.019	7.2	3.6
CH_4	0.032	12.0	0.75

The calculations are affected by the presence or absence of inerts or other extraneous materials which are carried through the system. This effect will make its appearance in the heat capacity summation. Thus in direct methanation, the presence of alkali will have some effect. In other instances, greater or lesser amounts of steam will have an effect, as will tars and oils produced. In order that the cases may be referenced to the same basis, however, these additional effects are not included in the subsequent calculations, but may be entered as adjustments.

The detail for the cases being considered is as follows.

DIRECT METHANATION

Reaction temperature: $\sim$ 1300-1400°F

$\Delta T \sim 1300°F$

Reaction:

1.020 C + 1.5 H$_2$O(g) + 0.020 O$_2$ →

0.5 CH$_4$ + 0.520 CO$_2$ + 0.5 H$_2$O(g)

$\Delta H_o = 0$

$\Sigma n_R(C_p)_R = 1.020 (3.0) + 1.5 (9.1) + 0.020 (7.2)$

$\qquad = 3.060 + 13.65 + 0.144 = 16.85$

$\Sigma n_P(C_p)_P = 0.5 (12.0) + 0.525 (11.0) + 0.5 (9.1)$

$\qquad = 6.0 + 5.775 + 4.55 = 16.325$

$\Delta H_T = 0 + (16.325 - 16.584)(1300) = -687.7$ Btu

The reaction is exothermic at temperature T. Therefore,

$\Sigma n_R(C_p)_R \Delta T = 16.854 (1300) = 21,910$ Btu

The above quantity will be referred to as the sensible heat summation.

The portion of the overall reaction supporting gasification with steam is external combustion with air:

0.020 C + 0.020 O$_2$ + 0.08 N$_2$ → 0.020 CO$_2$ + 0.08 N$_2$

The same temperature profile is assumed with the datum at standard conditions, and the same efficiencies of waste-heat utilization. This is implicit in the overall calculation above. While nitrogen as an inert does not affect the heat of reaction, there is an effect on the sensible heat summation. Therefore, as the adjusted value,

$\left[\Sigma n_P(C_p)_P \Delta T \right]_{adj} = 21,910 + 0.08 (7.2)(1300) = 22,658$ Btu

This figure is to be multiplied by $(\frac{1}{E_H} - E_H)$.

Appendix

INDIRECT METHANATION (O-BPG)

The steps are gasification with steam supported by oxygen, CO-shift and methanation.

Gasification:

Reaction temperature: may vary from 1200-1400°F through much of the bed, and up to 2200°F or more at the bottom of the bed, particularly for slagging operations

$\Delta T \sim 1500°F$

Reaction:

$$1.2\ C + 1.8\ H_2O(g) + 0.2\ O_2 \rightarrow 0.4\ CO + 0.8\ H_2 + 0.2\ CH_4 + 0.6\ CO_2 + 0.6\ H_2O(g)$$

$\Delta H_o = 0$

$\Sigma n_R(C_p)_R = 1.2\ (3.0) + 1.8\ (9.1) + 0.2\ (7.2)$

$= 3.6 + 16.38 + 1.4 = 21.42$

$\Sigma n_P(C_p)_P = 0.4\ (7.2) + 0.8\ (7.2) + 0.2\ (12.0) + 0.6\ (11.0) + 0.6\ (9.1)$

$= 2.88 + 5.76 + 2.4 + 6.6 + 5.46 = 23.10$

$\Delta H_T = 0 + (23.10 - 21.42)\ (1500) = 2520\ Btu$

The reaction at T is endothermic. Therefore, for the sensible heat summation,

$\Sigma n_P(C_p)_P \Delta T = 23.10\ (1500) = 34,650\ Btu$

CO-Shift:

Reaction temperature: $\sim 700\text{-}1000°F$

$\Delta T \sim 800°F$

Reaction:

$$0.4\ CO + 0.8\ H_2 + 0.1\ H_2O(g) \rightarrow 0.3\ CO + 0.9\ H_2 + 0.1\ CO_2$$

$\Delta H_o = 0.1\ (17,700) = -1,770\ Btu$

$$\Sigma n_R(C_p)_R = 0.4\,(7.2) + 0.8\,(7.2) + 0.1\,(9.1)$$
$$= 2.88 + 5.76 + 0.91 = 9.55$$
$$\Sigma n_R(C_p)_R = 0.4\,(7.2) + 0.8\,(7.2) + 0.1\,(9.1)$$
$$= 2.88 + 5.76 + 0.91 = 9.55$$
$$\Sigma n_P(C_p)_P = 0.3\,(7.2) + 0.9\,(7.2) + 0.1\,(11.0)$$
$$= 2.16 + 6.48 + 1.1 = 9.74$$
$$\Delta H_T = -1770 + (9.74 - 9.55)(800) = -1{,}618 \text{ Btu}$$

The reaction at T is exothermic. Therefore, the sensible heat summation is

$$\Sigma n_R(C_p)_R \Delta T = 9.55\,(800) = 7{,}640 \text{ Btu}$$

Methanation:

Reaction temperture: ~ 650-750°F

$\Delta T \sim 650°F$

Reaction:

$$0.3 \text{ CO} + 0.9 \text{ H}_2 \rightarrow 0.3 \text{ CH}_4 + 0.3 \text{ H}_2\text{O(g)}$$
$$\Delta H_o = 0.3\,(-88{,}700) = -26{,}610 \text{ Btu}$$
$$\Sigma n_R(C_p)_R = 0.3\,(7.2) + 0.9\,(7.2)$$
$$= 2.16 + 6.48 = 8.64$$
$$\Sigma n_P(C_p)_P = 0.3\,(12.0) + 0.3\,(9.1)$$
$$= 3.60 + 2.73 = 6.33$$
$$\Delta H_T = -26{,}610 + (6.33 - 8.64)(650) = -28{,}112 \text{ Btu}$$

The reaction at T is exothermic. Therefore, for the sensible heat summation.

$$\Sigma n_R(C_p)_R \Delta T = 8.64\,(650) = 5{,}616 \text{ Btu}$$

Total Sensible Heat Summation:

$$34{,}650 + 9{,}472 + 5{,}616 = 49{,}738 \text{ Btu}$$

Appendix

This figure will be multiplied by $(\frac{1}{E_H} - E_H)$.

INDIRECT METHANATION (External Combustion)

The steps are gasification supported by external combustion with air, CO-shift and methanation.

Gasification:

Reaction temperature: ~ 1800°F or higher

ΔT ~ 1700°F

Reaction:

$$1.3\ C + 1.9\ H_2O(g) + 0.3\ O_2 \rightarrow 0.7\ CO + 1.3\ H_2 + 0.3\ CO_2$$
$$+ 0.6\ H_2O(g) + 0.3\ CO_2$$

$\Delta H_0 = O$

$\Sigma n_R(C_p)_R = 1.3\ (3.0) + 1.9\ (9.1) + 0.3\ (7.2)$

$\qquad\qquad = 3.9 + 17.29 + 2.16 = 23.35$

$\Sigma n_P(C_p)_P = 0.7\ (7.2) + 1.3\ (7.2) + 0.3\ (11.0) + 0.6\ (9.1)$

$\qquad\qquad\qquad\qquad\qquad\qquad\qquad\qquad + 0.3\ (11.0)$

$\qquad\qquad = 5.04 + 9.36 + 3.3 + 5.46 + 3.3 = 26.46$

$\Delta H_T = 0 + (26.46 - 23.35)\ (1700) = 5{,}287$ Btu

The reaction at T is endothermic. Therefore, for the sensible heat summation,

$\Sigma n_R(C_p)_R \Delta T = 26.46\ (1700) = 44{,}982$ Btu

Correcting for the nitrogen in the air used for external combustion,

$\left[\Sigma n_P(C_p)_P \Delta T\right]_{adj} = 44{,}982 + 0.3\ (4)\ (7.2)\ (1700) = 59{,}670$ Btu

CO-Shift:

Conditions as before.

Reaction:

$$0.7\ CO + 1.3\ H_2 + 0.2\ H_2O(g) \rightarrow$$
$$0.5\ CO + 1.5\ H_2 + 0.2\ CO_2$$

$\Delta H_0 = 0.2\ (-17,700) = -3,540$ Btu

$\Sigma n_R(C_p)_R = 0.7\ (7.2) + 1.3\ (7.2) + 0.2\ (9.1)$

$\qquad = 5.04 + 9.36 + 1.82 = 16.22$

$\Sigma n_P(C_p)_P = 0.5\ (7.2) + 1.5\ (7.2) + 0.2\ (11.0)$

$\qquad = 3.6 + 10.8 + 2.2 = 16.6$

$\Delta H_T = -3,540 + (16.6-16.22)\ (800) = -3,236$ Btu

The reaction at T is exothermic. Therefore, for the sensible heat summation,

$$\Sigma n_R(C_p)_R \Delta T = 16.22\ (800) = 12,976\ \text{Btu}$$

Methanation:

Conditions as before.

Reaction:

$$0.5\ CO + 1.5\ H_2 \rightarrow 0.5\ CH_4 + 0.5\ H_2O(g)$$

$\Delta H_0 = 0.5\ (-88,700) = -44,350$ Btu

$\Sigma n_R(C_p)_R = 0.5\ (7.2) + 1.5\ (7.2)$

$\qquad = 3.6 + 10.8 = 14.4$

$\Sigma n_P(C_p)_P = 0.5\ (12.0) + 0.5\ (9.1)$

$\qquad = 6.0 + 4.55 = 10.55$

$\Delta H_T = -44,350 + (10.55-14.4)\ (650) = -46,853$ Btu

The reaction at T is exothermic. Therefore, for the sensible heat summation,

$$\Sigma n_R(C_p)_R\ \Delta T = 14.4\ (650) = 9,360\ \text{Btu}$$

Total Sensible Heat Summation:

$59,670 + 12,976 + 9,360 = 82,006$ Btu

Appendix

DIRECT HYDROGENATION

The steps are gasification supported by external combustion with air, CO-shift and hydrogasification.

Gasification:

As calculated in the preceding case, for the sensible heat summation,

$$\left[\Sigma n_P(C_p)_P \Delta T\right]_{adj} = 59{,}670 \text{ Btu}$$

CO-Shift:

Conditions as before.

Reaction:

$$0.7 \text{ CO} + 1.3 \text{ H}_2 + 0.7 \text{ H}_2\text{O(g)} \rightarrow 2 \text{ H}_2 + 0.7 \text{ CO}_2$$

$$\Delta H_o = 0.7 \,(-17{,}700) = -12{,}390 \text{ Btu}$$

$$\Sigma n_R(C_p)_R = 0.7 \,(7.2) + 1.3 \,(7.2) + 0.7 \,(9.1)$$

$$= 5.04 + 9.36 + 6.37 = 20.77$$

$$\Sigma n_P(C_p)_P = 2.0 \,(7.2) + 0.7 \,(11.0)$$

$$= 14.4 + 7.7 = 22.1$$

$$\Delta H_T = -12{,}390 + (22.1 - 20.77)(800) = -10{,}262 \text{ Btu}$$

The reaction at T is exothermic. Therefore, for the sensible heat summation,

$$\Sigma n_R(C_p)_R \Delta T = 20.77 \,(800) = 16{,}616 \text{ Btu}$$

Hydrogasification:

Reaction temperature: ~ 1200°F and higher

$\Delta T \sim 1200°F$

Reaction:

$$C + 2 \text{ H}_2 \rightarrow CH_4$$

$$\Delta H_o = -32{,}200 \text{ Btu}$$

$$\Sigma n_R(C_p)_R = 1.0 \,(3.0) + 2.0 \,(7.2)$$

$$= 3.0 + 14.4 = 17.4$$

$$\Sigma n_P(C_p)_P = 1.0 \,(12.0) = 12.0$$

$$\Delta H_T = -32{,}200 + (12.0 - 17.4)(1200) = -38{,}680 \text{ Btu}$$

The reaction at T is exothermic. Therefore, for the sensible heat summation,

$$\Sigma n_R (C_p)_R \Delta T = 17.4 (1200) = 20{,}880 \text{ Btu}$$

Total Sensible Heat Summation:

$$59{,}670 + 16{,}616 + 20{,}880 = 97{,}166 \text{ Btu}$$

SIZE REDUCTION

The subject of size reduction has been discussed in Chapter 2. Table 2.1 provides experimental data for the energy requirement for coals. For pulverization a value of 20 hp-hr/ton is assumed representative for producing a product for which 65-80% passes through a 200 mesh screen.

Transforming the above value,

$$20 (2545) = 50{,}900 \text{ Btu(elec)/ton}$$

At a 1/3 (or $\sim$ 30%) thermo-electric conversion efficiency, the requirement is in round numbers

$$\sim 150{,}000 \text{ Btu(thermal)/ton of coal}$$

or

$$\sim 75 \text{ Btu(t)/lb of coal}$$

If we are assuming a coal of 60% carbon, then the thermal energy input requirement is

$$\frac{75}{0.6} (12) = 1750 \text{ Btu(t)/lb-mole of carbon}$$

or, for the total feed,

$$n_C (1750)$$

In the case of the Lurgi gasifier, for instance, screened or sized coal is used instead of pulverized coal. Since, in any case, the power requirement is relatively small, the determination as made above will be used.

ACID GAS REMOVAL AND RECOVERY

The energy requirement for acid gas removal is based on the thermal or steam requirement for stripping CO_2, the principal acid gas compo-

Appendix

nent. This thermal requirement is in turn dependent upon the medium used for acid gas removal, as discussed in Chapter 10.

For the purposes here a diethanolamine solution is assumed. The heat of reaction or solution has been previously indicated as about 27,000 Btu/lb-mole of CO_2, and the steam requirement as about 90-100 lb of steam per mole of acid gas recovered. The heat of reaction will be used since it is a more definite quantity in terms of the thermal value.

According also to Kohl and Riesenfeld [in *Gas Purification*, Gulf Publishing Co., Houston (1974)], there is a pumping requirement of about 14 kwh(e) per lb-mole of acid gas removed and recovered. At a thermo-electric efficiency of $\sim 30\%$, this amounts to about 3,500 Btu(t)/lb-mole of CO_2.

Considering the amount of CO_2 involved, the energy requirements for acid gas removal are large. It is thus of considerable importance that waste-heat, if available, be used in one way or another in support of acid gas removal. While in instances reaction heat may be available, this source of heat may also be used elsewhere and deducted from the gross input requirements.

The single uniform source of low-level heat energy is from the condensation of excess reactant steam from gasification. This is represented by the quantity h(18,000) in Btu's.

By operating the gasifier at a necessary and sufficient pressure, this heat of condensation is available for use in the stripping of the acid gas from solution. This heat may be first used to generate steam for indirect heat transfer, or heat transfer may be made directly to the stripper.

The efficiency of waste heat utilization is in great part related to the capital investment in heat exchangers. As a practical limitation an efficiency of 90% is assumed for retrieval and 90% for input (or $\sim 80\%$ overall, for both retrieval and input). Therefore, based on the heat of solution, the net actual thermal input for acid gas removal and recovery becomes

$$\frac{g' [27,000]}{0.90} - h [18,000] (0.90) + g' (3500)$$

or

$$g' (33,500) - h (16,200) \text{ Btu}$$

where g' denotes the lb-moles of CO_2 produced internally.

Requirements for the further disposal of CO_2 and the conversion of H_2S to sulfur are regarded as covered by the contingency category.

COMPRESSION REQUIREMENTS

Compression of the product gases, or the compression of hydrogen in the case of hydrogasification, has an effect on the net process efficiency. Moreover, the degree of compression is not only related to the conversion processing but to the end-use of the product gas, as already explained.

Thus gas for pipeline transmission is customarily at ~ 1000 psi, gas for city-wide distribution is at ~ 100 psi, whereas plant gas may be required to be at only a few psig (or less). There is a decrease of an order of magnitude from one to the other.

There is an additional factor relevant to the separation and recovery of CO_2, and whether the CO_2 is to be compressed for utilization as a byproduct.

Rather than going into all the alternatives, the final product is referenced to 1000 psi, i.e. pipeline transmission pressure. This stipulation will involve compression in one way or another.

The formulas for reversible or isentropic compression — upon which most compressor calculations are based — are as follows:

$$d \ln T = \frac{k-1}{k} d \ln P$$

or

$$\frac{T_2}{T_1} = \left(\frac{P_2}{P_1}\right)^{\frac{k-1}{k}}$$

and

$$-dW_{theo} = n\, C_p\, dT$$

or

$$-W_{theo} = n\, C_p\, (T_2 - T_1)$$

where

- k = C_p/C_v, the ratio of the molar heat capacities at constant pressure and constant volume
- C_p = molar heat capacity at constant pressure
- n = moles of gas

Appendix

$-W_{theo}$ = theoretical work of compression in thermal units of mechanical energy. The minus sign denotes work done on the system.

For the purposes here the initial temperature is

$T_1 \sim 460 + 77 = 537°R$

An electro-mechanical efficiency of 80% is assumed (*Chemical Engineers' Handbook,* 3rd ed., p. 1261), and a thermo-electrical efficiency of 30%. Hence for the equivalent thermal energy input,

$$-W_{actual} = \frac{-W_{theo}}{0.80\,(0.30)}$$

A difficulty in applying the formulas is that k is a function of both temperature and pressure. It will therefore be assumed that values near standard conditions are the reference and that efficiency determinations are related to this datum. For the condition of $\sim$ 0°C and 1 atm,

	k	$\frac{k-1}{k}$
Air	$\sim$ 1.4	0.28
CH_4	$\sim$ 1.32	0.24
H_2	$\sim$ 1.4	0.28

In practice, intercooling is used between compression stages and/or following compression. This is in principle a source of low-grade heat, but no credit will be taken.

The compression inputs for the gaseous products or intermediates are as follows.

Direct Methanation:

The methane product after CO_2 separation is assumed at $\sim$ 20 psia. Hence

$$\left(\frac{1000}{20}\right)^{0.24} = 2.56$$

$T_2 = 537\,(2.56) = 1375\ °R\ \text{or}\ 915°F$

$-W_{theo} = 0.5\,(12.0)\,(1375-537) = 5{,}028\ \text{Btu(mech)}$

$-W_{actual} = \dfrac{5028}{0.80\,(0.30)} = 20{,}950\ \text{Btu(thermal)}$

Oxygen-Blown Pressure Gasification (O-BPG):

A gasifier pressure of ~ 350 psi was assumed. The same pressure will be used for the methanator product. Hence

$$\left(\frac{1000}{350}\right)^{0.24} = 1.29$$

$$T_2 = 537 (1.29) = 693°R \text{ or } 233°F$$

$$-W_{theo} = 0.5 (12.0) (693-537) = 936 \text{ Btu(mech)}$$

$$-W_{actual} = \frac{936}{0.80 (0.30)} = 3,900 \text{ Btu(thermal)}$$

External Combustion:

The gasifier and final product are assumed at ~ 20 psia. Hence, as calculated for direct methanation,

$$-W_{actual} = 20,950 \text{ Btu(thermal)}$$

Direct Hydrogenation:

While higher pressures (2000 psi plus) favor hydrogasification, recent technology (e.g., the Hygas process) indicates that lower pressures can be used; 1000 psi. Therefore, the problem reduces to that of compressing the hydrogen used to 1000 psi:

$$\left(\frac{1000}{20}\right)^{0.28} = 3.00$$

$$T_2 = 537 (3.00) = 1611°R \text{ or } 1151°F$$

$$-W_{theo} = 2.0 (7.1 (1611-537) = 15,250 \text{ Btu(mech)}$$

$$-W_{actual} = \frac{15250}{0.80 (0.30)} = 63,542 \text{ Btu(thermal)}$$

An argument can be made here for using pressure gasification (e.g., the Lurgi gasifier) to produce the hydrogen. There is, however, a trade-off between the energy input for oxygen and the energy input for compressing the hydrogen.

As to pressures greater than 1000 psi for hydrogasification, some of the compression input can be recouped by the use of expanders.

Appendix

ADDITIONAL PLANT ENERGY REQUIREMENTS

All other energy inputs are handled as a contingency item. In order to assign a number, the relation

$$n_C(5000) \quad \text{Btu(thermal)}$$

is used, which relates somewhat to plant complexity by introducing the quantity of carbon feed. The figure comes to roughly 10% of the total.

RESULTS

The calculations are entered in the following tables, with a graphical presentation of the results in Figures a, b and c.

Parameters of degree of conversion and efficiency of waste heat utilization are employed to put the determination on common bases.

If alkali (sodium carbonate) is to be added to the reacting system for direct methanation — as noted in Chapter 8 — the correction is small. For a coal of 60% carbon to which is added 10% sodium carbonate, the moles of Na_2CO_3 added per mole of carbon is

$$\frac{1}{0.60} \frac{(12)}{} \frac{0.10}{106} = 0.019 \text{ lb-mole } Na_2CO_3$$

The heat capacity is 26.4 Btu/lb-mole-°F. Thus

$$n\, C_p \Delta T = 0.019\,(26.5)\,1300 = 652 \text{ Btu}$$

This figure would be added to the above value of the sensible heat summation. Ash fluxing apparently does not occur at reaction conditions.

The nickel catalyst does not enter the calculation since it is essentially a fixed fluidized or ebullating bed.

Sulfur is not carried through the calculation. The amounts are small by comparison with the other materials; however, there can be an effect on catalyst performance as noted in previous chapters. Not only will reaction occur with the alkali, but the nickel catalyst may be poisoned by sulfur (sulfides). The verdict is still out, as has been pointed out, in part due to the alkali acting as a sulfur captor, in part due to the converson conditions, and in part since some sulfides are active hydrogenation catalysts (e.g., tungsten sulfide). In fact the activity of WS_2 is enhanced by the presence of H_2S. Furthermore nickel sulfide is a catalyst itself or in combination with other catalysts (e.g., WS_2). [Re: E. E. Donath, "Hydrogenation of Coal and Tar," *Chemistry of Coal Utilization,* Supplementary Volume, Wiley, New York (1963).]

These matters will not be pursued further at this point.

COMPARISIONS OF ENERGY INPUTS AND EFFICIENCY

TOTAL CARBON FEED / ENERGY INPUT OXYGEN PRODUCTION/AIR SUPPLY

	Theor. Gasif. a-c	Actual Gasif. (a-c)/E_{conv}	Theor. Comb c	Actual Comb. c/E_{comb} (E_{comb} = 2/3)	Total Carbon n_C		$c\Delta H_{O_2}$* O_2	Air	Divided by E_{O_2} = 2/3 O_2	Air
DIRECT METHANATION	1.00	1.00 (100%) 1.11 (90%) 1.25 (80%)	0.020	.030	1.030 1.14 1.28	(100%) (90%) (80%)		190		290**
INDIRECT METHANATION										
Oxygen-Blown Pressure Gasif.	1.00	1.00 (100%) 1.11 (90%) 1.25 (80%)	0.20	0.30	1.30 1.41 1.55	(100%) (90%) (80%)	18,200		27,300	
External Combustion	1.00	1.00 (100%) 1.11 (90%) 1.25 (80%)	0.30	0.45	1.45 1.56 1.70	(100%) (90%) (80%)		2,790		4,190**
DIRECT HYDROGENATION	2.00	2.00 (100%) 2.22 (90%) 2.50 (80%)	0.30	0.45	2.45 2.67 2.95	(100%) (90%) (80%)		2,790		4,190**

* ΔH_{O_2} = 91,000 Btu/lb-mole O_2 furnished as oxygen (Lurgi)
 = 9,300 Btu/lb-mole O_2 furnished as air

** Thermal equivalent for air compressor-blower

Appendix

STEAM (LATENT HEAT)

REACTION HEAT REQUIREMENTS

b	b/E_b (E_b=0.9)	h	hE_h (E_h=0)	$\frac{b}{E_b} - hE_n$	Net Latent Heat	(ΔH_o) overall	E_H	$(\Delta H_o)/E_H$ or $(\Delta H_o)E_H$	Sum of $\Sigma n_P(C_p)_P \Delta T$ and/or $\Sigma n_R(C_p)_R \Delta T$	$1/E_H$ minus E_H	Product
1.5	1.67	0.5	0	1.67	30,600	0	100%	0	(22,658)	0.0	0
							80%	0		0.45	10,196
							70%	0		0.73	16,540
							2/3	0		5/6	18,882
							60%	0		1.0	22,658
1.9	2.11	0.9	0	2.11	37,980	−28,400	100%	−28,400	47,906	0.0	0
							80%	−22,270		0.45	21,558
							70%	−19,880		0.73	34,971
							2/3	−18,933		5/6	39,922
							60%	−17,040		1.0	47,906
2.1	2.33	1.1	0	2.33	41,940	−47,590	100%	−47,590	82,006	0.0	0
							80%	−38,072		0.45	36,903
							70%	−33,313		0.73	59,864
							2/3	−31,727		5/6	68,338
							60%	−28,554		1.0	82,006
2.6	2.89	0.6	0	2.89	52,020	−44,590	100%	−44,590	97,166	0.0	0
							80%	−35,672		0.45	43,725
							70%	−31,213		0.73	70,931
							2/3	−29,727		5/6	80,972
							60%	−26,754		1.0	97,166

CALCULATION OF EFFICIENCY

Size Reduction n_C (1750)	Acid Gas Removal (Net)	Compression** to 1000 psig	Additional Requirements (n_C) 5000	Total Input (Btu) @ 100%	Overall Net Efficiency (Percent) @ 100%	Total Input (Btu) @ 90%	Overall Net Efficiency (Percent) @ 90%	Total Input (Btu) @ 80%	Overall Efficiency (Percent) @ 80%
1,803 (100%)	8,650	20,950	5,150 (100%)	67,443	70.5	68,185	63.3	69,130	56.0
1,995 (90%)			5,700 (90%)	77,639	64.6	78,385	58.0	79,326	51.2
2,240 (80%)			6,400 (80%)	83,983	61.0	84,725	54.7	85,670	48.3
				86,325	59.6	87,067	53.5	88,012	47.2
				90,101	57.4	90,843	51.5	91,788	45.4
2,275 (100%)	8,870	3,900	6,500 (100%)	58,425	60.0	59,168	55.0	60,113	49.7
2,468 (90%)			7,050 (90%)	85,663	47.5	86,406	43.5	87,351	39.2
2,713 (80%)			7,750 (80%)	101,916	40.1	102,659	36.7	103,694	33.0
				107,814	37.4	108,557	34.2	109,502	30.7
				117,691	32.9	118,434	30.0	119,379	26.9
2,538 (100%)	−1,070	20,950	7,250 (100%)	28,208	66.2	28,950	61.3	29,895	55.9
2,730 (90%)			7,800 (90%)	74,629	47.2	75,371	43.5	76,316	39.6
2,975 (80%)			8,500 (80%)	102,349	35.8	103,091	33.0	104,036	29.9
				112,409	31.6	113,151	29.2	114,096	26.4
				129,250	24.7	129,892	22.7	130,937	20.5
4,288 (100%)	23,700	63,542	12,250 (100%)	115,480	64.0	116,965	58.4	118,855	52.5
4,673 (90%)			13,350 (90%)	168,123	51.2	169,608	46.7	171,498	41.7
5,163 (80%)			14,750 (80%)	199,788	43.5	201,273	39.6	203,163	35.5
				211,315	40.7	212,800	37.1	214,690	33.2
				230,482	36.1	231,962	32.8	233,852	29.3

* Equals g(33,500)-h(16,200)
** May alternately be considered zero for gasification at 1,000 psi.

Appendix

HIGH-BTU GAS PRODUCTION
100% Conversion

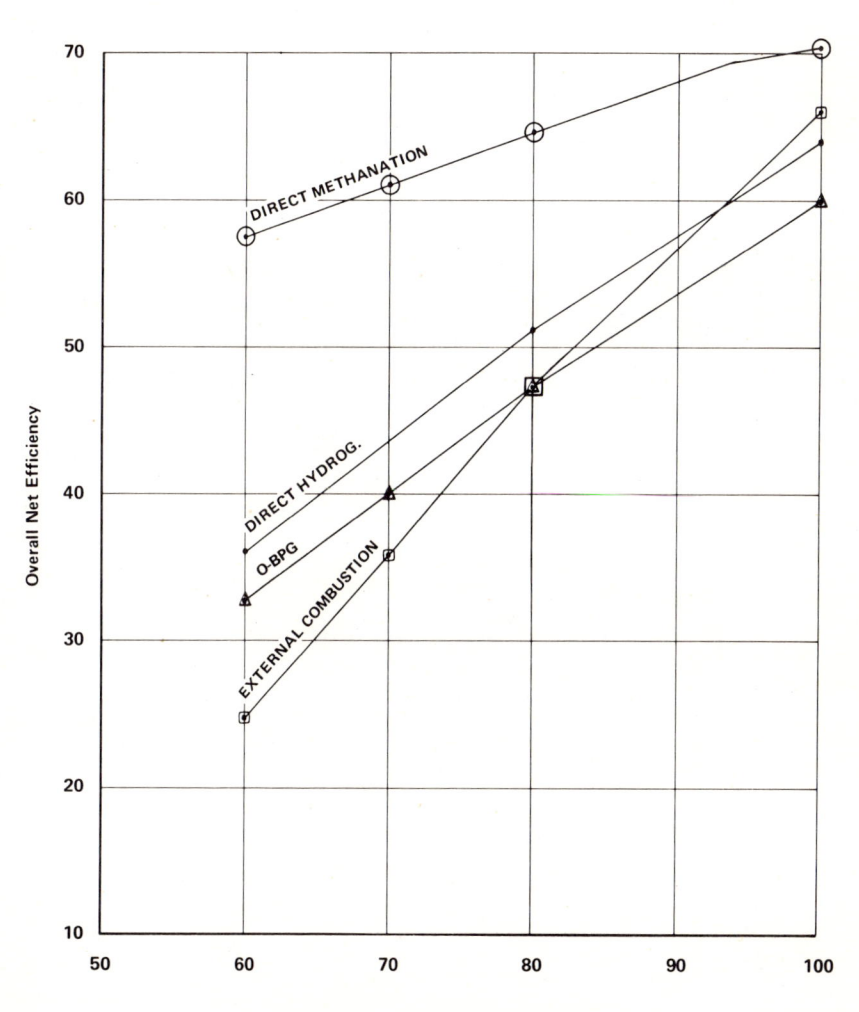

Overall Net Efficiency vs *Efficiency of Waste-Heat Utilization*

Curves: DIRECT METHANATION, DIRECT HYDROG., O-BPG, EXTERNAL COMBUSTION

HIGH-BTU GAS PRODUCTION
90% Conversion

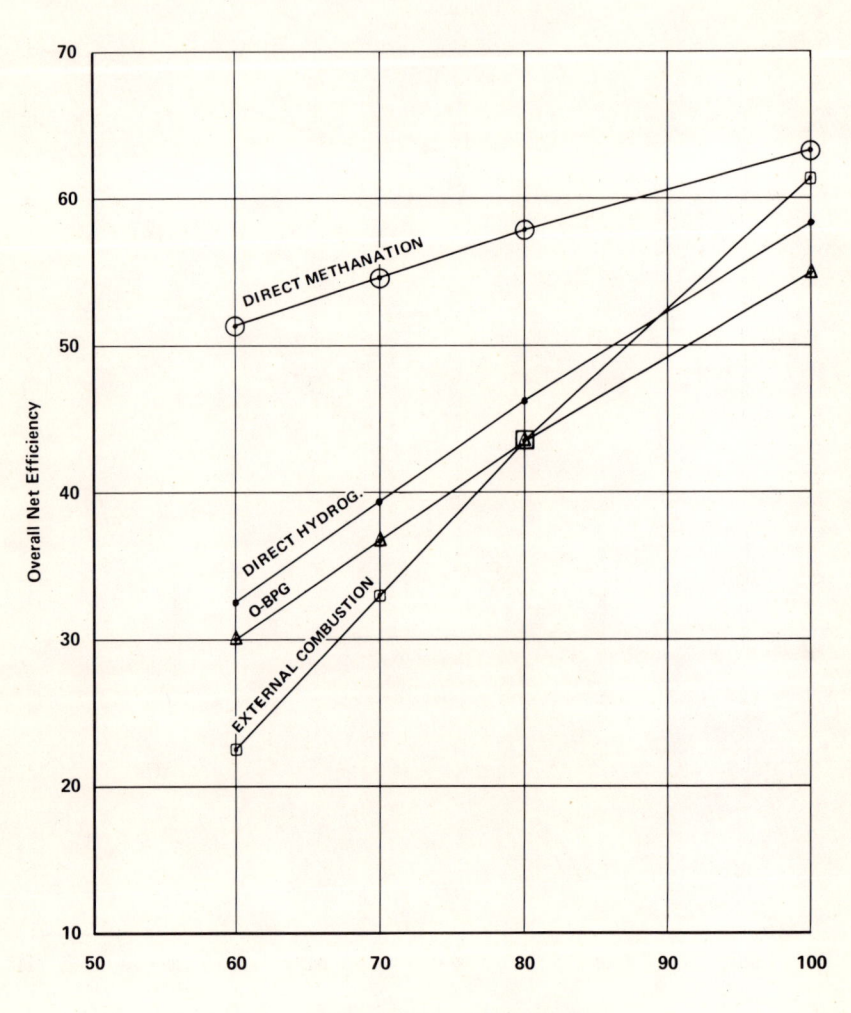

Appendix 423

HIGH-BTU GAS PRODUCTION
80% Conversion

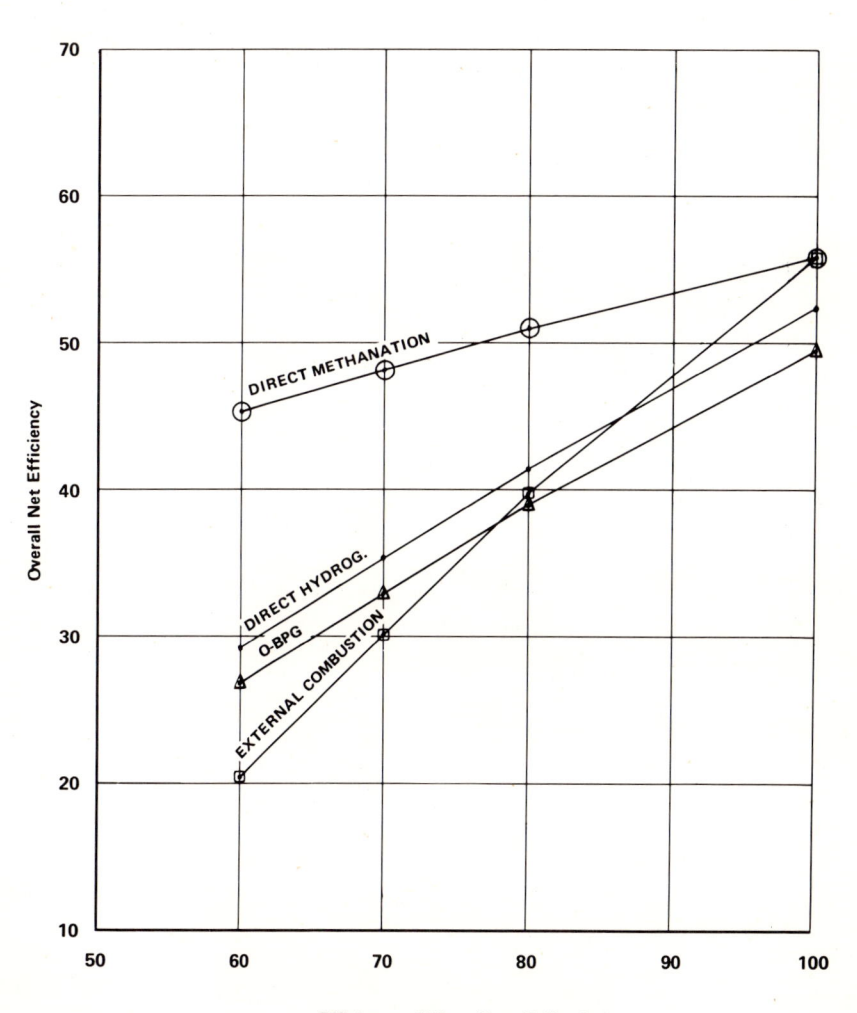

Appendix

V. PRODUCTION OF LOW- OR MEDIUM-BTU GAS

The gases produced from a gasifier may be utilized "as is" without further upgrading — that is, without removal of CO_2 (or H_2S) and/or without the follow-up CO-shift and methanation. These gases would comprise so-called low- or medium-Btu fuel gases and are preferably used "on site". Furthermore, in lieu of cooling the gases to ambient conditions, the gases may be burned hot with an attendant increase in efficiency. The course followed will depend upon whether the combustion products meet air-pollution standards as per the discussion in Chapter 10.

This route will be followed for several cases, based in part on information from the preceding sections.

The processes to be examined and compared are:

1. Direct methanation, e.g., the Hoffman process

2. Pressure gasification (Lurgi or equivalent)
    Oxygen-blown
    Air-blown

3. Gasification with external combustion

4. Entrained gasification (partial combustion)
    Oxygen-blown
    Air-blown

These categories were in part selected to correspond to the several systems analyzed in the report "Economics of Current and Advanced Processes for Fuel Gas Production," Electric Power Research Institute, Palo Alto, California, July, 1976. The second and fourth categories are similiar to those in the report.

ASSUMPTIONS

As stated before, standard conditions of 25°C (77°F) and one atm. are assumed for the datum, with pressure effects deemed negligible as far as the energy functions are concerned.

If the product is to be cold gas, the water present is condensed and removed (the gas will of course remain saturated unless dried). If the product is to be hot gas, all the water remains as vapor.

Heats of combustion for the gaseous components are:

	Gross	Net
CH_4	1013 Btu/SCF	913 Btu/SCF
H_2	325	275
CO	322	322

Rounded values are not used since the principal components vary.

No energy inputs for acid gas removal are entered.

For pressure gasification an allowance is given for energy recovery via an expander.

In the combustion calculations the gas is assumed to be used at essentially atmospheric pressure.

For the purpose of the calculation, the conversion is assumed at 100% for the several different parameters of waste-heat utilization: 60%, 2/3, 70%, 80% and 100%. This corresponds to the calculations of Section IV.

REACTION STOICHIOMETRY

Heat requirements and product characteristics for the several gasification systems are given or calculated as follows.

DIRECT METHANATION:

Reaction temperature: ~ 1300-1400°F

ΔT ~ 1300°F

Reaction:

$$1.020\ C + 1.5\ H_2O(g) + 0.020\ O_2 \rightarrow 0.5\ CH_4 + 0.5\ CO_2$$
$$+ 0.5\ H_2O(g) + 0.02\ CO_2$$

$\Delta H_0 = 0$

$\left[\Sigma n_R (C_p)_R \Delta T \right]_{adj} = 22{,}658$ Btu

Moles of product (moisture-free): $0.5 + 0.5 = 1.0$

Product Btu-rating (moisture-free):

556.5 Btu/SCF (gross)
456.5   (net)

Volume (moisture-free): 379 SCF

Appendix

Total heating value:

210,790 Btu (gross)
173,000   (net)

PRESSURE GASIFICATION (Lurgi or Equivalent):

Reaction temperature: ~ 1200-1400°F up to 2200°F or more

$\Delta T \sim 1500°F$

Reaction:

$$1.2\ C + 1.8\ H_2O(g) + \$\%2\ P_2 \rightarrow 0.4\ CO + 0.8\ H_2 + 0.2\ CH_4 + 0.6\ CO_2 + 0.6\ H_2O(g)$$

$\Delta H_0 = O$

Oxygen-Blown:

$\Sigma n_P(C_p)_P \Delta T = 34{,}650$ Btu

Moles of product (moisture-free): $0.4 + 0.8 + 0.2 + 0.6 = 2.0$

Product Btu-rating (moisture-free):

295.7 Btu/SCF (gross)
265.7   (net)

Volume (moisture-free):   758 SCF

Total heating value:

224,100 Btu (gross)
201,400   (net)

Air-Blown:

The reaction and conditions are assumed the same as for oxygen-blown, though some departure would occur in practice. The air introduces $0.2(4) = 0.8$ moles of nitrogen, which does not affect the heat of reaction but changes the sensible heat summation:

$$\left[\Sigma n_P(C_p)_P \Delta T\right]_{adj} = 34{,}650 + 0.8\ (7.2)\ (1500) = 43{,}290\text{ Btu}$$

Moles of product (moisture-free): $0.4 + 0.8 + 0.2 + 0.6 + 0.8 = 2.8$

Product Btu-rating (moisture-free):

   211.2 Btu/SCF (gross)
   189.8   (net)

Volume (moisture-free):   1,061 SCF

Total heating value:

   224,100 Btu (gross)
   201,400   (net)

Gasification with External Combustion:

Reaction temperature: ~ 1800°F or higher

ΔT ~ 1700°F

Reaction:

$$1.3\ C + 1.9\ H_2O(g) + 0.3\ O_2 \rightarrow 0.7\ CO + 1.3\ H_2 + 0.3\ CO_2$$
$$+ 0.6\ H_2O(g) + 0.3\ CO_2$$

$\Delta H_0 = 0$

$$\left[\Sigma n_P(C_p)_P \Delta T \right]_{adj} = 59{,}670\ Btu$$

Moles of product (moisture-free): 0.7 + 1.3 + 0.3 = 2.3

Product Btu-rating (moisture-free):

   281.7 Btu/SCF (gross)
   253.4   (net)

Volume (moisture-free):   871.7 SCF

Total heating value:

   245,600 Btu (gross)
   220,900   (net)

ENTRAINED GASIFICATION (Partial Combustion):

This case has not been considered in the determinations of the previous sections. As viewed here, gasification takes place at essentially atmospheric pressure.

Appendix

The conversion equation, as before, is

$$a\ C + b\ H_2O(g) + c\ O_2 \rightarrow$$
$$d\ CO + e\ H_2\ f\ CH_4 + g\ CO_2 + h\ H_2O(g)$$

The above is made up of the following equations, assuming lb-moles:

$C + H_2O(g) = CO + H_2$	56,500 Btu
$x\ [CO + H_2O(g) = H_2 + CO_2]$	$x\ [-17,700]$
$y\ [CO + 3\ H_2 = CH_4 + H_2O(g)]$	$y\ [-88,700]$
$z\ [C + O_2 = CO_2]$	$z\ [-169,300]$

Conditions are assumed such that no methane is produced, hence $y = 0$.

From the equivalence of Section III, assuming no moisture in the product gas,

$CO:\quad d = 1-x$

$CO_2:\quad g = x+z$

$H_2:\quad e = 1+x$

$H_2O:\quad h = b-1-x = 0$

where $a = 1+z$. Representative molar ratios in practice for partial combustion are as follows, which agree with equilibrium data for the temperature levels used:

$$\frac{CO}{CO_2} = \frac{1-x}{x+z} = \sim 3$$

$$\frac{CO}{H_2} = \frac{1-x}{1+x} = \sim 2$$

These values in general follow the equilibrium curves.

Solving for x and z,

$$x = -1/3 \qquad z = 7/9$$

Therefore,

$CO:\quad 4/3$

$CO_2:\quad 4/9$

$H_2:\quad 2/3$

$H_2O:\quad b - 1 + 1/3 = 0$ and $B = 2/3$

Coal Conversion

Also,

$$z = c = 7/9$$
$$a = 1+z = 16/9$$
$$a-c = 1$$

The standard heat of reaction is:

$$\Delta H_0 = 56{,}500 + x(-17{,}700) + z(-169.300)$$
$$= -69{,}278 \text{ Btu per } 16/9 \text{ lb-mole of C}$$
$$\text{or } -38{,}969 \text{ Btu per lb-mole of C}$$

Furthermore, the strongly exothermic reaction is autogenous — i.e., no external heat is necessary to fire the reaction. Note that significantly greater proportions of O_2 (or air) are used than in the other reactions, and that the standard heat of reaction is not set at zero.

Oxygen-Blown:

Reaction temperature: $\sim 1750\text{-}1800°F$

$\Delta T \sim 1700°F$

Reaction:

$$\frac{16}{9} C + \frac{2}{3} H_2O(g) + \frac{7}{9} O_2 \rightarrow \frac{4}{3} CO + \frac{2}{3} H_2 + \frac{4}{9} CO_2$$

$\Delta H_0 = -69{,}278$ Btu   (will also be exothermic at T)

$$\Sigma n_R(C_p)_R = \frac{16}{9}(3.0) + \frac{2}{3}(9.1) + \frac{7}{9}(7.2)$$
$$= 5.33 + 6.07 + 5.6 = 17.00$$

$$\Sigma n_R(C_p)_R \Delta T = 17.00 \, (1700) = 28{,}900 \text{ Btu}$$

Moles of product (moisture-free): 22/9

Product Btu-rating (moisture-free):

264.3 Btu/SCF (gross)
250.6   (net)

Volume (moisture-free): 926.4 SCF

Total heating value:

244,800 Btu (gross)
232,200   (net)

Appendix 431

Air-Blown:

The conversion is the same except that 7/9 (4) = 28/9 moles of nitrogen are introduced as inert. Therefore, for the sensible heat summation,

$$\left[\Sigma n_R (C_p)_R \Delta T\right]_{adj} = 28{,}900 + - (7.2)(1700) = 66{,}980 \text{ Btu}$$

Moles of product (moisture-free): 50/9

Product Btu-rating (moisture-free):

    116.3 Btu/SCF (gross)
    110.3   (net)

Volume (moisture-free): 2,105.6 SCF

Total heating value:

    244,800 Btu (gross)
    232,200   (net)

ENERGY INPUTS FOR OXYGEN OR AIR

The requirements are as established in Section IV except in the case of air-blown pressure gasification, which was not considered in Section IV.

Air-Blown Pressure Gasification:

For an operating pressure of 350 psi, the compression requirements are as follows:

$$\left(\frac{350}{14.7}\right)^{0.28} = 2.42$$

$T_2 = 537(2.42) = 1300°R \text{ or } 840°F$

$-W_{theo} = 5(0.20)(7.2)(1300-537) = 5{,}494 \text{ Btu(mech)}$

At 80% electro-mechanical efficiency and 30% thermo-electrical efficiency,

$$-W_{actual} = \frac{5494}{0.80(0.30)} = 22{,}892 \text{ Btu(thermal)}$$

EXPANSION

While no compression of product gases is involved, the gases produced in pressure gasification offer the option of flowing through an expander. Final presures are considered to be 1.5 atm. or 20 psia.

Pressure Gasification:

This pressure ratio gives

$$\frac{T_1}{T_2} = \left(\frac{350}{20}\right)^{0.28} = 2.23$$

where k is prorated to about 1.37 for the water-free product. Therefore, for most practical purposes, k ~ 1.4 and (k-1)/k ~ 0.28.

It is assumed that all the water vapor present is condensed prior to expansion and that the gases are reheated to a temperature T_1 such that the final expansion temperature is 77°F or 537°R. In practice, interstage heating may be necessary.

For the theoretical purposes of calculation, however,

$$T_1 = 537 (2.23) = 1198°R \text{ or } 738°F$$

This value is used as follows.

Oxygen-Blown:

On a moisture-free basis,

$$-W_{theo} = [0.4 (7.2) + 0.8 (7.2) + 0.2 (12.0) + 0.6 (11.0)] (537-1198)$$
$$= 17.64 (537-1198) = -11,660 \text{ Btu(mech) or } -9 \times 10^6 \text{ ft-lb}$$

At 80% mechanical efficiency for the expander (*Chemical Engineers' Handbook*, 3rd ed.),

$$-W_{actual} = -11,660 (0.80) = -9,328 \text{ Btu(mech)}$$

This is the actual mechanical or shaft work which would be produced.

This work may be performed directly (as on pumps or compressors) or may be converted to electricity, which in turn represents a thermal requirement. This latter route is chosen here as the more general basis. Using a 90% efficiency for the mechanical-electrical conversion and a 30% efficiency for the thermal electrical conversion,

$$-W_{actual} = \frac{-9328}{0.90 (0.30)} = -34,548 \text{ Btu(thermal)}$$

Appendix

The preheating requirement is 11,660 Btu, which at 90% efficiency of heat input is

$$\frac{11660}{0.9} = 12,956 \text{ Btu}$$

The net input is thus

$$-34,548 + 12,956 = -21,592 \text{ Btu}$$

Air-Blown:

The determination is the same except for the nitrogen present:

$$-W_{theo} = [17.64 + 0.3 (4) (7.2)] (537-1198)$$
$$= 23.40 (537-1198) = -15,467 \text{ Btu(mech)}$$
$$-W_{actual} = -15,467 (0.8) = -12,374 \text{ Btu(mech)}$$

or

$$-W_{actual} = \frac{-12374}{0.90 (0.30)} = -45,830 \text{ Btu(thermal)}$$

For preheating,

$$\frac{15467}{0.90} = 17,186 \text{ Btu(t)}$$

The net input is

$$-45,830 + 17,186 = -28,644 \text{ Btu(t)}$$

The theoretical oxygen or air requirement is divided by an efficiency $E_{O_2} = 2/3$ to obtain the actual requirement. While the oxygen all reacts in one way or another and does not show up in the product gas, the nitrogen is inert and shows up in the product gas. Therefore, in the case of the air-blown pressure gasifier, there is an adjustment for the additional nitrogen in the expander and preheater. This is the 50% excess over the theoretical. The adjustment is:

$$0.5 (0.2) (4) (7.2) (1198-537) \left[-\frac{0.8}{0.90 (0.30)} + \frac{1}{0.9} \right]$$
$$= 1903.68 [-2.963 + 1.111] = -3,526 \text{ Btu(t)}$$

The adjusted final value for the net input equivalent becomes

$$-28,644 - 3,526 = -32,170 \text{ Btu(t)}$$

ADDITIONAL REQUIREMENTS

This is a contingency item. The relation

$$n_C(3500) \quad Btu(t)$$

gives an input of roughly 10% of the total.

RESULTS

Overall net efficiency calculations are here based on net heating values since the principal components of the gas are not the same in all cases and since the moisture latent heat is not recovered in practice:

$$\text{Eff.} = \frac{\text{Product net heating value} - \Sigma \Delta H_{in}}{n_C(168{,}000)}$$

where

913 Btu/SCF = net heating value of methane

275 = net heating value of hydrogen

322 = net or gross heating value of carbon monoxide

The results are entered in the following tables and in Figure d.

Appendix

COMPARISON OF ENERGY INPUTS

TOTAL CARBON FEED

	Theo. Gasif. a-c	Actual Gasif. $(a-c)/E_{conv}$	Theo. Comb. c	Actual Comb. c/E_{comb} ($E_{comb} = 2/3$)	Total Carbon n_C
DIRECT METHANATION (~ Atm. Press.)	1.00	1.00 (100%) 1.11 (90%) 1.25 (80%)	0.020	0.030	1.030 (100%) 1.14 (90%) 1.28 (80%)
PRESSURE GASIFICATION (Lurgi or Equiv., 350 psi)	1.00	1.00 1.11 1.25	0.20	0.30	1.30 1.41 1.55
GASIFICATION WITH EXTERNAL COMBUSTION (~ Atm. Press.)	1.00	1.00 1.11 1.25	0.30	0.45	1.45 1.56 1.70
ENTRAINED GASIFICATION (~ Atm. Press.)	1.00	1.00 1.11 1.25	0.78	0.98*	1.98 2.09 2.23

* For air-entrained systems, low-temperature combustion is assumed to have a maximum efficiency of 80% (*Chemistry of Coal Utilization*, Supplementary Volume).

OXYGEN OR AIR PRODUCTION

	$c\Delta H_{O_2}^*$		$c\Delta H_{O_2}/E_{O_2}$ $(E_{O_2} = 2/3)$	
	Oxygen	Air	Oxygen	Air
DIRECT METHANATION (~ Atm. Press)	--	190	--	290**
PRESSURE GASIFICATION (Lurgi or Equiv., 350 psi)	18,200	22,892***	27,300	34,338
GASIFICATION WITH EXTERNAL COMBUSTION (~ Atm. Press.)	--	2,790	--	4,190
ENTRAINED GASIFICATION (~ Atm. Press.)	70,980	7,250	106,470	10,875

* H_{O_2} = 91,000 Btu/lb-mole of O_2 furnished as oxygen
  = 9,300 Btu/lb-mole of O furnishef as air (at 1.5 atm.)

** Thermal equiv. for air comp.-blower

*** Compressed to 350 psi

Appendix

STEAM (LATENT HEAT)

	b	b/E_b ($E_b = 0.9$)	h	hE_h ($E_h = 0$)	$b/E_b - hE_h$	Times 18000
DIRECT METHANATION (~ Atm. Press.)	1.5	1.67	0.5	0	1.67	30,600
PRESSURE GASIFICATION (Lurgi or Equiv., 350 psi)	1.9	2.11	0.9	0	2.11	37,980
GASIFICATION WITH EXTERNAL COMBUSTION (~ Atm. Press.)	2.1	2.33	1.1	0	2.33	41,940
ENTRAINED GASIFICATION (~ Atm. Press.)	0.67	0.74	0	0	0.74	13,320

REACTION HEAT REQUIREMENTS (GASIFICATION)

	ΔH_o	E_H	$\Delta H_o/E_H$ or $\Delta H_o E_H$	$\Sigma n_P(C_p)_P \Delta T$ or $\Sigma n_R(C_p)_R \Delta T$ Oxygen	Air	$1/E_H$ minus E_H	Product Oxygen	Air
DIRECT METHANATION (~ Atm. Press.)	0	100%	0	--	22,658	0.0	--	0
		80%	0			0.45		10,196
		70%	0			0.73		16,540
		2/3	0			5/6		18,882
		60%	0			1.0		22,658
PRESSURE GASIFICATION (Lurgi or Equiv., 350 psi)	0	100%	0	34,650	43,290	0.0	0	0
		80%	0			0.45	15,593	19,481
		70%	0			0.73	25,295	31,602
		2/3	0			5/6	28,875	36,075
		60%	0			1.0	34,650	43,290
GASIFICATION WITH EXTERNAL COMBUSTION (~ Atm. Press.)	0	100%	0	--	59,670	0.0	--	0
		80%	0			0.45		36,903
		70%	0			0.73		59,864
		2/3	0			5/6		68,338
		60%	0			1.0		82,006
ENTRAINED GASIFICATION (~ Atm. Press.)	−38,969	100%	−38,969	28,900	66,980	0.0	0	0
		80%	−31,175			0.45	13,005	30,141
		70%	−27,278			0.73	21,097	48,895
		2/3	−25,979			5/6	24,083	55,817
		60%	−23,381			1.0	28,900	66,980

OTHER ENERGY INPUTS

	Size Reduction $n_C(1750)$	Acid Gas Removal	Compression of Product — Oxygen	Compression of Product — Air	Additional Requirements $n_C(3500)$
DIRECT METHANATION (~ Atm. Press.)	1,803 (100%) 1,995 (90%) 2,245 (80%)	none	--	none	3,605 3,885 4,375
PRESSURE GASIFICATION (Lurgi or Equiv., 350 psi)	2,275 (100%) 2,468 (90%) 2,13 (80%)	none	−21,592*	−32,170**	4,650 4,935 5,425
GASIFICATION WITH EXTERNAL COMBUSTION (~ Atm. Press.)	2,538 (100%) 2,730 (90%) 2,975 (80%)	none	--	none	5,075 5,460 5,950
ENTRAINED GASIFICATION (~ Atm. Press.)	4,673 (100%) 4,865 (90%) 5,110 (80%)	none	none	none	6,930 7,315 7,805

* Expander, 350 psi to 20 psia
**Corrected for excess air (nitrogen)

PRODUCT (MOISTURE-FREE)

	Btu-Rating (Btu/SCF)		Total Volume (SCF)		Total Heating Value (Btu)	
	Oxygen	Air	Oxygen	Air	Oxygen	Air
DIRECT METHANATION (~ Atm. Press.)	--	556.5 (gross) 456.6 (net)	--	379	--	210,900 (gross) 173,000 (net)
PRESSURE GASIFICATION (Lurgi or Equiv., 350 psi)	295.7 265.7	211.2 (gross)* 189.8 (net)*	758	1061*	224,100 201,400	224,100 (gross) 201,400 (net)
GASIFICATION WITH EXTERNAL COMBUSTION (~ Atm. Press.)	--	281.7 (gross) 253.4 (net)	--	871	--	245,600 (gross) 220,900 (net)
ENTRAINED GASIFICATION (~ Atm. Press.)	264.3 250.6	116.3 (gross)* 110.3 (net)*	926.4	2105.6*	244,800 232,200	244,800 (gross) 232,200 (net)

* Not adjusted for excess air (nitrogen)

Appendix

CALCULATION OF EFFICIENCY FOR COLD GAS PRODUCTION

Basis: 100% conversion (steam-carbon reaction)
Net heating values used for gas product

	Total Input (Btu)		Overall Net Eff. (%)		Eff. of Waste-Heat Utilization (E_H)
	Oxygen	Air	Oxygen	Air	
DIRECT METHANATION (~Atm. Press)	--	36,298	--	79.0	100%
		46,494		73.1	80%
		52,838		69.4	70%
		55,180		68.1	2/3
		58,956		65.9	60%
PRESSURE GASIFICATION (Lurgi or Equiv., 350 psi)	48,338	44,798	70.0	71.7	100%
	63,931	64,579	62.9	62.6	80%
	73,633	76,400	58.5	57.2	70%
	77,213	80,873	56.9	55.2	2/3
	82,988	88,088	54.2	51.9	60%
GASIFICATION WITH EXTERNAL COMBUSTION (~Atm. Press)	--	90,646	--	53.5	80%
		113,607		44.0	70%
		122,081		40.6	2/3
		135,749		35.0	60%
ENTRAINED GASIFICATION (~Atms. Press.)	113,223	-3,171	42.0	70.8	100%
	125,222	17,628	35.8	64.5	80%
	129,617	29,617	32.2	60.9	70%
	129,497	33,902	30.9	59.6	2/3
	136,912	41,317	28.6	57.4	60%

PRODUCTION OF LOW- OR MEDIUM-BTU GAS

100% Conversion (Steam-Carbon Reaction)
Cold Gas Product—Net Heating Value

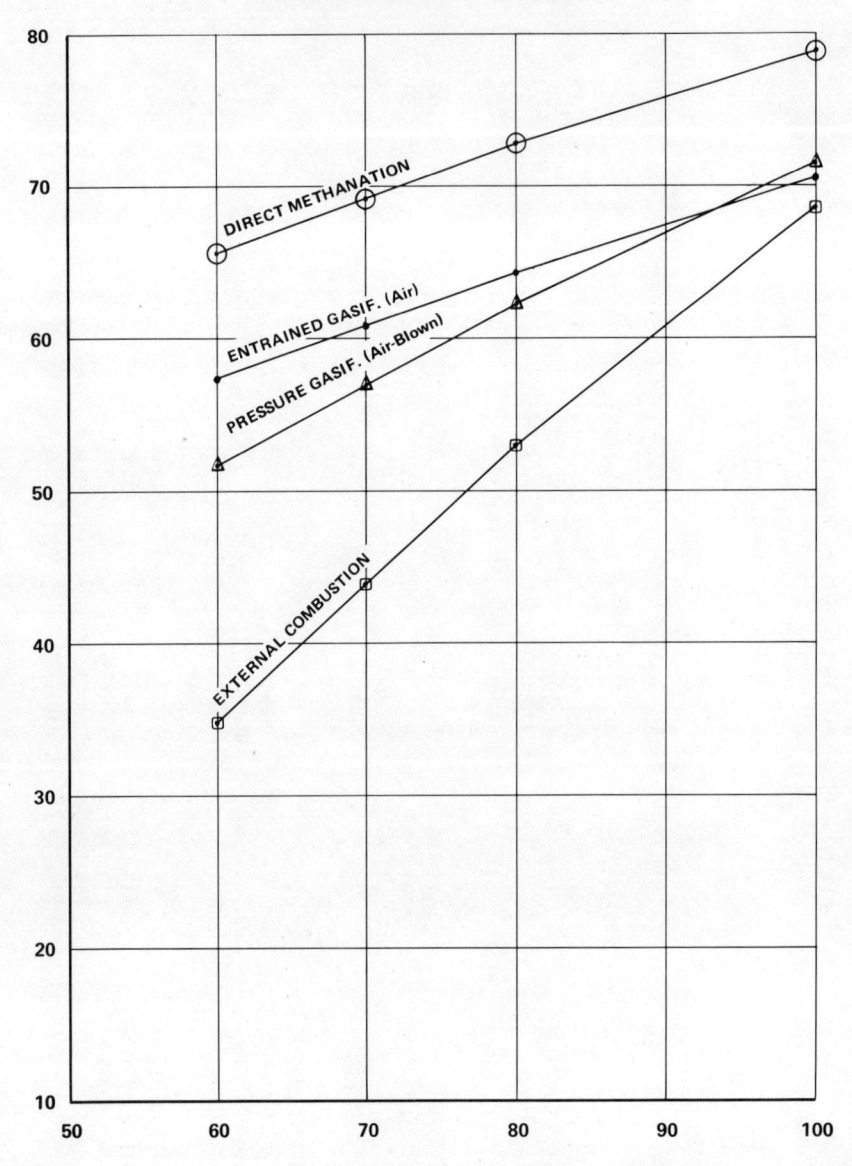

Efficiency of Waste-Heat Utilization

Appendix

COMBUSTION OF HOT GAS

The preceding determinations were made on the basis of a cold-gas product. If the gases are burned hot, where applicable, the Btu-rating may be adjusted upwards to allow for the sensible heat content of the hot gases.

At 100% efficiency of waste-heat utilization this adjustment will not affect the overall efficiency for production plus combustion if the combustion products are referenced to standard conditions as are the gasifier products.

At less than 100% efficiency of waste-heat utilization, however, a diminishing return sets in which will be more pronounced for either the cold gas or the hot gas, depending upon the relative efficacy of heat utilization of the gasifier system vs. the combustion system.

Other effects appear. Thus water vapor in the gasifier product is condensed if a cold gas product is produced. Otherwise it remains in the hot-gas product. Its presence as an inert affects the combustion characteristics of the gas.

From another viewpoint, the flame temperature for combustion of the hot gas is higher than for combustion of the cold gas — other things being equal — thus achieving higher heat transfer rates. The presence of an inert will tend to lower flame temperatures.

Specific heat release is a term coined to relate to furnace size and design. As employed here, the quantity is meant to be the heat of combustion divided by either the moles or volume of combustion products.

For several of the processes considered, the theoretical flame temperature and specific molar heat release are calculated using stoichiometric air for two conditions:

1) The product gas is moisture-free at standard conditions.

2) The product gas includes the excess moisture and is at 1000°F.

Gasification with pure oxygen is omitted. Therefore, the cases are confined to gasification using air, either directly (air-blown) or indirectly (with external combustion).

In pressure gasification, the hot gases would require throttling through an expander to meet Condition 2 while recovering pressure-volume energy. The equipment situation is doubtful. Otherwise the gases would be throttled in a Joule-Thomson expansion without the recovery of energy.

The results are tabulated in the following tables. The theoretical flame temperature T_f is calculated and entered as the difference ΔT_f, and will not necessarily agree with experimental values for a number of reasons.

As an additional note, if the gases are combusted hot, it is desirable to remove particulates and sulfur prior to combustion in order to minimize fouling of the heat transfer surfaces. Cyclones may be used to remove most of the solids carryover, whether ash or spent limestone is used to capture sulfur. Hot gas filters are also under development. Electrostatic precipitators can be used if the temperature is not too high. Other techniques attempt to minimize carryover, e.g., from fluidized beds, where the solids may or may not agglomerate (flux) — depending upon temperature, solids composition and residence times — and may be cycled out of the bed. The alternative is an add-on cleanup section for the combustion off-gases. The subject was discussed in some detail in Chapter 10.

Appendix

COMPARISON OF HOT-GAS VS. COLD-GAS COMBUSTION
MOLES OF PRODUCT GAS (AIR-SUPPORTED OR AIR-BLOWN)

	CO	H_2	CH_4	CO_2	H_2O*	Nitrogen theo / excess / total	Total Moles H_2O-free	Total Moles with H_2O	Stoichiometric Air O_2	Stoichiometric Air N_2
	d	e	f	g	h					
DIRECT METHANATION (~ Atm. Press.)			0.5	0.5	0.5		1.0	1.5	1.00	4.00
PRESSURE GASIFICATION (Lurgi of Equiv., 350 psi)	0.4	0.8	0.2	0.6	0.6	0.9 0.4** 1.2	3.2	3.8	1.00	4.00
GASIFICATION WITH EXTERNAL COMBUSTION (~ Atm. Press.)	0.7	1.3		0.3	0.6		2.3	2.9	1.00	4.00
ENTRAINED GASIFICATION (~ Atm. Press.)	4/3	2/3		4/9		28/9 7/9*** 35/9	57/9	57/9	1.00	4.00

* Is equal to zero for cold gas product
** 50% excess air used for gasification
*** 25% excess air used for gasification

ADJUSTED NET HEATING VALUE FOR HOT PRODUCT GAS

	Total Heating Value (net) Btu	$\Sigma n_i(C_p)_i$	Times ΔT where $\Delta T = 1000-77 = 923°F$	Adjusted Net Heating Value Total Btu	Btu per Mole	Btu per SCF
DIRECT METHANATION (~Atm. Press.)	173,000	16.05	14,800	187,800	125,200	330
PRESSURE GASIFICATION (Lurgi or Equiv., 350 psi)	201,400	31.74	29,300	230,700	60,700	160
GASIFICATION WITH EXTERNAL COMBUSTION (~Atm. Press.)	220,900	23.16	21,400	242,300	83,600	221
ENTRAINED GASIFICATION (~Atm. Press.)	232,200	54.68	50,500	282,700	44,600	118

Appendix

COMBUSTION PRODUCTS AND FLAME TEMPERATURE

	CO_2	H_2O Cold Gas	H_2O Hot Gas	N_2	Total Cold Gas	Total Hot Gas	$\Sigma n_i(C_p)_i$ Cold Gas	$\Sigma n_i(C_p)_i$ Hot Gas	$\Delta T_f(°F)$ Cold Gas	$\Delta T_f(°F)$ Hot Gas
DIRECT METHANATION (~Atm. Press.)	1.0	1.0	1.5	4.0	6.0	6.5	48.90	53.45	3538	3514
PRESSURE GASIFICATION (Lurgi or Equiv., 350 psi)	1.0	1.2	1.8	5.2	7.4	8.0	59.36	64.82	3393	3559
GASIFICATION WITH EXTERNAL COMBUSTION (~Atm. Press.)	1.0	1.3	1.9	4.0	6.3	6.9	51.63	57.09	4279	4244
ENTRAINED GASIFICATION (~Atm. Press.)	16/9	2/3	2/3	71/9	93/9	93/9	82.38	82.38	2819	3432

SPECIFIC MOLAR HEAT RELEASE

	Net Heating Value		Total Moles		Heat Release per Mole Comb. Prod.		Total Moles of Carbon	Heat Release per Mole of Carbon	
	Cold Gas	Hot Gas	Cold Gas	Hot Gas	Cold Gas	Hot Gas		Cold Gas	Hot Gas
DIRECT METHANATION (~ Atm. Press)	173,000	187,800	6.0	6.5	28,834	28,892	1.03	167,961	182,330
PRESSURE GASIFICATION (Lurgi or Equiv., 350 psi)	201,400	230,700	7.4	8.0	27,216	28,838	1.30	154,923	177,461
GASIFICATION WITH EXTERNAL COMBUSTION (~ Atm. Press)	220,900	242,300	6.3	6.9	35,063	35,116	1.45	152,345	167,103
ENTRAINED GASIFICATION (~ Atm. Press.)	232,200	282,700	93/9	93/9	22,471	27,358	1.98	117,273	142,778

Appendix

Combustion of Pure Components:

The combustion of the individual components provide a point of reference for the combustion of methane:

$$CH_4 + 2 O_2 \xrightarrow{8 N_2} CO_2 + 2 H_2O(g)$$

The net heat of combustion is 913 Btu/SCF or 346,027 Btu/lb-mole. The theoretical flame temperature difference is

$$\Delta T_f = \frac{346027}{11.0 + 2(9.1) + 8(7.2)} = 3986°F$$

The specific molar heat release is

$$\frac{346027}{1 + 2 + 8} = 31{,}457 \text{ Btu/lb-mole}$$

For the combustion of carbon,

$$C + O_2 \xrightarrow{4 N_2} CO_2$$

Based on a heat of combustion of ~14,000 Btu/lb, the flame temperature difference is

$$\Delta T_f = \frac{16800}{11.0 + 4(7.2)} = 4221°F$$

The specific molar heat release is

$$\frac{168000}{1 + 4} = 33{,}600 \text{ Btu/lb-mole}$$

For hydrogen,

$$H_2 + \frac{1}{2} O_2 \xrightarrow{2 N_2} H_2O(g)$$

The net heat of combustion is 275(379 = 104,225 Btu/lb-mole. Therefore,

$$\Delta T_f = \frac{104225}{9.1 + 2(7.2)} = 4435°F$$

The specific heat release is

$$\frac{104225}{1 + 2} = 34{,}742 \text{ Btu/lb-mole}$$

For carbon monoxide,

$$CO + \tfrac{1}{2} O_2 \xrightarrow{2 N_2} CO_2$$

The net heat of combustion is 332(379) = 122,038 Btu/lb-mole. Therefore,

$$\Delta T_f = \frac{122038}{11.0 + 2(7.2)} = 4804°F$$

The specific heat release is

$$\frac{122038}{1+2} = 40{,}679 \text{ Btu/lb-mole}$$

CONCLUSIONS:

In the first three cases there is little advantage in combustion of the hot gases compared to combustion of the cold gases, provided the overall heat recovery and utilization is suitably efficient. For entrained gasification or partial combustion, there is more clearly an advantage in combustion the hot producer gas providing environmental standards can be met.

Interestingly, higher flame temperatures are achieved for the third case, gasification supported by the external combustion of air. This is probably due to the fact that fewer combustion products are formed relative to the heat release.

Heat release based on the total carbon feed is in the same order as the overall efficiency.

In last analysis, however, environmental considerations may dictate that the product gases be cooled, partially condensed and scrubbed prior to combustion in order to minimize particulate and sulfur emissions. While sulfur may be captured at gasifier temperatures by the addition of alkali or alkaline earth materials, particulate carryover may still exist, calling for add-on scrubbers. The question becomes one of whether to scrub prior to combustion or to scrub after combustiuon. If hot gas cleanup can be developed and utilized, however, the situation changes.

Subject Index

Abrasion, *see* attrition
Absorption, 345 ff.
    with extractive reflux, 349
Acceptors, 343
Acetic acid, 74, 75
Acid gas removal and recovery, 345 ff., 412 ff.
    steam load, 331
Acids, organic, 224
Activated carbon, 356
Activation, 242 ff., 272 ff.
    by oxidation, 242, 275
    by reduction, 243, 244, 274
    times, 243
Active sites or centers, 272
Activity, 8, 114, 233, 243, 265 ff.
Adams platinum oxide, 229
Addition reactions, 73 ff.
Additives, 113
Adsorbents, 355
Adsorption, 8, 104
    of metals, 78
    of reactants, 104, 266, 267, 272
Adsorptive methods, 355
Agglomerating-Ash process, 45, 214
Agglomeration, 16, 18, 72, 85
Air, 4
    excess, 100
    primary, 133
    secondary, 133
Air-blowing, 6, 154, 197
Air-blown gasifiers, 154, 197, 295, 372 ff., 424 ff.
Air-blown pressure gasification, 431
Air/fuel ratio, 188
    see excess air
Air humidification, 132
Air pollution, 229
Alcohols, 224, 228, 229
Aldehydes, 224
Alkali carbonates, 8
    as catalysts, 208 ff., 211
    as promotors, 238, 239
    for acid gas removal, 345 ff.
    for SO$_2$ removal, 339 ff.

Alkali content, 100
Alkanolalkylamines, 356
Alkylamines, 356
Alumina as catalyst, 210, 226, 229
    as support, 238
Aluminosilicates, 195
Aluminum catalysts, 210
Aluminum chlorides as catalyst, 65, 192, 229
Aluminum oxide (alumina), 166
    as catalyst, 195, 229
Aluminum silicates, 167
Amines, 347 ff.
Ammonia, 64, 70, 106, 308
    for acid gas removal, 346
    removal, 360
    synthesis, 244
Ammonia synthesis catalyst, 244
Ammoniacal compounds, 346
Ammonium molybdate, 186
Analysis of coal
    basis, 49
    ultimate, 48
    western coal, 375
Aniline, 61
Anodic sites, 273 ff.
Aqueous liquor, 69, 70, 308
Aqueous medium, 131
Aqueous suspension, 126
Arc, *see* electric arc
Aromatic compounds, hydrogenation of, 229
Aromaticity, 64, 200
Aromatics, 228
Aromatization, 229
Ash, 48, 99 417
    as catalyst, 177, 192, 210
    deposits, 99
    fly ash, 330
        see particulates
    fusibility, 49
    fusion, 84, 99
    sulfur in, 64, 282
Ash resistivity, 338
    see electrostatic precipitators

Asphalt, 197, 199
Asphaltenes, 63, 197, 199
Asphaltic materials, 196
Asphalts, 63
ASTM analysis, 48, 51
ASTM classification, 49-51
ASTM curve
Astro-flow process, 176
Atgas process, 41
Atmospheric convection cooling, 333
Atmospheric oxidation, 86
Attrition, 234, 278
Authothermic operation, 15, 214
Autogenous conversion, 430

Bacteria, 53, 269
Baghouses, 339, 444
Barium carbonate
　as promotor, 239
Barium oxide, 114
BASF-Flesch-Demig gasifier, 298
Basic oxygen furnace
　see BOF
Batch reactors, 9
Battelle process, 211, 280
BCR Gasif, 332, 335
　see Bi-Gas gasif.
Benzene, 63, 380
Bi-Gas gasifier, 303, 332, 335
　see BCR Gasif.
Bi-Gas process, 17, 39
Binding agents, 239
Biological systems, 269
Bisulfite process, 339
Bitumens, 62, 197, 199
Bituminous coals
　extraction of, 63
　hydrolysis of, 76
Blast furnaces, 288
Blue gas, see water gas
BOF (basic oxygen furnace), 169, 310 ff.
Boilers, 133 ff.
Boric acid catalyst, 229
Boron, 236
Boron oxide, 167
Boudouard reaction, 163, 169, 313
Briquettes, 114
British Gas Corporation (BGC) gasifier, 305
Bromides, 117

Bromine, 74
Brown coal, 51, 65, 76
Boundary layer, 113
Btu-rating, 12, 48, 49, 176, 426 ff.
Bulk velocity, 113
Burning of coal particle, 110
Burning time, 104, 112

Caking, 72, 85
　Also see agglomeration
Calcium carbonate
　as promotor, 239
　for SO_2 removal, 343
Calcium catalysts, 210
Calcium oxide, 167
　as catalyst, 210
Calcium sulfate, 339
Calcium sulfate (gypsum) sludge, 330
Camphor, 114
Carbide formation, 167
Carbides, 234, 235, 236, 237, 244
Carbon
　active, 65
　as catalyst, 293
　free, 200
　reactions of, 106 ff.
Carbon balances, 383 ff.
Carbon black, 97
　as catalyst, 293
Carbon deposition, 227, 242
Carbon dioxide
　byproduct, 285, 329, 299, 330
　conversion to CO, 313
　　see Boudouard reaction
　effect on catalyst reduction, 243
　emissivity, 98
　removal, 345 ff.
　system, with carbonate-sulfide - H_2S, 282
Carbon disulfide, 63
Carbon formation, 116
Carbon/hydrogen ratio, 5
Carbon monoxide, 153 ff., 287 ff.
　concentration, 177
　decomposition, 293
　from CO_2, 313
　　see Boudouard reaction
　generation, 287, 294 ff.
　　maximizing, 315
　in situ product, 178

Subject Index

in catalyst activation, 242, 244
low concentrations, 321
oxidation of, 121
properties, 292
reactions, 292 ff.
removal and recovery, 317
separation, 317
sources, 292
system, oxide-sulfide - H_2S-H_2, 282
Carbonate-sulfide-H_2S-CO_2 system 282
Carbonates
   as catalysts, 208 ff.
   as promotors, 239
Carbonitrides, 236
Carbonization, 11, 68 ff.
Carbonization of catalysts, 242
   effect of hydrogen, 242
   effect of sulfur, 282
Carbonyls, 272
Carriers, 238, 239
   see supports
   effect on activity, 239
Catalysis, 7, 8, 113, 208, 265 ff.
   in combustion, 162
   in pyrolysis, 65
   in reactions with steam, 206 ff.
Catalysis, *see also* heterogeneous catalysis and homogeneous catalysis
Catalyst, effect on product species, 268, 269
Catalyst activation
   see activation
Catalyst activity, 233, 238 ff., 272 ff.
   see activation
Catalyst attrition, 278
Catalyst carbonization
   see carbonization of catalysts
Catalyst composition, 233 ff.
Catalyst durability, 242, 278
Catalyst heterogeneity
   see heterogeneity
Catalyst life, 242, 248
Catalyst poisoning
   see poisoning of catalysts
Catalyst preparation, 238 ff.
Catalyst properties, 233 ff.
Catalyst regeneration, 272 ff.
Catalyst supports, 234
Catalysts, 5, 7, 8, 9, 185, 186, 192, 194, 195
   impregnated, 187

phases, 235
slurries, 17
Cathodic sites, 273 ff.
Caustic soda as catalyst, 208
Cementite, 234
Cenospheres, 105
Ceramic materials as supports, 234
Chain breaking, 108
Channel process, 97
Char, 64, 69, 70, 196
Chemical conversion, 3
Chemicals, 6
Chlorinated compounds, 116
Chlorination, 73
Chlorine, 160
Chlorine trifluoride
Chromium, 228, 229
Chromium oxide catalyst, 241
Classification
Circulating water, 331
Claus conversion, 359
Claus process, 346
Clay, 195
   active, 65
   fireclay, 210, 226
CO-shift reaction, 6, 115, 178, 226, 372 ff.
CO_2-Acceptor process, 42, 214, 332, 335
Coal, physical and chemical changes, 47 ff.
   analysis, 48
   characteristics, 48
   coalification, 48, 54
   high-sulfur, 89
   moisture, 48
   rank, 49
   structure, 47, 63
Coal/air transport, 113
Coal analysis, *see* analysis of coal
Coal conversion processes, listing of, 14
Coal dust, 108
Coal/fuel oil slurries, 126
Coal-paste, 193
Coal quality, 84
Coal rank, 49
Coal/steam reactions, 288 ff.
Coal/water systems, 126 ff.
Coalcon Hydrocarbonization process, 28
Coalification, 48, 54, 65
Coal-steam reaction, 6, 7
Cobalt carbides, 234, 237

Subject Index

Cobalt catalysts, 210, 234
   in Fischer-Tropsch synthesis, 225 ff.
Cobalt-molybdate catalysts, 194
Cobalt oxide, 166
   as catalyst, 210
Cobalt sulfide as catalyst, 229
Cobalt-thorium catalyst, 241
COED process, 25, 69, 332, 335
COGAS gasifier, 306
COGAS process, 44, 214
Coke or coking, 11, 16, 63, 65, 68, 69, 72, 85, 114, 288, 314
   ovens, 70
      Also see char, carbonization
Colloidal suspensions, 197
Combustibility, 84
Combustion, 3, 4, 11, 81, 134
   coal particle, 110 ff.
   mode, 90, 103, 109
   rate, 110, 118
      see burning time
   surface, 103
Combustion mechanisms, 102 ff., 160
Combustion of coal/water slurries, 127
Combustion systems, 133
Comminution, 55
Complexes, 104, 274
Composition of products, CO-H reactions, 249
Compression, 359, 414
Concurrent flow, 10, 307
Conductors, 270
Consol process, 29, 196, 198
Consumptive water, 331
Continuous operations, 9
Convection, 97, 134, 333
Conversion processes, listing, 14
Cooling water, 331 ff.
   see circulating water
   makeup, 333
Copper
   as promotor, 239
Copper aluminum chloride, 321
Copper ammonium salts, 178, 293, 294, 317
Copper salts
   reactivity of CO, 294
Corrosion cells, 273
Countercurrent flow, 10, 307
Cracking, 6
m-Cresol, 61

Critical temperature, 88
Cryogenic processing, 211, 212, 315, 322
Crystal growth
   see grain growth
Cyanides, 106, 308, 360
Cyclic operations, 9
Cyclone burner, 343
Cyclone furnace, 129
Cyclone separators, 343, 444

"d" shell
Dalton's law of constant composition, 236, 265
D.C.-A.C. conversion, 21
De-ashed process, 196
Decarbonization
   see carbonization of catalysts
   effect of sulfur, 282
Dehydrogenation, 200, 229
Depolymerization, 193
   see polymerization
Deposition of carbon, 227, 242
Density, 54, 58
Desorption, 8
   of products, 104, 266, 267
Dew point, 115
Diamagnetism, 235
Diatomaceous earth, 238
Dielectric constant, 55
Diethanolamine (DEA), 347 ff.
Diethylene glycol, 359
Differential thermal analysis, 66
Diffusion, 97, 102, 103, 104
Diffusion, *also see* mass transfer
Diffusion-controlling, 82, 115, 118, 154
Diffusion flames, 103
Diglycolamine (DGA), 347
Diisopropanolamine (DIPA), 347
Diluents
   Also see inerts
   inactive, 116
Dimethyl ether, 347
Direct combustion, *see* combustion
Direct conversion, 7, 269 ff., 277 ff., 367 ff.
Direct gasification, 207
Direct heating, 69
Direct hydrogenation, 3, 4, 6, 14, 371 ff.
Direct methanation
   see direct conversion
Dispersed phase, 62

Subject Index

Disposal of residues, 330
Distillates, 12
Distillation (effect), 4, 134, 177, 205
   azeotropic, 73, 349
   steam, 18, 72
   vacuum, 73
Dolomite, 238, 343
   calcined dolomite
      see CO_2-Acceptor process
Double-alkali process, 342
   see bisulfite
Downshot furnaces, 134
Dropping beds, 9
Dry scrubber, 343
Drying, 58, 359
Dulong's formula, 49, 376

Ebullating bed, 9, 278
Efficiency
   comparison of, 420 ff., 441 ff.
   definition, 368, 398, 399, 434
   for cooling water, 335 ff.
   of combustion, 94, 125, 443
   of conversion, 366 ff.
      see net energy analysis
   of oxygen utilization, 387, 397
   of waste-heat utilization, 420 ff., 441 ff.
Electric arc, 18
Electric field, 166
Electrical conductivity, 55
Electrochemical phenomena, 265, 273
Electrodes, 21
Electrofluidic gasification, 214, 290
Electro-magneto-chemical phenomena, 166
Electrostatic precipitators, 338, 339, 444
Electrothermal gasification, 15
Elemental sulfur, 330, 359
Elongational viscosity, 199
   see Trouton viscosity
Emf, 275
Emissions, 11
Emissivity, 98
Endothermicity, 6, 7, 213
   effect of moisture, 120
Energetics of combustion, 90 ff., 156
Energy, 1
   conversion, 18
Energy losses in gas scrubbers, 344

Enthalpy, 55
Entrained gasification, 428 ff.
Entrained systems, 9,. 137 ff., 142, 143, 307
   catalysts, 255
Equations of conversion, 90, 156
Equilibrium, 4, 8, 83, 206, 267, 268
   in combustion, 156
   reaction, 157, 206
Equilibrium, see also heterogeneous equilibrium and homogeneous equilibrium
Equilibrium constants, 8, 267
Esters, 224
Ethane formation, 226, 237
Ethanolamines, 352 ff.
Ether carbonitrile, 356
Ethylene formation, 226, 237
Evaporation losses, 337
Excess air, 4, 100
Exothermicity, 6, 7, 87
   sign conventions, 91
Expansion, 432, 443
Experimental systems, 117 ff., 251 ff.
Explosion of methane, 96
Explosion velocity, 108, 123
External combustion in support of gasification, 388 ff.
Extraction of metals, 78
Extractive distillation, 350, 351
Extractive reflux, 350, 351
Extracts, see solvents
Exxon Catalytic Coal Gasification process, 211, 212, 280
Exxon Gasification process, 46
Exxon Liquefaction process, 34

Falling particle, 18, 216
   Also see dropping beds
Fast-dropping beds, see dropping beds
Fatty acids, 229
Feeders, 134
   fluidized, 18
Ferric oxide, 191, 208, 226
   see iron catalysts
   see steam-iron process
Ferrites, 234
Ferromagnetism, 235
Ferrous sulfate, 191
Ferrous sulfide, 191

Subject Index

Fireclay, 210, 226
Fire-tube boilers, 133
Fischer-Tropsch catalysts, 272
   see Group VIII transition metals (*Also see* Chapter 7)
Fischer-Tropsch process, 5, 12, 260
Fischer-Tropsch synthesis, 11, 224 ff.
Fixed beds, 9, 10, 69, 91, 169, 230
   see slowly-moving beds
   catalysts, 18, 251
Fixed carbon, 48, 52
Flame propagation, 107
Flame reactions, 83
Flame speed, 113, 116
Flame temperature, 98, 118, 315, 443
Flames
   diffusion, 103
   premixed, 103
Flammability limits, 108
Flashback, 113, 142
Flow
   Eulerian (continuum), 117
   Lagrangean (point-particle), 117
Flow reactors, 9
   catalytic, 194
Fluidization, 185, 251
   advantages, 216
Fluidized beds, 9, 10, 17, 18, 69, 188, 251, 278, 307
Fluidized catalyst, 245, 278
Fluidized catalytic cracking (FCC), 382
Fluidized combustion, 12, 117, 136 ff., 344
Fluidized pyrolysis, 69, 71
Fluidized velocity, 142
Fluor process, 347
Fly ash, 330
   see particulates
Formic acid, 293
Fouling, 99
Fracturing, 173, 176
Free-burning, 85
Free-carbon content, 200
   of iron catalysts, 244
Free energies, 270
Friability, 89
Frontfeed furnace, 134
Fuel beds, 135
Fuel cells, 21, 22
Fuels, 6
Fugacity, see activity
Fuller's earth, 195

Furnace performance, 123
Furnaces, 133 ff.
Fused alkalis, 76
Fused iron catalysts, 167
Fusion of catalysts
   see sintering of catalysts

Galvanic cells, 273
Garrett process, 26, 382
Gas-phase reactions, 103
Gas producers, 295 ff.
Gas recycle, 17, 279
Gasification, 15, 295 ff.
   see name of gasifier
   see oxygen-blown, air-blown, external combustion
Gasifiers, 295 ff.
   see name of gasifier
   air-blown, 295 ff., 424 ff., 431
   liquid yields, 308
      see tars and oils
   oxygen-blown, 295, 372 ff., 424 ff.
Geothite, 245
Giammarco-Vetrocoke process, 347
Grain growth, 240, 242, 276
β-graphite, 156
Grates, 9, 134
Grind, 84
Grindability, 54, 58
Group VIII transition elements, 7, 165, 233, 234, 265, 272, 277
Gypsum, 339
   sludge, 330

Hägg carbide, 234, 235, 237
Hardgrove grindability
Hardness, 54
H-Coal process, 16, 32, 194, 332, 333, 335
Heat balance, 92, 95
Heat capacity, 54, 353
   table, 405
Heat input, 93
Heat of combustion, 270, 391
   calculation of, 49, 376
      see Dulong's formula
Heat of formation, 352, 391
Heat of reaction, 81, 153, 157, 224, 352, 383 ff., 391
   effect of temperature, 402 ff.
Heat of solution, 352 ff.
Heat output, 93

Subject Index

Heat release, 443
Heat removal, 6, 253, 403 ff.
   see efficiency of waste-heat utilization
Heating value, 49, 58, 434
   of cold gas, 440
   of hot gas, 444
Heavy oil let down (HOLD), 193
Helium, 16
Hematite, 235, 244
Heterogeneity, 83, 238, 265, 275
Heterogeneous catalysis, 8, 265 ff.
Heterogeneous equilibrium, 8, 267 ff.
Heterogeneous interaction, 118
Heterogeneous reactions, 8, 103, 265 ff., 267, 268, 269, 273
High-Sulfur coals, 89
Higher alcohols, 228
Hoffman process, 280 ff., 371 ff.
   see direct methanation or direct conversion
H-Oil process, 16, 194
Hole theory, 265
Homogeneous behavior
   ignition, 105
   reactions, 267
Homogeneous catalysis, 8, 267 ff., 272, 273
Homogeneous equilibrium, 8, 269 ff.
Homogeneous process, 269
Homogeneous reactions, 266 ff.
Hot-gas cleanup, 313, 344, 444
Hot-gas recycle, 17
Hot potassium carbonate solutions, 347
Hot precipitator, 339
Humic materials (acids), 43, 62, 74, 75
Humidification, 132
Hydrane process, 16, 35
Hydrates, as catalysts, 208
Hydraulic transport, 127
Hydrocarbonization, 16, 28
Hydrocarbons, 5, 224, 229
   aliphatic, solubility in, 63
   aliphatic, olefinic, aromatic, naphthenic, 64
Hydrocol process, 230, 257
Hydrocracking, 6, 229
Hydrogasification, 16, 183, 371 ff.
Hydrogen, 7, 153, 183, 193, 287 ff.
   effect of in steam-carbon reaction, 207
   generation, 287, 288 ff.
   in catalyst activation, 242
   oxidation of, 121
   reactions of, 106
   system, oxide-sulfide-H_2S-CO, 282
Hydrogen cyanide, 106, 294, 308
   removal, 360
Hydrogen fluoride, 211
Hydrogen sulfide, 17, 64, 192
   conversion, 358
   effect on catalyst activity, 241
   removal, 241, 245 ff.
   system, carbonate-sulfide-CO_2, 282
   system, oxide-sulfide-H_2-CO, 282
Hydrogen sulfide-sulfur-sulfur dioxide equilibrium, 173
Hydrogenation, 3, 4, 6, 11, 16, 75, 76, 185 ff., 192, 293
   dry, 16, 185 ff.
   liquid-phase, 16, 184, 189 ff.
   mild, 195 ff.
   vapor-phase, 183, 184, 194 ff.
   yields, 195
Hydrogenation activity, 233
Hydrolysis, 75, 124
Hydrorefining, 241
Hydrous oxides, 243
Hygas gasifier, 302
Hygas process, 16, 17, 31, 188, 290, 332, 335

Ignition, 106, 109, 134
   range, 107
   spontaneous, 88
   temperature, 107
   zone, 107
IGT Gasification, 332, 335
   see Hygas
Impregnated catalysts, 187, 211, 239
In situ gasification, 173
   carbon monoxide from, 178
   channelization, 173
   leakage, 177
   low-Btu gas from, 177
   superheated steam in, 177
In situ processes, 18
Incandescence, 81
Index of rank, 52
Indirect heat transfer, 14, 214
Indirect hydrogenation, 3, 5, 14, 371
Indirect methanation, 371 ff.

Inerts, 115, 116
    Also see diluents
Inorganic acids as catalysts, 211
Inorganic sulfur, 53
Interface, 83
Internal pressure, 60
Interstial compounds, 236
Inverse process, 307
Iodine, 74
Iodine number, 62
Ionization potential, 273, 274
Iridium, 233
Iron carbides, 235, 236, 238
Iron catalysts, 7, 166, 167, 178, 210, 211, 234, 237, 239, 2444, 245, 251
    fused, 167, 244
    in Fischer-Tropsch synthesis, 225 ff.
    in hydrogenation, 186
    reduction, 244
Iron-copper catalysts, 239
Iron disulfide, 88
Iron ores as catalysts, 244
Iron oxide as catalyst, 65, 210, 277
    see iron catalysts
Iron oxide composition, 243
Iron oxide process, 16
    see steam-iron process
Iron oxide systems, 237, 240, 243
Iron sulfide
    as catalyst, 195
Iron system, 235
IRSID-CAFL processes, 311
Isomerization, 229

Kellogg Molten-Salt process, 43, 124 332, 335
Ketones, 224
Kiesulguhr, 238
Kindling temperature, 107
Kogasin, 225
Koppers-Totzek gasifier, 12, 298, 300

Latent heat requirements, 401
    see Dulong's formula
Lautamasse, 244
Leonardite, 78
Lignite, 65, 78, 84
Liquid fuels, 11
Liquid carrier, 183
Liquid-phase hydrogenation, 183, 184, 189 ff.

Lime as catalyst, 208
Limestone, 339
Limonite, 245
Low-BTU gas, 153, 169, 345, 425 ff.
Low-temperature oxidation, 87
Low-temperature reactions, 85
Low-temperature separations
    see cryogenic processing
Lurgi gasifier, 12, 17, 208, 215, 298, 372, 385, 399, 427
    efficiency, 399
Lurgi process, 36

Macerals, 53
Maghemite, 235, 238
Magnesium catalyst, 241
Magnesium oxide, 167
Magnesium salts, 114
Magnetic field, 21
Magnetic susceptibility, 55, 235
Magnetite, 18, 234, 236, 244, 245
Magnetohydrodynamics, *see* MHD
Mass transfer, 8
Mechanical properties, 55
Medium-Btu gas, 345, 425 ff.
Metal carbonyls, 272
Metal catalysts in steam-carbon reactions, 210
Metal naphthenates, 186
Metal oxides, 99
    as catalysts, 208
Metals of Group VIII, 233
Metallic couples as catalysts, 208
Metallic oxides
    sulfur removal, 358
Metallic sulfides, 192
Methanation, 6, 17, 225, 371 ff.
    liquid-phase, 17
    yields, 226, 371 ff.
Methane, 11, 17, 106, 186, 206, 377
    explosion of, 96
    production, 371 ff.
Methanol, 21, 223, 228, 380
    absorbent for acid gases, 347
        see Rectisol process
Methyldiethanolamine (MDEA), 347
MHD, 19, 20, 21, 105
Microbiological systems, 54, 269
Middle oil, 193
Mild hydrogenation, 195 ff.
Mill scale, 244, 245

Subject Index

Mode of combustion, 90, 103, 109
Mode of contact, 153
Moisture, 87, 93, 116, 119 ff., 124, 443
  disadvantages, 125
  effect on combustion, 119 ff., 132
  effect on CO production, 315
Moisture content
  of coals, 84, 115
  of gases, 443
Moisture in coals, 48, 58
  analysis, 49, 59
  combined, 58
  drying, 60
  effect on combustion, 58, 60, 119 ff.
  effect on density, 58
  effect on heating value, 58
  effect on preparation, 58
  free, 58
Molecular sieves, 359
Molten salt
  see Kellogg process
  see Rockgas process
  see hot-gas cleanup
Molybdenum, 228, 229, 241
Molybdenum compounds
  as catalysts, 191, 194, 241
Molybdenum naphthenate, 186
Molybdenum oxide
  as catalyst, 166, 195, 228, 230
Molybdenum sulfide
  as catalyst, 229, 241
Monoethanolamine (MEA), 347 ff.
Montan wax, 62
Moving bed, 9, 382
Moving burden, see moving bed
Multiphase systems, 10

Naphtha conversion, 234
Naphthalene, 61
Naphthenates, 186
Natural gas reforming, 7
  see reforming
Net energy analysis, 366 ff.
  definition, 367 ff.
Net energy yield, 335
Newtonian behavior
  see non-Newtonian substances
NGPRP
  see Northern Great Plains Resources Program

Nickel carbide, 234
Nickel carbonyl, 272
Nickel catalysts, 7, 17, 178, 210, 234
  in Fischer-Tropsch sysnthesis, 225 ff.
Nickel-Kieselguhr, 229
Nickel-Manganese catalyst, 241
Nickel oxides, 275, 276
  system, sulfide - H_2S-H_2-CO, 282
Nickel sulfide as catalyst, 195, 229, 417
  system, oxide - H_2S-H_2-CO, 282
Nitric acid as reactant, 74
Nitrides, 234, 236
Nitrogen, 115
  addition of, 115
  in coal, 48
Nitrogen compounds, 64
Nitrogen oxides
  absorption of, 358
Nitrous oxide, 116
Nonconductors, 276
Nonluminous gases, 98
Non-Newtonian substances, 199
Non-stoichiometric character of transition elements, 165
Northern Great Plains Resources Program (NGPRP), 299, 331, 336
Nuclear explosives, 176
Nuclear heat, 16, 214

Occidental process, see Garrett process
Octane, 377
OG process, 311
Oil recycle, 17
Oil shale, 335
Oils, 11, 64, 70, 308
  fuels, 70
  recycle, 308
Oily constituents, 197, 199
Oily phase, 62
Olefins, 229
One-stage conversion, see direct conversion
Operating conditions
  hydrogenation, 190
  indirect hydrogenation, 245 ff.
Operating variables, effect of, 245 ff.
Organic acids, 75
Organic sulfur, 53, 241
Organosol, 62
Osmium, 233
Overall equilibrium constant, 267

Subject Index

Oxidation, 81, 115, 131
  atmospheric, 78, 86
  by water, 76
  carbon monoxide, 121
  hydrogen, 121
  low-temperature, 74, 87
  of catalysts,
    see activation
  pretreatment, 65
  residues, 63
  zone, 91
Oxidation levels in catalysts, 238, 275
Oxidation zone, 91
Oxide catalysts, 236
Oxides
  as catalysts, 234, 272 ff.
  film thickness, 167
  hydrous, 243
  metal, 99
Oxo process, 227
Oxygen, 6, 87
  efficiency of use, 387, 399
  for heat, 207
  production, 400
Oxygen-blown gasifiers, 295 ff., 425 ff.
Oxygen-blown pressure gasification, 385 ff.
Oxygenated compounds, 5, 224, 380
Ozone, 74

Palladium, 233
Palladium-on-calcium, 200
Paraffins, 229
Paramagnetism, 235
Partial combustion, 3, 4, 153 ff., 294 ff., 427 ff.
Partial oxidation
  see partial combustion
Particulates, 11, 338, 444
  emission standard, 338
Paste slugs, 127
Peat, 49, 62, 66
Permeability, 173
Petrography, 53 ff.
Petroleum asphalts, 63
Phases, catalyst, 235
Phenanthroline, 200
Phenol, 61
Phenols, 308
  removal, 360

Phosgene, 293
Phosphoric acid as catalyst, 65, 211, 229
Physical properties, 54 ff.
  of coal, 54 ff.
  of carbon monoxide, 292
Plasmas, 18, 21
Plasticity in coals, 68, 72
Platinum, 21, 229, 233, 234, 276
Platinum group catalysts, 229
Pneumatic or gas transport, 127
Poisoning of catalysts, 234, 241, 275, 282, 417
Pollution control, 11, 338
Polyethylene glycol, 347
Polymerization, 60, 61, 193
  of olefins, 229
Porosity, 173
Potassium carbonate
  as catalyst, 208 ff., 211, 244
  as promotor, 239
  solutions, 347
Potassium dichromate as reactant, 74
Potassium hydroxide as reactant, 77
Potassium oxide, 244, 245
Potassium permanganate, 63, 74, 229
Pott-Broche process, 62, 195, 198
Precipitation of catalyst, 240
Premixed flames, 103
Preparation, 55
Pressure effect of
  in CO-H_2 reactions, 242, 243, 247
  in hydrogenation, 187, 190
  in steam reactions, 207, 209
  methane yield, 17, 227
  on catalyst durability, 242
Pressure-drop, 344
Pressure gasification, 427
  see oxygen-blown pressure gasification
Primary air, 133
Process cooling water, 331 ff.
  see cooling water or circulating water
Producer gas, 11, 12, 153, 169, 294 ff., 425 ff.
Product compositions, CO-H_2 reactions, 249
  effect of catalyst, 269
Product species, effect of catalyst, 268, 269
Project Rio Blanco, 176
Project Seacoke, 196

Subject Index

Project Thunderbird, 174
Project Wagonwheel, 176
Promotors, 234
Propylene carbonate, 347
Pulverized coal, 84, 110, 130, 298
Pulverized fuels, 133
   see pulverized coal
Pyridine, 61
Pyrites, 88, 89
   oxidation, 88, 244
Pyrites as catalysts, 244
Pyrolysis, 4, 6, 11, 14, 64 ff., 381
   during firing, 72
   effect of moisture, 72

Quench-cooler, 268
Quenching, 18, 268

Radiation, 97, 134
Random motion, 10
Raney nickel, 17, 229
Rank, 49 ff., 84
   effect of, 63, 185, 191
Rate, 4
Rate constant, 109
Reacting systems, 9 ff.
Reaction conditions
   Fischer-Tropsch, 224 ff., 231, 232
   hyrogenation, 185, 194
   steam, 215, 406 ff.
Reaction equilibrium,
   effect of catalyst, 268
   see equilibrium
Reaction heat requirements, 402 ff.
Reaction rate
   effect of catalyst, 268
Reaction time, 248
   see residence time
Reactions
   gas-phase, 103
   of carbon, 106 ff.
   of gaseous hydrocarbons, 83
Reactions of carbon monoxide and hydrogen, 223 ff., 274
   products, 249
Reactions with air or oxygen, 81 ff.
Reactivity, 8, 156
Reactors, 9 ff.
Recirculation, 98, 99
Rectisol process, 347

Redox process, 288
Reduction, 75
   of catalysts, *see* activation zone, 91
Reduction zone, 91
Reforming, 21, 82, 106, 227, 275
Removal of heat
   see heat removal
Repolymerization, 193
   see polymerization
Reserves, 1, 2
Residence time, 188, 190, 245, 248
Resins, 62, 199
Retention time, 189
Reversion to homogeneous equilibrium, 268
Rhodium, 233
RMProcess, 246
Rochdale process, 213
Rockgas molten salt process, 312, 313
Rubidium carbide, 237
Rubidium catalysts, 225, 232, 233

Saponification number, 62
SASOL plant, 12, 230, 259, 332, 335
Scavengers, 241, 279
Scrubbers
   see wet scrubbers
Seaboard process, 347
Secondary air, 133
Selective absorption, 349
Selective removal, 345
Selexol process, 347
Semiconductors, 211, 273, 274, 276
Semi-continuous operations, 9
Sensible heat losses
   see efficiency of combustion
Sensible heat summation, 406
Separation of coal, 55
Shale oil, 335
Shear viscosity, 199
Siderite, 234, 244
Sign conventions, 91
Silica
   as carrier, 238
   as catalyst, 210, 226
Silica gel, 238
   as adsorbent, 359
Silicates as catalysts, 208
Silicon dioxide, 167

Single-alkali process, 341
Single-stage conversion, see direct conversion
Sintering of catalysts, 234, 239, 242
Sites, 272
  see active sites
Size reduction, 54-56, 412
  energy requirements, 56, 412
Sized coal, 298
Slag or slagging, 16, 153, 216, 308
  producer, 308
  zone, 308
Slaked lime, 339
Slowly-moving beds, 9, 10, 169
  see fixed beds
Slurries, 11, 126, 131
  recovery from, 131
Slurry pipelining, 126, 285
Sodium acid phosphate, 167
Sodium bisulfite process, 339
Sodium carbonate
  as catalyst, 65, 186, 208 ff., 210, 211, 226, 277, 417
  as promotor, 239
Sodium chloride as catalyst, 65
Sodium compounds in catalysis, 210, 211
Sodium hydroxide as catalyst, 65
Sodium hydroxide as reactant, 74, 75, 76
Sodium oxide, 114
Sodium vanadate, 356
Softening point, 49, 99
Solvation, 183
Solvent extraction, see solvents
Solvents, 4, 14, 60 ff., 66
  pyrolizing action, 77
Solvent Refined Coal process, see SRC process
Solvents, acid gas removal, 347
Soot blowers, 100
Soot formation, 96
Sorbents, 343, 355
  see adsorbents
Space time
  see residence time
Space velocity, 153, 246
  effect of, 206
Specific gravity, 54
Specific heat, 54
Specific heat release, 443

Spencer or Pittsburg-Midway process, 196, 332, 335
  see de-ashed process
  see SRC process
Spheres, 109
Splitting, 194
Spontaneous combustion, 85, 88, 89
Spontaneous ignition, 88
Sprayed catalyst, 239
Spray drying, 239
SRC (solvent refined coal) process, 30, 196, 198
Stannous chloride as catalyst, 186
Steam, 124
  effect of, 172, 189
  excess, see proportions
  proportions, 207, 371, 385, 387
  reactions, 205 ff.
  stripping, 72
  temperature control, 308
Steam-carbon reaction, 120
Steam gasification, 6, 205 ff., 287 ff., 371 ff.
Steam gasification systems, 214
Steam-hydrogen mixtures, 17
Steam-iron process, 16, 289, 290, 291
Steam-jet ignition, 98
Steam load, acid gas removal and recovery, 331
Steam oxidation of catalysts, 242, 275
Steam reforming, see reforming
Steel converters, 310, 315
  see steel, production of
Steel, production of
  see blast furnaces, BOF, Redox process
Steel turnings, 18, 244
Steeply-sloping seams, 175
Stoichiometry of conversion, 375 ff., 383 ff.
Storage, 84
Stratoblast service, 176
Stretford process, 359
Stripping, 355
Substitution reactions, 73 ff.
Sulfates, 53, 64, 282
  as catalysts, 208
Sulfides, 53, 64, 192
  as catalysts, 195, 208, 229, 417
  system, carbonate - H_2S-CO_2, 282

Subject Index

Sulfur, 11, 17, 64, 173
    effect on catalyst carbonization, 282
    effect on catalysts, 241
    elemental, 330, 359
    in ash, 64, 282
    in coal analysis, 48
    inorganic, 53
    organic, 53, 241
    removal, 444
        see hydrogen sulfide removal
    tolerance of catalysts, 241, 242, 279, 282, 417
Sulfur, alkali capture during combustion, 344
Sulfur, emission standard, 338
Sulfur dioxide, 173
    removal of, 338 ff.
Sulfur oxides
    removal of, 338 ff.
Sulfur poisoning, 234, 282, 417
    see poisoning
    see sulfur tolerance of catalyst
Sulfur removal, 241, 344
    see sulfur
Sulfur tolerance of catalysts, 241, 242, 276, 279, 282, 417
Sulfur trioxide, 115
Sulfuric acid, 211, 229
Supports, 234, 238
Surface area of catalysts, 240
Surface combustion, 103
Surface effects, 108
Surface oxides, 82
Surface reactions, 8, 104, 266, 267 ff., 273
Surfaces, 83
Surfactants, 78
Suspended beds, 10
Suspensions, 126
Synol process, 227
Synthane gasifier, 304
Synthane process, 17, 40
Synthesis gas, 5, 6, 7, 11, 206, 226
Synthetic bitumens, 199
Synthin, 225
Synthine process, 225
Synthoil, 225
Synthoil process, 33
Synthol, 225

Talc, 167

Tars, 11, 64, 70, 195, 308
    fuels, 70
    recycle, 308
Temperature
    in steam gasification, 207
Temperature, effect of, 187, 190
    in CO-H_2 reactions, 246
    on CO formation, 4, 82, 153, 315
Temperature control, 6, 17
Tetralin, 61
Theoretical flame temperatures, 98
Thermo-chemical processes, 1
Thermochemistry, 90 ff.
Thermodynamic efficiencies, 395 ff.
Thermodynamic properties,
    of coal, 54 ff.
    of carbon monoxide, 292
Thermionic conversion, 19
Thermoelectric conversion, 19
Thermograms, 66
Thermogravimetric balance, 66
Thorium catalyst, 241
Thorium oxide, 228, 229, 242
Three-zone theory, 92
Tin as catalyst, 186
Tin compounds as catalyst, 186, 191
Titanium oxide as catalyst, 191
Topping, 21
TOSCOAL process, 27, 214, 382
Town gas, 11, 12, 154
Trace elements, 49
Transition elements, see Group VIII transition elements
Transport, 113, 126
    hydraulic, 127
    pneumatic, 127
Transport reactors, 10
Triethanolamine (TEA), 347 ff.
Trouton viscosity, 199
True-boiling-point curve, 67
TRW process, 218, 280
Tube furnaces, 134
Tungsten oxide, 167
Tungsten sulfide as catalyst, 195, 229, 241, 242, 417
Two-zone theory, 92

Ulmins, 53
Ultrasonics, 105, 339
    sludge removal, 339
Uranium, 78

Vacuum carbonate process, 347
Valency difference, 275
   see oxidation level
Vanadium catalysts, 210
Vapor-phase catalysis, 272
   see homogeneous catalysis
Vapor-phase hydrogenation, 183, 184, 194 ff.
Vapor pressure, 58
Variables, 169
Vibrators, 339
Viscosity, 199
Viscosity-temperature-susceptibility (VTS), 199
Volatile residue, 52
Volatiles, 48, 52
Vortex scrubber, 345

Washing, 55
Waste disposal
   see disposal
Waste-heat utilization, 403 ff.
   see heat removal, efficiency of waste-heat utilization
Water, 83, 168, 299
Water content, 155
   see moisture content
Water dew point, 115
Water gas, 4, 6, 206
Water-gas reactions, 205
Water-gas shift, see CO-shift
Water rate, 331
Water requirements for coal conversion, 331 ff.
Water-tube boilers, 133
Water vapor, 83, 108, 123
   effect on catalyst activity, 242
   effect on catalyst reduction, 243
   effect on combustion, 120 ff.
Waxes
   from extraction, 62
   from Fischer-Tropsch synthesis, 229
Weather resistance of bitumens, 199
Weathered coal, 78, 86
Wellman-Galusha gasifier, 12, 298, 301
Wet scrubbers, 339, 345
Wettability
   to tar or pitch, 84
   to water, 84
Winkler gasifier, 12, 18, 215, 216, 298, 299

Winkler process, 37
Wood, 62

Zeolites
   as adsorbents, 357
   as catalysts, 192
Zinc chloride as ctalyst, 65, 192
Zinc oxide
   as catalyst, 195, 228, 229
Zirconia, 21